physics

for scientists and engineers

Paul A. Tipler

physics

and engineers
for scientists

Fourth Edition

Volume 3

Modern Physics: Quantum Mechanics, Relativity, and The Structure of Matter

W.H. FREEMAN AND COMPANY/WORTH PUBLISHERS

Physics for Scientists and Engineers
Fourth Edition, Volume 3
Paul A. Tipler

Copyright © 1999 by W.H. Freeman and Company
Copyright © 1990, 1982, 1976 by Worth Publishers, Inc.
All rights reserved
Manufactured in the United States of America
Library of Congress Catalog Card Number: 98-60168
Volume 1 (Chapters 1–21) paperback ISBN: 1-57259-491-8
Volume 1 (Chapters 1–21) hardcover ISBN: 1-57259-812-3
Volume 2 (Chapters 22–35) paperback ISBN: 1-57259-492-6
Volume 2 (Chapters 22–35) hardcover ISBN: 1-57259-813-1
Volume 3 (Chapters 36–41) paperback ISBN: 1-57259-490-X
Volume 3 (Chapters 36–41) hardcover ISBN: 1-57259-814-X
Volumes 1 and 2, ISBN: 1-57259-614-7
Volumes 1, 2, and 3, ISBN: 1-57259-615-5

Printing: 1 2 3 4 5 02 01 00 99 98

Executive Editors: Anne C. Duffy and Susan Finnemore Brennan
Development Editors: Steven Tenney and Morgan Ryan, with Richard Mickey
Marketing Managers: Kimberly Manzi and John Britch
Design: Malcolm Grear Designers
Art Director: George Touloumes
Production Editor: Margaret Comaskey
Production Manager: Patricia Lawson
Layout: Fernando Quiñones and Lee Mahler
Picture Editor: Elyse Rieder
Graphic Arts Manager: Demetrios Zangos
Three-dimensional art by DreamLight Incorporated
Illustrations: DreamLight Incorporated and Mel Erikson Art Services
Composition: Compset, Inc.
Separations: Creative Graphic Services
Printing and Binding: R. R. Donnelley and Sons
Cover Image: Sand atop a vertically driven shaker table spontaneously
forms a roughly sinusoidal outline. Image by Max Aguilera-Hellweg.

Illustration credits begin on page IC-1 and constitute
an extension of the copyright page.

W.H. Freeman and Company
41 Madison Avenue
New York, NY   10010   U.S.A.

For Claudia

preface

In this fourth edition I have worked toward four goals:

1. To help students increase their experience and ability in problem solving
2. To make the reading of the text easier and more fun for students
3. To bring the presentation of physics up to date to reflect the importance of the role of quantum theory
4. To make the text more flexible for the instructor in a wide variety of course formats

Enhanced Problem Solving

To help students learn how to solve problems, the number of worked *Examples* that correspond to intermediate-level problems has been greatly increased. Especially notable is a new two-column side-by-side example format that has been developed to better display the text and equations in worked examples. Care has been taken to show the students a logical method of solving problems. Examples begin with strategies, and often diagrams, in a *Picture the Problem* prologue. When possible, the first step gives an equation relating the quantity asked for to other quantities. This is usually followed by a statement of the general physical principle that applies. For example, this step may be "Apply Newton's second law" or "Use conservation of energy." Examples usually conclude with *Remarks* that discuss the problem and solution, and in many cases there are additional *Check the Result* sections that teach the student how to check the answer, as well as *Exercises* that present additional related problems, which students can solve on their own.

Also new are innovative, interactive types of examples, each labeled *Try it yourself.* In these, students are told in the left column how to proceed with each step of the problem-solving process, but in the right column are given only the answer. Thus, students are guided through the problem, but must independently work through the actual derivations and calculations.

A *Problem-Solving Guide* appears at the end of each chapter in the form of a summary of the worked examples in the chapter. The Problem-Solving Guide is designed to help students recognize types of problems and find the right conceptual strategy for solving them. Here again, general principles such as applying Newton's second law or the conservation of energy are emphasized.

Concluding each chapter is a selection of approximately one hundred *Problems*. The problems are grouped by type, which may or may not coincide with the section titles in the chapter. Each problem is designated easy, intermediate, or challenging. Qualitative questions and problems are integrated with quantitative problems within each group, in the hope that this organization will elevate the stature of qualitative problems in the minds of students (and instructors). At the back of the book, *Answers* are given to the odd-numbered problems. Preceding the answers for each chapter is a *Problem Map* that charts which odd-numbered intermediate-level problems correspond with worked examples in the text. Complete solutions to every other odd-numbered problem, worked out in the two-column example format, are available in the *Solutions Manual for Students.*

I do not believe that students can be given too much help in solving problems. Students learn best when they are successful at the tasks they are given. The hierarchy of worked examples, "Try it yourself" examples, Problem-Solving Guide, and Problem Map gives the student and the instructor maximum flexibility by leading the student through progressive levels of independence. "Try it yourself" problems take students step by step through a problem without doing the math for them. The Problem-Solving Guide gives an overview of the techniques that have been demonstrated in the chapter. The Problem Map shows students who are having difficulty where help may lie in the chapter but gives no other assistance.

Student Interest

Much effort has gone into making the written text more lively and informal. Students build their understanding of physics on the physics they've already learned, each concept serving as a building block that will provide the foundation for further inquiry. Over one hundred enthusiastic student reviews indicate that the changes in the fourth edition will successfully reach the widest range of students and will help them to enjoy learning and doing physics rather than focusing on the difficulty of the subject. To further stimulate the interest of students, supplemental, brief *"Exploring ..."* sections offer essays on various topics of interest to science and engineering undergraduates.

Modern Physics in the Introductory Course

Although quantum theory revolutionized the way we describe the physical world more than 70 years ago, we have been slow to integrate it into our introductory physics courses. To make physics more relevant to today's students, the mass–energy relationship and energy quantization sections are included in the conservation of energy chapter, and the quantization of angular momentum is discussed in the chapter on the conservation of angular momentum. These ideas are then used throughout the text, for example, in Chapter 19 to explain the failure of the equipartition theorem.

In addition, two optional chapters, "Wave–Particle Duality and Quantum Physics" (Chapter 17) and "The Microscopic Theory of Electrical Conduction" (Chapter 27), have been written so that instructors who choose to do so can integrate them into a two-semester course along with the usual topics in classical physics. These chapters offer something completely new—support for professors who choose to introduce quantum physics earlier in the course. Chapter 17 on the wave–particle duality of nature is the concluding chapter in Part II, immediately following the chapter on superposition and standing waves. This chapter introduces the idea of the wave–particle duality of light and matter and uses the frequency quantization of standing waves, just studied in the previous chapter, to introduce energy quantization of confined systems. Many students have heard of quantum theory and are curious about it. Having just studied frequency quantization that arises in standing waves, students can easily grasp energy quantization from standing electron waves, once they have seen from diffraction and interference patterns that electrons have wave properties. Because there is little time to cover even the usual material in the introductory course, some instructors are reluctant to consider adding even one more chapter such as Chapter 17. I would argue that quantum physics is at least as important as many of the other topics we teach.

Chapter 27 on the quantum explanation for electrical conduction is positioned so that it can be covered immediately after the discussion of electric current and dc circuits. The classical model of conduction is developed, concluding with the relation between resistivity and the average speed v_{av} and mean free path λ of electrons. The classical and quantum interpretations of v_{av} and λ are then discussed using the particle-in-a-box problem, discussed in the optional Chapter 17, to introduce the Fermi energy. Simple band theory is discussed to show why materials are conductors, insulators, or semiconductors. My hope in offering these optional chapters is that, given the choice, instructors will take advantage of the means to incorporate simple quantum theory into their elementary physics course.

Flexibility

To accommodate professors in a wide variety of course formats and to respond to the preferences of previous users of this text, there has been some revision in the order of material. With this new edition, instructors can give their students a brief exposure to modern physics integrated with the classical topics, or they can choose to skip the optional chapters on quantum physics entirely, perhaps returning to them in the final part of the course when this material is traditionally taught. To make room for these optional quantum chapters, some traditional material may be deleted from the course. To aid the instructor, material that can be skipped without jeopardizing coverage in other sections has been placed in optional sections. There are also two optional chapters in addition to Chapters 17 and 27. Chapter 12, "Static Equilibrium and Elasticity," and Chapter 21, "Thermal Properties and Processes," gather material that instructors sometimes choose to skip over or offer as added reading. The "optional" labeling of sections and chapters enables the instructor to pick and choose among topics with confidence that no material in nonoptional sections depends on previous coverage of an optional topic. Optional sections and chapters are clearly marked by gray borders down the side of the page. Some optional material, such as numerical methods and the use of complex numbers to solve the driven oscillator equation, is presented in "Exploring ..." essays.

Acknowledgments

Many people have contributed to this edition. I would like to thank everyone who used the earlier editions and offered comments and suggestions.

Gene Mosca, James Garland, Robert Lieberman, and Murray Scureman provided detailed reviews of nearly every chapter. Gene Mosca also wrote the student study guide along with Ron Gautreau. Robert Leiberman and Brooke Pridmore class-tested parts of the book, and assisted in obtaining student reviews and feedback. Howard McAllister was instrumental in the development of a standard approach to problem solving in the examples.

Many new problems were provided by Frank Blatt and Boris Korsunsky. Frank Blatt wrote the solutions manuals and offered many helpful suggestions. Jeff Culbert helped to enliven the problem sets with his story problems. Several of the graphs at the ends of the examples were provided by Robert Hollebeek.

I received invaluable help in manuscript checking from Murray Scureman, Thor Stromberg, and Howard Miles, and in checking problems and solutions from Thor Stromberg, Howard Miles, Robert Detenbeck, Daniel G. Tekleab, Jeannette Myers, Scott Sinawi, John Pratte, Yuriy Zhestkov, Huidong Guo, Fred Watts, Ilon Joseph, Monwhea Jeng, Harry Chu, and Roy Wood. Any errors remaining are of course my responsibility.

I would particularly like to thank the more than one hundred students who read and studied from various chapters and provided detailed and valuable comments. Many instructors have provided extensive and invaluable reviews of one or more chapters. They have all made fundamental contributions to the quality of this revision. I would therefore like to thank:

Michael Arnett, *Iowa State University*
William Bassichis, *Texas A&M*
Joel C. Berlinghieri, *The Citadel*
Frank Blatt, *Retired*
John E. Byrne, *Gonzaga University*
Wayne Carr, *Stevens Institute of Technology*
George Cassidy, *University of Utah*
I. V. Chivets, *Trinity College, University of Dublin*
Harry T. Chu, *University of Akron*
Jeff Culbert, *London, Ontario*
Paul Debevec, *University of Illinois*
Robert W. Detenbeck, *University of Vermont*
Bruce Doak, *Arizona State University*
John Elliott, *University of Manchester, England*
James Garland, *Retired*
Ian Gatland, *Georgia Institute of Technology*
Ron Gautreau, *New Jersey Institute of Technology*
David Gavenda, *University of Texas at Austin*
Newton Greenburg, *SUNY Binghamton*
Huidong Guo, *Columbia University*
Richard Haracz, *Drexel University*
Michael Harris, *University of Washington*
Randy Harris, *University of California at Davis*
Dieter Hartmann, *Clemson University*
Robert Hollebeek, *University of Pennsylvania*
Madya Jalil, *University of Malaya*
Monwhea Jeng, *University of California, Santa Barbara*
Ilon Joseph, *Columbia University*
David Kaplan, *University of California, Santa Barbara*

John Kidder, *Dartmouth College*
Boris Korsunsky, *Northfield Mt. Hermon School*
Andrew Lang (graduate student), *University of Missouri*
David Lange, *University of California, Santa Barbara*
Isaac Leichter, *Jerusalem College of Technology*
William Lichten, *Yale University*
Robert Lieberman, *Cornell University*
Fred Lipschultz, *University of Connecticut*
Graeme Luke, *Columbia University*
Howard McAllister, *University of Hawaii*
M. Howard Miles, *Washington State University*
Matthew Moelter, *University of Puget Sound*
Eugene Mosca, *United States Naval Academy*
Aileen O'Donughue, *St. Lawrence University*
Jack Ord, *University of Waterloo*
Richard Packard, *University of California*
George W. Parker, *North Carolina State University*
Edward Pollack, *University of Connecticut*
John M. Pratte, *Clayton College & State University*
Brooke Pridmore, *Clayton State College*
David Roberts, *Brandeis University*
Lyle D. Roelofs, *Haverford College*
Larry Rowan, *University of North Carolina at Chapel Hill*
Lewis H. Ryder, *University of Kent, Canterbury*
Bernd Schuttler, *University of Georgia*

Cindy Schwarz, *Vassar College*
Murray Scureman, *Amdahl Corporation*
Scott Sinawi, *Columbia University*
Wesley H. Smith, *University of Wisconsin*
Kevork Spartalian, *University of Vermont*
Kaare Stegavik, *University of Trondheim, Norway*
Jay D. Strieb, *Villanova University*
Martin Tiersten, *City College of New York*
Oscar Vilches, *University of Washington*
Fred Watts, *College of Charleston*
John Weinstein, *University of Mississippi*
David Gordon Wilson, *MIT*
David Winter, *Columbia University*
Frank L. H. Wolfe, *University of Rochester*
Roy C. Wood, *New Mexico State University*
Yuriy Zhestkov, *Columbia University*

Focus Group Participants

Cherry Hill, New Jersey, July 15, 1997

John DiNardo, *Drexel University*
Eduardo Flores, *Rowan College*
Jeff Martoff, *Temple University*
Anthony Novaco, *Lafayette College*
Jay Strieb, *Villanova University*
Edward Whittaker, *Stevens Institute of Technology*

Denver, Colorado, August 15, 1997

Edward Adelson, *Ohio State University*
David Bartlett, *University of Colorado at Boulder*
David Elmore, *Purdue University*
Colonel Rolf Enger, *United States Air Force Academy*
Kendal Mallory, *University of Northern Colorado*
Samuel Milazzo, *University of Colorado at Colorado Springs*
Anders Schenstrom, *Milwaukee School of Engineering*
Daniel Schroeder, *Weber State University*
Ashley Schultz, *Fort Lewis College*

Student Reviewers

For this edition we invited the input of student reviewers at all stages of manuscript development. A number of the student reviews were blind submissions. The reviews of the following students were especially helpful:

Jesper Anderson, *Haverford College*
Anthony Bak, *Haverford College*
Luke Benes, *Cornell University*
Deborah Brown, *Northwestern University*
Andrew Burgess, *University of Kent, Canterbury*
Sarah Burnett, *Cornell University*
Sara Ellison, *University of Kent, Canterbury*
Ilana Greenstein, *Haverford College*
Sharon Hovey, *Northwestern University*
Samuel LaRoque, *Cornell University*
Valerie Larson, *Northwestern University*
Jonathan McCoy, *Haverford College*
Aaron Todd, *Cornell University*
Katalin Varju, *University of Kent, Canterbury*
Ryan Walker, *Haverford College*
Matthew Wolpert, *Haverford College*
Julie Zachiariadis, *Haverford College*

I would also like to thank the reviewers of previous editions, whose contributions are part of the foundation of this edition:

Walter Borst, *Texas Technological University*
Edward Brown, *Manhattan College*
James Brown, *The Colorado School of Mines*
Christopher Cameron, *University of Southern Mississippi*
Roger Clapp, *University of South Florida*
Bob Coakley, *University of Southern Maine*
Andrew Coates, *University College, London*
Miles Dresser, *Washington State University*
Manuel Gómez-Rodríguez, *University of Puerto Rice, Río Piedras*
Allin Gould, *John Abbott College C.E.G.E.P., Canada*

Dennis Hall, *University of Rochester*
Grant Hart, *Brigham Young University*
Jerold Izatt, *University of Alabama*
Alvin Jenkins, *North Carolina State University*
Lorella Jones, *University of Illinois, Urbana-Champaign*
Michael Kambour, *Miami-Dade Junior College*
Patrick Kenealy, *California State University at Long Beach*
Doug Kurtze, *Clarkson University*
Lui Lam, *San Jose State University*
Chelcie Liu, *City College of San Francisco*
Robert Luke, *Boise State University*
Stefan Machlup, *Case Western Reserve University*
Eric Matthews, *Wake Forest University*
Konrad Mauersberger, *University of Minnesota, Minneapolis*
Duncan Moore, *University of Rochester*
Elizabeth Nickles, *Albany College of Pharmacy*
Harry Otteson, *Utah State University*
Jack Overley, *University of Oregon*
Larry Panek, *Widener University*
Malcolm Perry, *Cambridge University, United Kingdom*

Arthur Quinton, *University of Massachusetts, Amherst*
John Risley, *North Carolina State University*
Robert Rundel, *Mississippi State University*
John Russell, *Southeastern Massachusetts University*
Michael Simon, *Housatonic Community College*
Jim Smith, *University of Illinois, Urbana-Champaign*
Richard Smith, *Montana State University*
Larry Sorenson, *University of Washington*
Thor Stromberg, *New Mexico State University*
Edward Thomas, *Georgia Institute of Technology*
Colin Thomson, *Queens University, Canada*
Gianfranco Vidali, *Syracuse University*
Brian Watson, *St. Lawrence University*
Robert Weidman, *Michigan Technological University*
Stan Williams, *Iowa State University*
Thad Zaleskiewicz, *University of Pittsburgh, Greensburg*
George Zimmerman, *Boston University*

Finally, I would like to thank everyone at Worth and W. H. Freeman Publishers for their help and encouragement. I was fortunate to work with two talented developmental editors. Steve Tenney worked on the beginning phases of the book and is responsible for many of the innovative ideas, such as the example format, summary format, problem-solving guide, and problem map. Morgan Ryan worked on the final stages, including the entire art program, and made significant improvements in the entire book. I am grateful also for the contributions of Kerry Baruth, Anne Duffy, Margaret Comaskey, Elizabeth Geller, Yuna Lee, Sarah Segal, Patricia Lawson, and George Touloumes.

Berkeley, California
December 1997

Paul Tipler

supplements

For Students

Study Guide

Volume 1 (Chapters 1–21) ISBN: 1-57259-511-6
Volumes 2 and 3 (Chapters 22–41) ISBN: 1-57259-512-4

Each chapter contains a description of key ideas, potential pitfalls, true-false questions that test essential definitions and relations, questions and answers that require qualitative reasoning, and problems and solutions.

Solutions Manual for Students

Volume 1 (Chapters 1–21) ISBN: 1-57259-513-2
Volumes 2 and 3 (Chapters 22–41) ISBN: 1-57259-524-8

The *Solutions Manual for Students* provides answers to every other odd end-of-chapter problem, presented in the same format and with the same level of detail as the *Instructor's Solutions Manual* (see below).

Tipler PLUS⊕ CD-ROM

The *Tipler PLUS⊕* CD-ROM is specifically designed to complement the learning process started in the text. On the CD-ROM, students will find a wealth of features to enhance the learning process. Interactive solution-builder exercises based on additional example problems build problem-solving skills. Video clips of lab demonstrations and applied physics bring main objectives to life. Animated quizzes based on the 3D graphics in the text test concepts from each chapter. And Web links lead the student to the sprawling world of physics on the Web. The student version of *Tipler PLUS⊕*, like the instructor's version, can be updated via the Web.

For Instructors

Instructor's Solutions Manual

Volume 1 (Chapters 1–21) ISBN: 1-57259-514-0
Volumes 2 and 3 (Chapters 22–41) ISBN: 1-57259-515-9

Complete solutions to all problems in the text are worked out in the same two-column format as the examples.

Test Bank

Approximately 3500 multiple-choice questions span all sections of the text. Each question is identified by topic and noted as factual, conceptual, or numerical. ISBN: 1-57259-517-5

Computerized Test-Generation System

A database comprises the questions in the *Test Bank*. Instructors can custom design their tests with the *Computerized Test Bank*. For Windows, ISBN: 1-57259-519-1; for Macintosh, ISBN: 1-57259-520-5

Instructor's Resource Manual

Demonstrations and a film and video cassette guide are included. ISBN: 1-57259-516-7

Transparencies

Approximately 150 full-color acetates of figures and tables from the text are included, with type enlarged for projection. Volume 1, ISBN: 1-57259-521-3; Volumes 2 and 3, ISBN: 1-57259-674-0

Instructor's Tipler PLUS⊕ CD-ROM

The instructor's version of the *Tipler PLUS⊕* CD-ROM includes everything on the student CD as well as syllabus-making software in one easy-to-navigate environment. Just indicate what part of the book you are teaching and when and *Tipler PLUS⊕* will link your syllabus to a wealth of CD-ROM and Web content. You can click the update button for new Web links, exercises, and updated content or create your own annotated study and lecture aids. *Tipler PLUS⊕* can even create an e-mail list of your students and fellow instructors. In addition to the syllabus-maker software and the material on the student CD-ROM, the instructor's CD-ROM also features selected items from the *Study Guide* and the *Instructor's Resource Manual*.

about the author

Paul Tipler was born in the small farming town of Antigo, Wisconsin, in 1933. He graduated from high school in Oshkosh, Wisconsin, where his father was superintendent of the Public Schools. He received his B.S. from Purdue University in 1955 and his Ph.D. at the University of Illinois in 1962, where he studied the structure of nuclei. He taught for one year at Wesleyan University in Connecticut while writing his thesis, then moved to Oakland University in Michigan, where he was one of the original members of the physics department, playing a major role in developing the physics curriculum. During the next 20 years, he taught nearly all the physics courses and wrote the first and second editions of his widely used textbooks *Modern Physics* (1969, 1978) and *Physics* (1976, 1982). In 1982, he moved to Berkeley, California, where he now resides, and where he wrote *College Physics* (1987) and the third edition of *Physics* (1991). In addition to physics, his interests include music, hiking, and camping, and he is an accomplished jazz pianist and poker player.

The author as a student, 1954

For over 20 years, the formula has been

Tipler = Quality

Tipler *Physics for Scientists and Engineers,* continues to be the best resource a student c have for learning physics. Dynamic features like these guide the student to mastery . . .

EXAMPLES

- Text includes a greater number of intermediate-level worked examples.

- Each example has a **"Picture the Problem"** section that teaches students how to solve the problem conceptually before solving it mathematically. By learning how to find and organize the relevant information in a problem, students learn to think like a physicist.

- A major innovation is the potent **two-column side-by-side format** for the solutions to examples. Concepts are explained on the left, and the math is presented on the right. This format allows students to make the connections between the equation and what it means.

- Most examples conclude with a **"Remark"** that supplies additional information, discussion of common errors, and advice on solving problems as a physicist would.

Example 6-8

You ski downhill on waxed skis that are nearly frictionless. (*a*) What work is done on you as you ski a distance *s* down the hill? (*b*) What is your speed on reaching the bottom of the run? Assume the length of the ski run is *s*, its angle of incline is θ, and your mass is *m*. The height of the hill is then $h = s \sin \theta$.

Picture the Problem We assume that you are a particle. Two forces act on you: gravity, $m\vec{g}$, and the normal force exerted by the hill, $\vec{F}_n$ (Figure 6-15*a*). Only gravity does work on you, because the normal force is perpendicular to the hill, and hence has no component in the direction of your motion. The work–kinetic energy theorem with $v_i = 0$ gives the final speed *v*.

Figure 6-15*b* shows a free-body diagram for you on skies. The net force is $mg \sin \theta$, which is the component of the weight in the direction of the displacement Δs.

Figure 6-15a **Figure 6-15b**

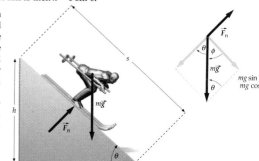

(*a*)1. The work done by gravity as you traverse the slope is $m\vec{g} \cdot \vec{s}$: $W = m\vec{g} \cdot \vec{s} = mgs \cos \phi = mgs \sin \theta$

2. From Figure 6-15*a*, the angle θ is related to *h* and *s*: $\sin \theta = \dfrac{h}{s}$

3. Substitute *h* for $s \sin \theta$: $W = mgh$

(*b*) Apply the work–kinetic energy theorem to find the final speed *v*: $W = mgh = \dfrac{1}{2} mv^2 - 0 \quad \text{or} \quad v = \sqrt{2gh}$

Remarks $mg \sin \theta = mg \cos \phi$ is the component of the weight in the direction of the displacement. This is the component that does work on you. The final speed is independent of the angle θ, and the same as if the skier had dropped vertically a height *h* with acceleration *g*. If θ were smaller, the skier would travel a greater distance to drop the same vertical distance *h*, but the component of the force of gravity in the direction of motion would be less. The two effects cancel, and the work done by gravity is *mgh* independent of the angle of the slope. Figure 6-16 shows that for a hill of arbitrary shape, the work done by the earth on the skier is *mgh*.

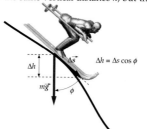

Figure 6-16 Skier skiing down a hill of arbitrary shape. The work done by the ear during a displacement $\Delta \vec{s}$ is $m\vec{g} \cdot \Delta \vec{s} = mg \Delta s \cos \phi = mg \Delta h$, where Δh is the vertical distance dropped. The total work done by the earth when the skier skis down a vertical distance *h* is $W = \int_0^s m\vec{g} \cdot d\vec{s} = mg \int_0^s \cos \phi \, ds = mg \int_0^h dh = mgh$, independent of the shape of the hill.

- When appropriate, **"Check the Result"** sections teach students how to check their own work.

- Many examples are followed by one or more related **exercises**. Answers are given, but it is up to the student to relate the exercise to the worked-out example.

Check the Result The component of *A* along *B* is A cos ϕ = ($\sqrt{13}$ m) cos 70.6° = 1.2 m.

Exercise (*a*) Find $\vec{A} \cdot \vec{B}$ for $\vec{A} = (3 \text{ m})\hat{i} + (4 \text{ m}) \hat{j}$ and $\vec{B} = (2 \text{ m}) \hat{i} + (8 \text{ m}) \hat{j}$. (*b*) Find *A*, *B*, and the angle between $\vec{A}$ and $\vec{B}$ for these vectors. (*Answers* (*a*) 38 m², (*b*) A = 5 m, B = 8.25 m, ϕ = 23°)

Example 2-15

A car is speeding at 25 m/s ($\approx$ 90 km/h $\approx$ 56 mi/h) in a school zone. A police car starts from rest just as the speeder passes and accelerates at a constant rate of 5 m/s^2. (*a*) When does the police car catch the speeding car? (*b*) How fast is the police car traveling when it catches up with the speeder?

Picture the Problem To determine when the two cars will be at the same position, we write the positions x_s of the speeder and x_p of the police car as functions of time and solve for the time t when $x_s = x_p$.

(*a*) 1. Write the position functions for the speeder and the police car:
$$x_s = v_s t \quad \text{and} \quad x_p = \tfrac{1}{2} a_p t^2$$

    2. Set $x_s = x_p$ and solve for the time t:
$$v_s t = \tfrac{1}{2} a_p t^2; \quad t = 0 \quad \text{(initial condition)}$$
$$t = \frac{2v_s}{a_p} = \frac{2(25 \text{ m/s})}{5 \text{ m/s}^2} = 10 \text{ s}$$

(*b*)   The velocity of the police car is given by $v = v_0 + at$ with $v_0 = 0$:
$$v_p = a_p t = (5 \text{ m/s}^2)(10 \text{ s}) = 50 \text{ m/s}$$

Remark The final speed of the police car in (*b*) is exactly twice that of the speeder. Since the two cars covered the same distance in the same time, they must have had the same average speed. The speeder's average speed, of course, is 25 m/s. For the police car to start from rest and have an average speed of 25 m/s, it must reach a final speed of 50 m/s.

Exercise How far have the cars traveled when the police car catches the speeder? (*Answer* 250 m)

Remark In Figure 2-13 the solid lines depict the speeder and the police car in this example. The dashed lines are variations on the example. The smaller acceleration depicted by the lower dashed line means the police car takes longer to reach the speeder. In the higher dashed line, the acceleration is the same as in the example, but the police car does not start accelerating until 4 s after the speeder passes by.

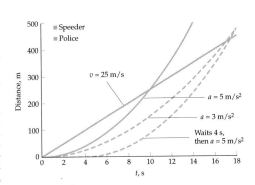

Figure 2-13

Example 6-7   *try it yourself*

A particle is given a displacement $\Delta \vec{s} = 2 \text{ m } \hat{i} - 5 \text{ m } \hat{j}$ along a straight line. During the displacement, a constant force $\vec{F} = 3 \text{ N } \hat{i} + 4 \text{ N } \hat{j}$ acts on the particle (Figure 6-14). Find (*a*) the work done by the force, and (*b*) the component of the force in the direction of the displacement.

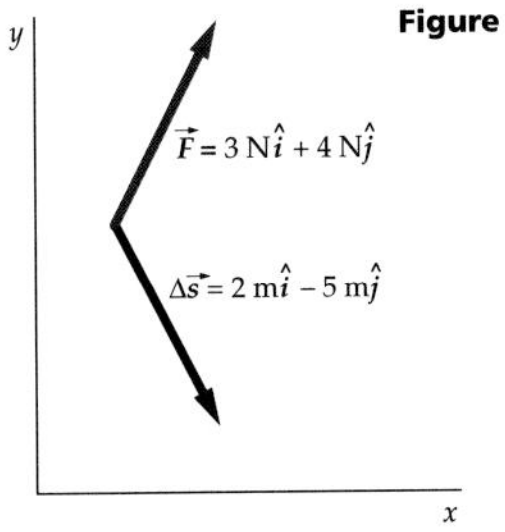

Picture the Problem The work W is found by computing $W = \vec{F} \cdot \Delta \vec{s} = F_x \Delta x + F_y \Delta y + F_z \Delta z$. Since $\vec{F} \cdot \Delta \vec{s} = F \cos \phi |\Delta \vec{s}|$, we can find the component of $\vec{F}$ in the direction of the displacement from

$$F \cos \phi = \frac{(\vec{F} \cdot \Delta \vec{s})}{|\Delta \vec{s}|} = \frac{W}{|\Delta \vec{s}|}$$

Figure 6-14

$\vec{F} = 3 \text{ N}\hat{i} + 4 \text{ N}\hat{j}$

$\Delta \vec{s} = 2 \text{ m}\hat{i} - 5 \text{ m}\hat{j}$

Cover the column to the right and try these on your own before looking at the answers.

Steps	Answers
(*a*) Compute the work done W.	$W = -14 \text{ N·m}$
(*b*) 1. Compute $\Delta \vec{s} \cdot \Delta \vec{s}$ and use your result to find the distance $|\Delta \vec{s}|$.	$|\Delta \vec{s}| = \sqrt{29} \text{ m}$
2. Compute $F \cos \phi = W/|\Delta \vec{s}|$.	$F \cos \phi = -2.60 \text{ N}$

Remark The component of the force in the direction of the displacement is negative, so the work done is negative.

Exercise Find the magnitude of $\vec{F}$, and the angle ϕ between $\vec{F}$ and $\Delta \vec{s}$. (*Answer* $F = 5 \text{ N}, \phi = 121°$)

exploring

Numerical Methods: Euler's Method

If a particle moves under the influence of a *constant* force, its acceleration is constant and we can find its velocity and position from the constant-acceleration formulas in Chapter 2. But consider a particle moving through space where the force on it, and therefore its acceleration, depends on its position and velocity. The velocity and acceleration of the particle at one instant determine its position and velocity at the next instant, which then determines its acceleration at that instant. The actual position, velocity, and acceleration of an object all change continuously with time. We can approximate this by replacing the continuous time variations with small time steps of duration Δt. The simplest approximation is to assume constant acceleration during each step. This approximation is called **Euler's method**. If the time interval is sufficiently short, the change in acceleration during the interval will be small and can be neglected.

Let x_0, v_0, and a_0 be the known position, veloc-

$$x_2 = x_1 + v_1 \Delta t$$

In general, the connection between the position and velocity at time t_n and time $t_{n+1} = t_n + \Delta t$ is given by

$$v_{n+1} = v_n + a_n \Delta t \qquad 3$$

and

$$x_{n+1} = x_n + v_n \Delta t \qquad 4$$

To find the velocity and position at some time t, we therefore divide the time interval $t - t_0$ into a large number of smaller intervals Δt and apply Equations 3 and 4, beginning at the initial time t_0. This involves a large number of simple, repetitive calculations that are easily done on a computer. The technique of breaking the time interval into small steps and computing the acceleration, velocity, and position at each step using the values from the previous step is called numerical integration.

Drag Forces

To illustrate the use of numerical methods, let us consider a problem in which a sky diver is dropped from rest at some height under the influences of gravity and a drag force that is proportional to the square of the speed. We will find the velocity v and the distance traveled x as functions of time.

The equation describing the motion of an object of mass m dropped from rest is Equation 5-7 with $n = 2$:

$$\sum F_y = mg - bv^n = ma_y$$

Summary

1. Work, kinetic energy, potential energy, and power are important derived dynamic quantities.
2. The work–kinetic energy theorem is an important relation derived from Newton's laws applied to a particle.
3. The dot product of vectors is a mathematical definition that is useful throughout physics.

Topic	Remarks and Relevant Equations	
1. Work		
Constant force	The work done by a constant force is the product of the component of the force in the direction of motion and the displacement of the force:	
	$W = F \cos \theta\, \Delta x = F_x\, \Delta x$	6-1
Variable force	$W = \int_{x_1}^{x_2} F_x\, dx = $ area under the F_x-versus-x curve	6-9
Force in three dimensions	$W = \int_1^2 \vec{F} \cdot d\vec{s}$	6-14
Units	The SI unit of work and energy is the joule (J):	
	$1\,\text{J} = 1\,\text{N·m}$	6-2
2. Kinetic Energy	$K = \dfrac{1}{2} mv^2$	6-6

PROBLEMS

- Types of problems are denoted by color swatches: **yellow** denotes conceptual problems and a **gray** band indicates optional or exploring sections.

- The difficulty level is denoted by bullets.

- Qualitative problems are included in context with related quantitative problems.

Problems

In a few problems, you are given more data than you actually need; in a few other problems, you are required to supply data from your general knowledge, outside sources, or in-formed estimates.

	Conceptual Problems
	Problems from Optional and Exploring sections

•	Single-concept, single-step, relatively easy
••	Intermediate-level, may require synthesis of concepts
•••	Challenging, for advanced students

Conditions for Equilibrium

1 • True or false:

(a) $\Sigma\vec{F} = 0$ is sufficient for static equilibrium to exist.
(b) $\Sigma\vec{F} = 0$ is necessary for static equilibrium to exist.
(c) In static equilibrium, the net torque about any point is zero.
(d) An object is in equilibrium only when there are no forces acting on it.

2 • A seesaw consists of a 4-m board pivoted at the center. A 28-kg child sits on one end of the board. Where should a 40-kg child sit to balance the seesaw?

3 • In Figure 12-23, Misako is about to do a push-up. Her center of gravity lies directly above point P on the floor, which is 0.9 m from her feet and 0.6 m from her hands. If her mass is 54 kg, what is the force exerted by the floor on her hands?

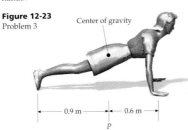

Figure 12-23
Problem 3

Center of gravity

0.9 m — 0.6 m

P

4 • Juan and Bettina are carrying a 60-kg block on a 4-m board as shown in Figure 12-24. The mass of the board is 10 kg. Since Juan spends most of his time reading cookbooks, whereas Bettina regularly does push-ups, they place the block 2.5 m from Juan and 1.5 m from Bettina. Find the force in newtons exerted by each to carry the block.

Figure 12-24 Problem 4

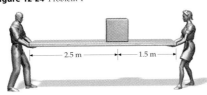

2.5 m — 1.5 m

Figure 12-25 Problem 5

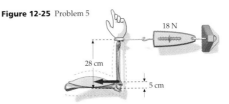

18 N

28 cm

5 cm

the pivot point. If the scale reads 18 N when she exerts her maximum force, what force is exerted by the biceps muscle?

6 • A crutch is pressed against the sidewalk with a force $\vec{F}_c$ along its own direction as in Figure 12-26. This force is balanced by a normal force $\vec{F}_n$ and a frictional force $\vec{f}_s$. (a) Show that when the force of friction is at its maximum value, the coefficient of friction is related to the angle θ by $\mu_s = \tan\theta$. (b) Explain how this result applies to the forces on your foot when you are not using a crutch. (c) Why is it advantageous to take short steps when walking on ice?

Figure 12-26 Problem 6

θ

$\vec{F}_c$

$\vec{f}_s$

$\vec{F}_n$

The Center of Gravity

7 • True or false: The center of gravity is always at the geometric center of a body.

8 • Must there be any material at the center of gravity of an object?

9 • If the acceleration of gravity is not constant over an object, is it the center of mass or the center of gravity that is the pivot point when the object is balanced?

10 • Two spheres of radius R rest on a horizontal table with their centers a distance $4R$ apart. One sphere has twice the weight of the other sphere. Where is the center of gravity of this system?

11 • An automobile has 58% of its weight on the front wheels. The front and back wheels are separated by 2 m.

General Problems

66 • If the net torque about some point is zero, must it be zero about any other point? Explain.

67 • The horizontal bar in Figure 12-52 will remain horizontal if

(a) $L_1 = L_2$ and $R_1 = R_2$.
(b) $L_1 = L_2$ and $M_1 = M_2$.
(c) $R_1 = R_2$ and $M_1 = M_2$.
(d) $L_1 M_1 = L_2 M_2$.
(e) $R_1 L_1 = R_2 L_2$.

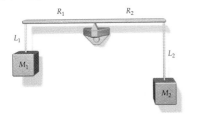

R_1 R_2

L_1

L_2

M_1

M_2

68 • Which of the following could not have units of N/m^2?

(a) Young's modulus
(b) Shear modulus
(c) Stress
(d) Strain

69 •• Sit in a chair with your back straight. Now try to stand up without leaning forward. Explain why you cannot do it.

70 • A 90-N board 12 m long rests on two supports, each 1 m from the end of the board. A 360-N block is placed on the board 3 m from one end as shown in Figure 12-53. Find the force exerted by each support on the board.

Figure 12-53
Problem 70

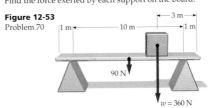

1 m — 10 m — 1 m

3 m

90 N

$w = 360$ N

contents in brief

Volume 1

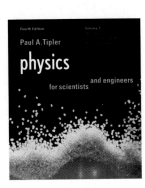

PART IV electricity and magnetism

Volume 2

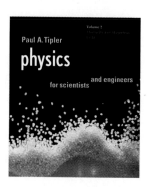

PART V light

Appendix

PART **VI** modern physics: quantum mechanics, relativity, and the structure of matter

Volume 3

contents

PART VI — modern physics: quantum mechanics, relativity, and the structure of matter

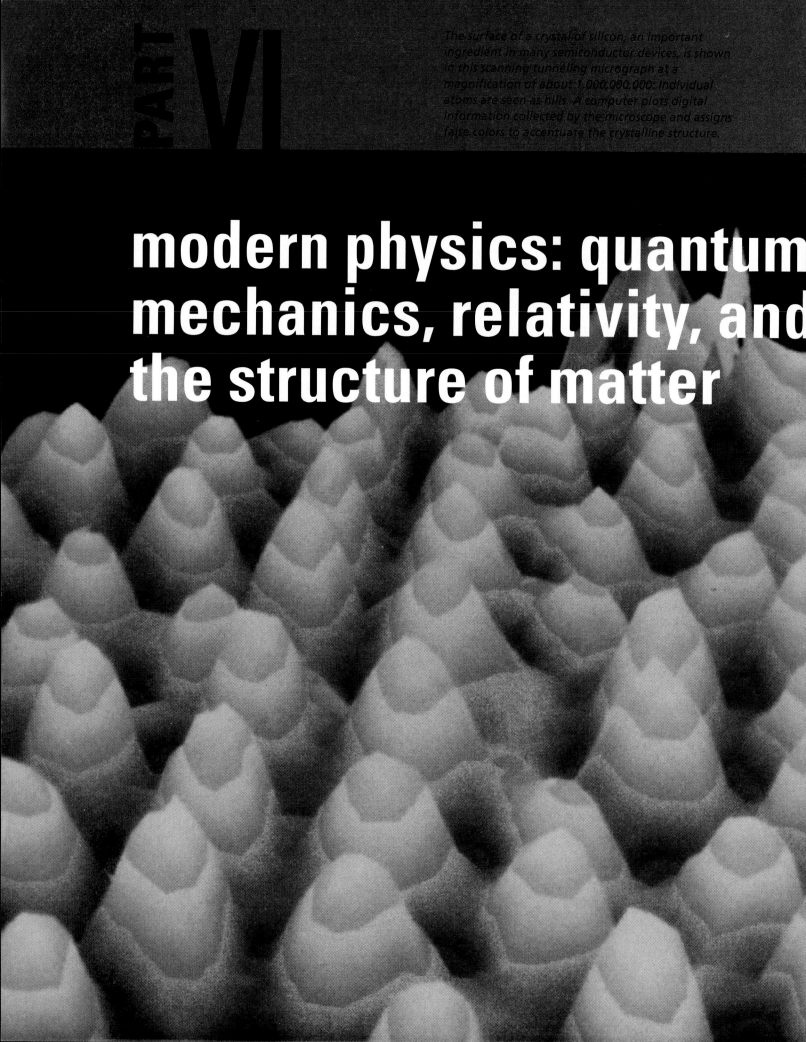

PART VI

modern physics: quantum mechanics, relativity, and the structure of matter

Applications of the Schrödinger Equation

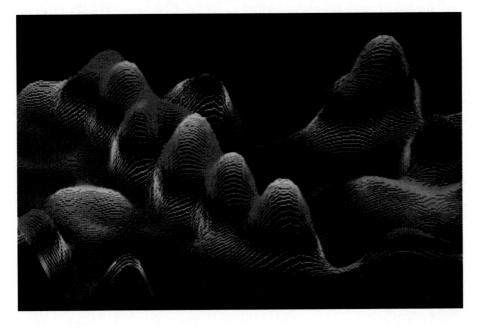

False-color scanning tunneling micrograph of a segment of a DNA molecule magnified about one and a half million times. A sample of double-stranded DNA was dissolved in a salt solution and deposited on graphite prior to being imaged in air by the microscope. The row of orange-yellow peaks corresponds to the ridges of the DNA double helix.

In Chapter 17 we found that electrons and other "particles" have wave properties and are described by a wave function $\Psi(x, t)$. The probability of finding the particle in some region of space is proportional to the square of the wave function. We mentioned that the wave function is a solution of the Schrödinger equation, and we discussed some solutions qualitatively without reference to the equation itself. In particular, we showed how the standing-wave conditions lead to quantization of energy for a particle confined to a one-dimensional box.

This chapter is a continuation of the material introduced in Chapter 17. We discuss the Schrödinger equation and apply it to the particle in the box problem and to several other situations in which a particle is confined to a region of space to illustrate how boundary conditions lead to energy quantization. We then show how the Schrödinger equation leads to barrier penetration, and discuss the extension of the Schrödinger equation to more than one dimension and to more than one particle. You should review the material in Chapter 17 before moving on to the following discussion.

36-1   The Schrödinger Equation

Like the classical wave equation (Equation 15-9b), the Schrödinger equation is a partial differential equation in space and time. Like Newton's laws of motion, it cannot be derived. Its validity, like that of Newton's laws, lies in its agreement with experiment. In one dimension, the Schrödinger equation is[†]

$$-\frac{\hbar^2}{2m}\frac{\partial^2\Psi(x,t)}{\partial x^2} + U\Psi(x,t) = i\hbar\frac{\partial\Psi(x,t)}{\partial t}$$

36-1

Time-dependent Schrödinger equation

where U is the potential-energy function. Equation 36-1 is called the **time-dependent Schrödinger equation.** Unlike the classical wave equation, it relates the second space derivative of the wave function to the *first* time derivative of the wave function, and it contains the imaginary number $i = \sqrt{-1}$. The wave functions that are solutions of this equation are not necessarily real. $\Psi(x,t)$ is not a measurable function like the classical wave functions for sound or string waves. The probability of finding a particle in some region of space dx is certainly real though, so we must modify slightly the equation for probability density given in Chapter 17 (Equation 17-14). We take for the probability of finding a particle in some region dx

$$P(x,t)\,dx = |\Psi(x,t)|^2\,dx = \Psi^*\,\Psi\,dx$$

36-2

where Ψ^*, the complex conjugate of Ψ, is obtained from Ψ by replacing i by $-i$ wherever it appears.[‡]

When the potential energy U does not depend on time (for example, when a particle's wave function corresponds to a standing wave), the time-dependent Schrödinger equation can be simplified by writing the wave function in the form

$$\Psi(x,t) = \psi(x)e^{-i\omega t}$$

36-3

The right side of Equation 36-1 is then

$$i\hbar\frac{\partial\Psi(x,t)}{\partial t} = i\hbar(-i\omega)\psi(x)e^{-i\omega t} = \hbar\omega\psi(x)e^{-i\omega t} = E\psi(x)e^{-i\omega t}$$

where $E = \hbar\omega$ is the energy of the particle. Substituting $\psi(x)e^{-i\omega t}$ into Equation 36-1 and canceling the common factor $e^{-i\omega t}$ we obtain an equation for $\psi(x)$ called the **time-independent Schrödinger equation**

$$-\frac{\hbar^2}{2m}\frac{d^2\psi(x)}{dx^2} + U(x)\psi(x) = E\psi(x)$$

36-4

Time-independent Schrödinger equation

where we have written U as $U(x)$ to emphasize the fact that there is no time dependence in the equation.

The calculation of the allowed energy levels in a system involves only the time-independent Schrödinger equation, whereas finding the probabilities of transition between these levels requires the solution of the time-dependent equation. In this book we will be concerned only with the time-independent Schrödinger equation.

The solution of Equation 36-4 depends on the form of the potential-energy function $U(x)$. When $U(x)$ is such that the particle is confined to some region of space, only certain discrete energies E_n give solutions ψ_n that can satisfy the normalization condition (Equation 17-15):

$$\int_{-\infty}^{\infty}|\psi|^2\,dx = 1$$

[†] Although we simply state the Schrödinger equation, Schrödinger himself had a vast knowledge of classical wave theory that led him to this equation.

[‡] Every complex number can be written in the form $z = a + bi$, where a and b are real numbers and $i = \sqrt{-1}$. The complex conjugate of z is $z^* = a - bi$, so $z^*z = (a + bi)(a - bi) = a^2 + b^2 = |z|^2$. Complex numbers are discussed more fully in Appendix D.

The complete time-dependent wave functions are then given, from Equation 36-3, by

$$\Psi_n(x, t) = \psi_n(x)e^{-i\omega_n t} = \psi_n(x)e^{-i(E_n/\hbar)t} \qquad \text{36-5}$$

We will illustrate the use of the time-independent Schrödinger equation by solving it for the problem of a particle in a box. The potential energy for a one-dimensional box from $x = 0$ to $x = L$ is shown in Figure 36-1. It is called an **infinite square-well potential** and is described mathematically by

$$U(x) = 0, \qquad 0 < x < L$$

$$U(x) = \infty, \qquad x < 0 \quad \text{or} \quad x > L \qquad \text{36-6}$$

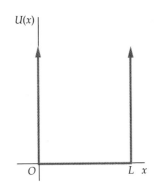

Figure 36-1 The infinite square-well potential energy. For $x < 0$ and $x > L$, the potential energy $U(x)$ is infinite. The particle is confined to the region in the well $0 < x < L$.

Inside the box, the potential energy is zero, whereas outside the box it is infinite. Since we require the particle to be in the box, we have $\psi(x) = 0$ everywhere outside the box. We then need to solve Schrödinger's equation inside the box subject to the condition that $\psi(x)$ must be zero at $x = 0$ and at $x = L$.

Inside the box, Schrödinger's equation is

$$-\frac{\hbar^2}{2m}\frac{d^2\psi(x)}{dx^2} = E\psi(x)$$

or

$$\frac{d^2\psi(x)}{dx^2} = -\frac{2mE}{\hbar^2}\psi(x) = -k^2\psi(x) \qquad \text{36-7}$$

where

$$k^2 = \frac{2mE}{\hbar^2} \qquad \text{36-8}$$

The general solution of Equation 36-7 can be written as

$$\psi(x) = A\sin kx + B\cos kx \qquad \text{36-9}$$

where A and B are constants. At $x = 0$, we have

$$\psi(0) = A\sin(k0) + B\cos(k0) = 0 + B$$

The boundary condition $\psi(x) = 0$ at $x = 0$ thus gives $B = 0$, and Equation 36-9 becomes

$$\psi(x) = A\sin kx \qquad \text{36-10}$$

The wave function is thus a sine wave with the wavelength λ related to the wave number k in the usual way, $\lambda = 2\pi/k$. The boundary condition $\psi(x) = 0$ at $x = L$ restricts the possible values of k and therefore the values of the wavelength λ, and (from Equation 36-8) the energy $E = \hbar^2 k^2/2m$. We have

$$\psi(L) = A\sin kL = 0 \qquad \text{36-11}$$

This condition is satisfied if kL is π or any integer times π, that is, if k is restricted to the values k_n given by

$$k_n = n\frac{\pi}{L}, \qquad n = 1, 2, 3, \ldots \qquad \text{36-12}$$

Substituting this result into Equation 36-8 and solving for E gives us the allowed energy values:

$$E_n = \frac{\hbar^2 k_n^2}{2m} = \frac{\hbar^2}{2m}\left(n\frac{\pi}{L}\right)^2 = n^2\left(\frac{h^2}{8mL^2}\right) = n^2 E_1 \qquad \text{36-13}$$

$$E_1 = \frac{h^2}{8mL^2} \qquad \text{36-14}$$

where we have used $\hbar = h/2\pi$. Equation 36-14 is the same as Equation 17-19, which we obtained by fitting an integral number of half-wavelengths into the box.

For each value of n there is wave function $\psi_n(x)$ given by

$$\psi_n(x) = A_n \sin \frac{n\pi x}{L}$$

36-15

which is the same as Equation 17-22 with the constant $A_n = \sqrt{2/L}$ determined by normalization.[†]

36-2   A Particle in a Finite Square Well

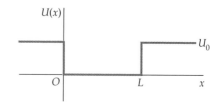

Figure 36-2 The finite square-well potential energy.

The quantization of energy that we found for a particle in an infinite square well is a result that follows from the general solution of the Schrödinger equation for any particle confined to some region of space. We will illustrate this by considering the qualitative behavior of the wave function for a slightly more general potential-energy function, the finite square well shown in Figure 36-2. This potential-energy function is described mathematically by

$$U(x) = U_0, \qquad x < 0$$

$$U(x) = 0, \qquad 0 < x < L$$

36-16

$$U(x) = U_0, \qquad x > L$$

This potential-energy function is discontinuous at $x = 0$ and $x = L$, but it is finite everywhere. The solutions of the Schrödinger equation for this type of potential-energy function depend on whether the total energy E is greater or less than U_0. We will not discuss the case of $E > U_0$, except to remark that in that case the particle is not confined and any value of the energy is allowed; that is, there is no energy quantization. Here we assume that $E < U_0$.

Inside the well, $U(x) = 0$, and the time-independent Schrödinger equation is the same as for the infinite well (Equation 36-7):

$$\frac{d^2\psi(x)}{dx^2} = -\frac{2mE}{\hbar^2}\,\psi(x) = -k^2\psi(x)$$

where $U(x)$ is the potential energy and E is the total energy, with

$$k^2 = \frac{2mE}{\hbar^2}$$

The general solution is of the form

$$\psi(x) = A \sin kx + B \cos kx$$

In this case, $\psi(x)$ is not required to be zero at $x = 0$, so B is not zero. Outside the well, the time-independent Schrödinger equation is

$$\frac{d^2\psi(x)}{dx^2} = \frac{2m}{\hbar^2}(U_0 - E)\psi(x) = \alpha^2\psi(x)$$

36-17

where

$$\alpha^2 = \frac{2m}{\hbar^2}(U_0 - E) > 0$$

36-18

The wave functions and allowed energies for the particle can be found by solving Equation 36-17 for $\psi(x)$ outside the well and then requiring that $\psi(x)$ and $d\psi(x)/dx$ be continuous at the boundaries $x = 0$ and $x = L$. The solution of Equation 36-17 is not difficult (for positive values of x, it is of the form $\psi(x) = Ce^{-\alpha x}$), but applying the boundary conditions involves much tedious

---

[†] See Equation 17-15.

algebra and is not important for our purpose. The important feature of Equation 36-17 is that the second derivative of $\psi(x)$, which is related to the curvature of the wave function, has the same sign as the wave function ψ. If ψ is positive, $d^2\psi/dx^2$ is also positive and the wave function curves away from the axis as shown in Figure 36-3a. Similarly, if ψ is negative, $d^2\psi/dx^2$ is negative and ψ again curves away from the axis as shown in Figure 36-3b. This behavior is very different from that inside the well, where ψ and $d^2\psi/dx^2$ have opposite signs so that ψ always curves toward the axis like a sine or cosine function. Because of this behavior outside the well, for most values of the energy E in Equation 36-17, $\psi(x)$ becomes infinite as x approaches $\pm\infty$; that is, most wave functions $\psi(x)$ are not well behaved outside the well. Though they satisfy the Schrödinger equation, such functions are not proper wave functions because they cannot be normalized. The solutions of the Schrödinger equation are well behaved (that is, approach 0 as $|x|$ becomes very large) only for certain values of the energy. These energy values are the allowed energies for the finite square well.

Figure 36-3 (a) A positive function with positive curvature. (b) A negative function with negative curvature.

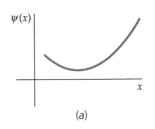

(a)

(b)

Figure 36-4 shows a well-behaved wave function with a wavelength λ_1 inside the well corresponding to the ground-state energy. The behavior of the wave functions corresponding to nearby wavelengths and energies is also shown. Figure 36-5 shows the wave functions and probability distributions for the ground state and first two excited states. From this figure, we can see that the wavelengths inside the well are slightly longer than the corresponding wavelengths for the infinite well (Figure 17-14), so the corresponding energies are slightly less than those for the infinite well. Another feature of the finite-well problem is that there are only a finite number of allowed energies. For very small values of U_0, there is only one allowed energy.

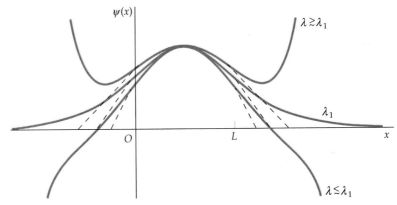

Figure 36-4 Functions satisfying the Schrödinger equation with wavelengths near the wavelength λ_1 corresponding to the ground-state energy $E_1 = \hbar^2/2m\lambda_1^2$ in the finite well. If λ is slightly greater than λ_1, the function approaches infinity, like the function in Figure 36-3a. At the critical wavelength λ_1, the function and its slope approach zero together. If λ is slightly less than λ_1, the function crosses the x axis while the slope is still negative. The slope then becomes more negative because its rate of change $d^2\psi/dx^2$ is now negative. This function approaches negative infinity as x approaches infinity.

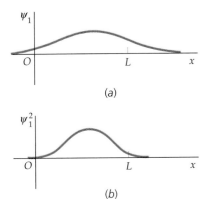

(a)

(b)

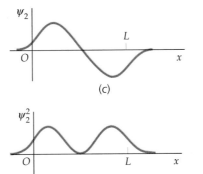

(c)

(d)

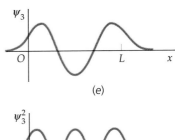

(e)

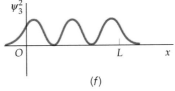

(f)

Figure 36-5 Graphs of the wave functions $\psi_n(x)$ and probability distributions $\psi_n^2(x)$ for $n = 1, 2,$ and 3 for the finite square well. Compare these graphs with those of Figure 17-14 for the infinite square well, where the wave functions are zero at $x = 0$ and $x = L$. The wavelengths here are slightly longer than the corresponding ones for the infinite well, so the allowed energies are somewhat smaller.

Note that the wave function penetrates beyond the edges of the well at $x = L$ and $x = 0$, indicating that there is some small probability of finding the particle in the region in which its total energy E is less than its potential energy U_0. This region is called the *classically forbidden region* because the kinetic energy, $E - U_0$, would be negative when $U_0 > E$. Since negative kinetic energy has no meaning in classical physics, it is interesting to speculate on the result of an attempt to observe the particle in the classically forbidden region. It can be shown from the uncertainty principle that if an attempt is made to localize the particle in the classically forbidden region, such a measurement introduces an uncertainty in the momentum of the particle corresponding to a minimum kinetic energy that is greater than $U_0 - E$. This is just great enough to prevent us from measuring a negative kinetic energy. The penetration of the wave function into a classically forbidden region does have important consequences in barrier penetration, which will be discussed in Section 36-4.

Much of our discussion of the finite-well problem applies to any problem in which $E > U(x)$ in some region and $E < U(x)$ outside that region, as we see in the next section.

36-3   The Harmonic Oscillator

The potential energy for a particle of mass m attached to a spring of force constant K is

$$U(x) = \tfrac{1}{2}Kx^2 = \tfrac{1}{2}m\omega_0^2 x^2 \qquad \text{36-19}$$

where $\omega_0 = \sqrt{K/m}$ is the natural frequency of the oscillator. Classically, the object oscillates between $x = +A$ and $x = -A$. Its total energy is $E = \tfrac{1}{2}m\omega_0^2 A^2$, which can have any positive value, or zero.

This potential-energy function, shown in Figure 36-6, applies to any system undergoing small oscillations about a position of stable equilibrium. For example, it could apply to the oscillations of the atoms of a diatomic molecule such as H_2 or HCl oscillating about their equilibrium separation. In the region $-A \leq x \leq A$ between the classical turning points, the total energy is greater than the potential energy and the Schrödinger equation can be written

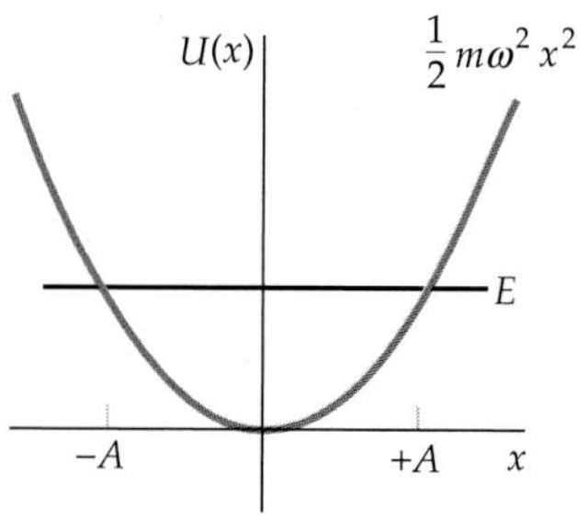

$$\frac{d^2\psi(x)}{dx^2} = -k^2\psi(x) \qquad \text{36-20}$$

Figure 36-6 Harmonic oscillator potential.

where $k^2 = 2m[E - U(x)]/\hbar^2$ now depends on x. The solutions of this equation are no longer simple sine or cosine functions because the wave number $k = 2\pi/\lambda$ now varies with x; but since $d^2\psi/dx^2$ and ψ have opposite signs, ψ will always curve toward the axis and the solutions will oscillate.

Outside the classical turning points ($|x| > A$), the potential energy is greater than the total energy and the Schrödinger equation is similar to Equation 36-17:

$$\frac{d^2\psi(x)}{dx^2} = +\alpha^2\psi(x) \qquad \text{36-21}$$

except that here $\alpha^2 = (2m/\hbar^2)[U(x) - E] > 0$ depends on x. For $|x| > A$, $d^2\psi/dx^2$ and ψ have the same sign, so ψ will curve away from the axis and there will be only certain values of E for which solutions exist that approach zero as x approaches infinity.

For the harmonic oscillator potential energy function, the Schrödinger equation is

$$-\frac{\hbar^2}{2m}\frac{d^2\psi(x)}{dx^2} + \frac{1}{2}m\omega_0^2 x^2\psi(x) = E\psi(x)$$ 36-22

Wave Functions and Energy Levels

Rather than pursue a general solution to the Schrödinger equation for this system, we simply present the solution for the ground state and the first excited state.

The ground-state wave function $\psi_0(x)$ is found to be a Gaussian function centered at the origin:

$$\psi_0(x) = A_0 e^{-ax^2}$$ 36-23

where A_0 and a are constants. This function and the wave function for the first excited state are shown in Figure 36-7.

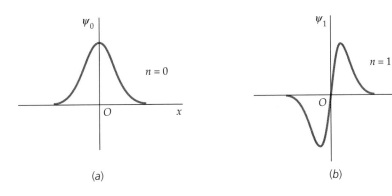

(a)   (b)

Figure 36-7 (a) Ground-state wave function for the harmonic oscillator potential. (b) Wave function for the first excited state of the harmonic oscillator potential.

Example 36-1

Verify that $\psi_0(x) = A_0 e^{-ax^2}$ is a solution of the Schrödinger equation for the harmonic oscillator.

Picture the Problem We take the first and second derivative of ψ with respect to x and substitute into Equation 36-22. Since this is the ground-state wave function, we write E_0 for the energy E.

1. Compute $d\psi_0/dx$:

$$\frac{d\psi_0(x)}{dx} = -2axA_0 e^{-ax^2}$$

2. Compute $d^2\psi_0/dx^2$:

$$\frac{d^2\psi_0(x)}{dx^2} = -2aA_0 e^{-ax^2} + 4a^2x^2A_0 e^{-ax^2}$$

3. Substitute these derivatives into the Schrödinger equation:

$$-\frac{\hbar^2}{2m}\left(-2aA_0 e^{-ax^2} + 4a^2x^2A_0 e^{-ax^2}\right) + \frac{1}{2}m\omega_0^2 x^2A_0 e^{-ax^2}$$

$$= E_0 A_0 e^{-ax^2}$$

4. Cancel the common factor $A_0 e^{-ax^2}$:

$$-\frac{\hbar^2}{2m}\left(-2a + 4a^2x^2\right) + \frac{1}{2}m\omega_0^2 x^2 = E_0$$

5. The equation in step 4 must hold for all x. Substitute $x = 0$:

$$E_0 = \frac{\hbar^2 a}{m}$$

6. Substitute this result into the equation in step 4: $-\dfrac{\hbar^2}{2m}(4a^2x^2) + \dfrac{1}{2}m\omega_0^2 x^2 = 0$

7. The coefficients of the x^2 terms must equal zero: $-\dfrac{\hbar^2}{2m}(4a^2) + \dfrac{1}{2}m\omega_0^2 = 0$

8. Solve for a:

$$a^2 = \dfrac{m^2\omega_0^2}{4\hbar^2}$$

$$a = \dfrac{m\omega_0}{2\hbar}$$

6. Substitute this result into the equation for E_0 in step 5: $E_0 = \dfrac{\hbar^2 a}{m} = \dfrac{1}{2}\hbar\omega_0$

Remark We have shown that the given function satisfies the Schrödinger equation for any value of A_0 as long as the energy E_0 is given by $E_0 = \tfrac{1}{2}\hbar\omega_0$. In general, when we have an equation such as $Ax^3 + Bx^2 + Cx + D = 0$, which must hold for *all values* of x, each coefficient must be zero. That is, $A = B = C = D = 0$.

We see from this example that the ground-state energy is given by

$$E_0 = \dfrac{\hbar^2 a}{m} = \dfrac{1}{2}\hbar\omega_0 \qquad\qquad 36\text{-}24$$

The first excited state has a node in the center of the potential well, just as with the particle in a box.[†] The wave function $\psi_1(x)$ is

$$\psi_1(x) = A_1 x e^{-ax^2} \qquad\qquad 36\text{-}25$$

where $a = m\omega_0/2\hbar$, as in Example 36-1. This function is also shown in Figure 36-7. Substituting $\psi_1(x)$ into the Schrödinger equation as was done for $\psi_0(x)$ in Example 36-1 yields the energy of the first excited state,

$$E_1 = \tfrac{3}{2}\hbar\omega_0$$

In general, the energy of the nth excited state of the harmonic oscillator is

$$E_n = (n + \tfrac{1}{2})\hbar\omega_0 \qquad 36\text{-}26$$

as indicated in Figure 36-8. The fact that the energy levels are evenly spaced by the amount $\hbar\omega_0$ is a peculiarity of the harmonic oscillator potential. As we saw in Chapter 17, the energy levels for a particle in a box, or for the hydrogen atom, are not evenly spaced. The precise spacing of energy levels is closely tied to the particular form of the potential-energy function.

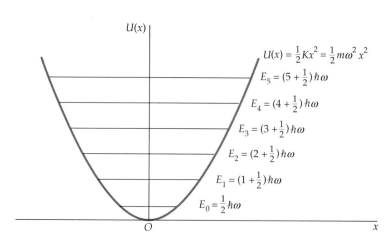

Figure 36-8 Energy levels in the harmonic oscillator potential.

[†] Each higher-energy state has one additional node in the wave function.

36-4   Reflection and Transmission of Electron Waves: Barrier Penetration

In Sections 36-2 and 36-3, we were concerned with bound-state problems in which the potential energy is larger than the total energy for large values of $|x|$. In this section, we consider some simple examples of unbound states for which E is greater than $U(x)$. For these problems, $d^2\psi/dx^2$ and ψ have opposite signs, so $\psi(x)$ curves toward the axis and does not become infinite at large values of $|x|$.

Step Potential

Consider a particle of energy E moving in a region in which the potential energy is the step function

$$U(x) = 0, \quad x < 0$$
$$U(x) = U_0, \quad x > 0$$

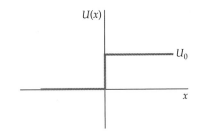

Figure 36-9 Step potential. A classical particle incident from the left, with total energy $E > U_0$, is always transmitted. The change in potential energy at $x = 0$ merely provides an impulsive force that reduces the speed of the particle. A wave incident from the left is partially transmitted and partially reflected because the wavelength changes abruptly at $x = 0$.

as shown in Figure 36-9. We are interested in what happens when a particle moving from left to right encounters the step.

The classical answer is simple. To the left of the step, the particle moves with a speed $v = \sqrt{2E/m}$. At $x = 0$, an impulsive force acts on the particle. If the initial energy E is less than U_0, the particle will be turned around and will then move to the left at its original speed; that is, the particle will be reflected by the step. If E is greater than U_0, the particle will continue to move to the right but with reduced speed given by $v = \sqrt{2(E - U_0)/m}$. We can picture this classical problem as a ball rolling along a level surface and coming to a steep hill of height h given by $mgh = U_0$. If the initial kinetic energy of the ball is less than mgh, the ball will roll part way up the hill and then back down and to the left along the lower surface at its original speed. If E is greater than mgh, the ball will roll up the hill and proceed to the right at a lesser speed.

The quantum-mechanical result is similar when E is less than U_0. Figure 36-10 shows the wave function for the case $E < U_0$. The wave function does not go to zero at $x = 0$ but rather decays exponentially, like the wave function for the bound state in a finite square-well problem. The wave penetrates slightly into the classically forbidden region $x > 0$, but it is eventually completely reflected. This problem is somewhat similar to that of total internal reflection in optics.

For $E > U_0$, the quantum-mechanical result differs markedly from the classical result. At $x = 0$, the wavelength changes abruptly from $\lambda_1 = h/p_1 = h/\sqrt{2mE}$ to $\lambda_2 = h/p_2 = h/\sqrt{2m(E - U_0)}$. We know from our study of waves that when the wavelength changes suddenly, part of the wave is reflected and part is transmitted. Since the motion of an electron (or other particle) is governed by a wave equation, the electron will be sometimes transmitted and sometimes reflected. The probabilities of reflection and transmission can be calculated by solving the Schrödinger equation in each region of space and comparing the amplitudes of the transmitted and reflected waves with that of the incident wave. This calculation and its result are similar to finding the fraction of light reflected from an air–glass interface. If R is the probability of reflection, called the **reflection coefficient,** this calculation gives

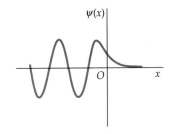

Figure 36-10 When the total energy E is less than U_0, the wave function penetrates slightly into the region $x > 0$. However, the probability of reflection for this case is 1, so no energy is transmitted.

$$R = \frac{(k_1 - k_2)^2}{(k_1 + k_2)^2} \qquad\qquad 36\text{-}27$$

where k_1 is the wave number for the incident wave and k_2 is that for the transmitted wave. This result is the same as that in optics for the reflection of light at normal incidence from the boundary between two media having different indexes of refraction n (Equation 33-11). The probability of transmission T, called the **transmission coefficient,** can be calculated from the reflection coefficient, since the probability of transmission plus the probability of reflection must equal 1:

$$T + R = 1 \qquad\qquad\qquad \text{36-28}$$

Example 36-2

A particle of energy E_0 traveling in a region in which the potential energy is zero is incident on a potential barrier of height $U_0 = 0.2E_0$. Find the probability that the particle will be reflected.

Picture the Problem We need to calculate the wave numbers k_1 and k_2 and use them to calculate the reflection coefficient R from Equation 36-27. The wave numbers are related to the kinetic energy K by $K = p^2/2m = \hbar^2 k^2/2m$.

1. The probability of reflection is the reflection coefficient:

$$R = \frac{(k_1 - k_2)^2}{(k_1 + k_2)^2}$$

2. Calculate k_1 from the initial kinetic energy E_0:

$$\frac{\hbar^2 k_1^2}{2m} = E_0$$

$$k_1 = \sqrt{2mE_0/\hbar^2} = 1.41\sqrt{mE_0/\hbar^2}$$

3. Relate k_2 to the final kinetic energy K_2:

$$\frac{\hbar^2 k_2^2}{2m} = K_2 = E_0 - U_0 = E_0 - 0.2E_0 = 0.8E_0$$

4. Solve for k_2:

$$k_2 = \sqrt{2m(0.8E_0)/\hbar^2} = 1.26\sqrt{mE_0/\hbar^2}$$

5. Substitute these values to calculate R:

$$R = \frac{(k_1 - k_2)^2}{(k_1 + k_2)^2} = \frac{(1.41 - 1.26)^2}{(1.41 + 1.26)^2} = 0.00316$$

Remark The probability of reflection is only 0.3%. This probability is small because the barrier height reduces the kinetic energy by only 20%. Since k is proportional to the square root of the kinetic energy, the wave number and therefore the wave length is changed by only 10%.

Exercise Express the index of refraction n of light in terms of the wave number k, and show that Equation 33-11 for the reflection of light at normal incidence is the same as Equation 36-27.

In quantum mechanics, a localized particle is represented by a wave packet, which has a maximum at the most probable position of the particle. Figure 36-11 shows a wave packet representing a particle of energy E incident on a step potential of height U_0, which is less than E. After the encounter, there are two wave packets. The relative heights of the transmitted packet and reflected packet indicate the relative probabilities of transmission and reflection. For the situation shown here, E is much greater than U_0, and the probability of transmission is much greater than that of reflection.

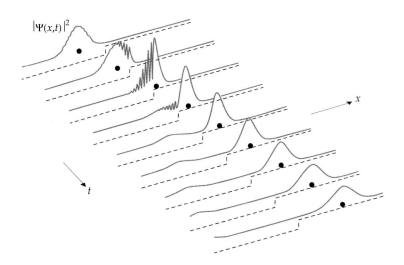

Figure 36-11 Time development of a one-dimensional wave packet representing a particle incident on a step potential for $E > U_0$. The position of a classical particle is indicated by the dot. Note that part of the packet is transmitted and part is reflected.

Barrier Penetration

Figure 36-12*a* shows a rectangular potential barrier of height U_0 and width a given by

$$U(x) = 0, \qquad x < 0$$

$$U(x) = U_0, \qquad 0 < x < a$$

$$U(x) = 0, \qquad x > a$$

We consider a particle of energy E, which is slightly less than U_0, that is incident on the barrier from the left. Classically, the particle would always be reflected. However, a wave incident from the left does not decrease immediately to zero at the barrier but will instead decay exponentially in the classically forbidden region $0 < x < a$. Upon reaching the far wall of the barrier ($x = a$), the wave function must join smoothly to a sinusoidal wave function to the right of the barrier as shown in Figure 36-12*b*. This implies that there is some probability of the particle (which is represented by the wave function) being found on the far side of the barrier even though, classically, it should never pass through the barrier. For the case in which the quantity $\alpha a = \sqrt{2ma^2(U_0 - E)/\hbar^2}$ is much greater than 1, the transmission coefficient is proportional to $e^{-2\alpha a}$:

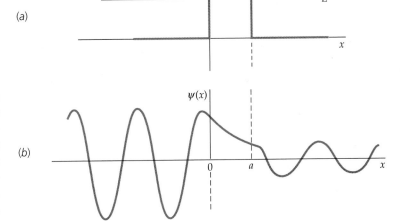

Figure 36-12 (*a*) Rectangular potential barrier. (*b*) Penetration of the barrier by a wave with total energy less than the barrier energy. Part of the wave is transmitted by the barrier even though, classically, the particle cannot enter the region $0 < x < a$ in which the potential energy is greater than the total energy.

$$T \propto e^{-2\alpha a} \qquad\qquad 36\text{-}29$$

Transmission through a barrier

with $\alpha = \sqrt{2m(U_0 - E)/\hbar^2}$. The probability of penetration of the barrier thus decreases exponentially with the barrier thickness a and with the square root of the relative barrier height $(U_0 - E)$.

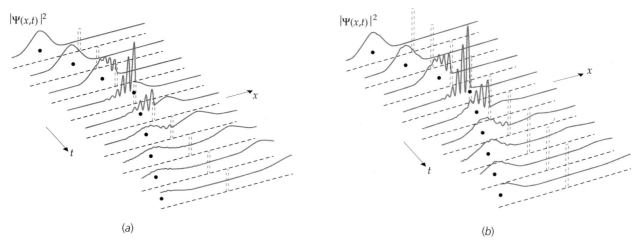

(a)

(b)

Figure 36-13 Barrier penetration. (*a*) A wave packet representing a particle incident on a barrier of height just slightly greater than the energy of the particle. For this particular choice of energies, the probability of transmission is approximately equal to the probability of reflection, as indicated by the relative sizes of the transmitted and reflected packets. (*b*) The same particle incident on a barrier of height much greater than the energy of the particle. A very small part of the packet tunnels through the barrier. In both drawings, the position of a classical particle is indicated by a dot.

Figure 36-13*a* shows a wave packet incident on a potential barrier of height U_0 that is considerably greater than the energy of the particle. The probability of penetration is very small, as indicated by the relative sizes of the reflected and transmitted packets. In Figure 36-13*b* the barrier is just slightly greater than the energy of the particle. In this case the probability of penetration is about the same as the probability of reflection. Figure 36-14 shows a particle incident on two potential barriers of height just slightly greater than the energy of the particle.

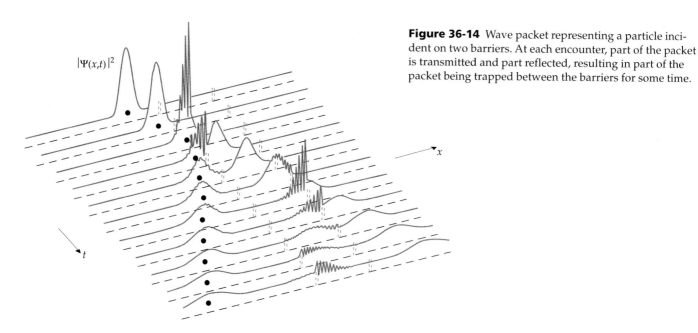

Figure 36-14 Wave packet representing a particle incident on two barriers. At each encounter, part of the packet is transmitted and part reflected, resulting in part of the packet being trapped between the barriers for some time.

As we have mentioned, the penetration of a barrier is not unique to quantum mechanics. When light is totally reflected from a glass–air interface, the light wave can penetrate the air barrier if a second piece of glass is brought within a few wavelengths of the first. This effect can be demonstrated with a laser beam and two 45° prisms (Figure 36-15). Similarly, water waves in a ripple tank can penetrate a gap of deep water (Figure 36-16).

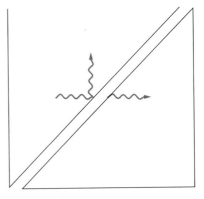

Figure 36-15 Penetration of an optical barrier. If the second prism is close enough to the first, part of the wave penetrates the air barrier even when the angle of incidence in the first prism is greater than the critical angle.

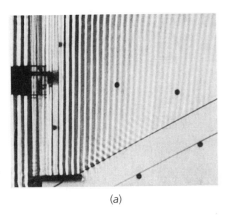
(a)

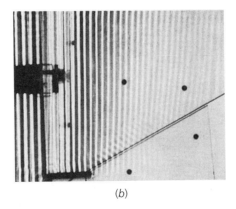
(b)

Figure 36-16 Penetration of a barrier by water waves in a ripple tank. In (a) the waves are totally reflected from a gap of deeper water. When the gap is very narrow, as in (b), a transmitted wave appears.

The theory of barrier penetration was used by George Gamow in 1928 to explain the enormous variation in the half-lives for α decay of radioactive nuclei. (Alpha particles are helium nuclei emitted from larger atoms in radioactive decay; they consist of two protons and two neutrons tightly bound together.) In general, the smaller the energy of the emitted α particle, the longer the half-life. The energies of α particles from natural radioactive sources range from about 4 to 7 MeV, whereas the half-lives range from about 10^{-5} second to 10^{10} years. Gamow represented a radioactive nucleus by a potential well containing an α particle as shown in Figure 36-17. Without knowing very much about the nuclear force that is exerted on the α particle within the nucleus, he represented it by a square well. Just outside the well, the α particle with its charge of $+2e$ is repelled by the nucleus with its charge $+Ze$, where Ze is the remaining nuclear charge. This force is represented by the Coulomb potential energy $+k(2e)(Ze)/r$. The energy E is the measured kinetic energy of the emitted α particle, because when it is far from the nucleus its potential energy is zero. After the α particle is formed inside the radioactive nucleus, it bounces back and forth inside the nucleus, hitting the barrier at the nuclear radius R. Each time it strikes the barrier, there is some small probability of its penetrating and appearing outside the nucleus. We can see from Figure 36-17 that a small increase in E reduces the relative height of the barrier $U - E$ and also its thickness. Because the probability of penetration is so sensitive to the barrier thickness and relative height, a small increase in E leads to a large increase in the probability of transmission and therefore to a shorter lifetime. Gamow was able to derive an expression for the half-life as a function of E that is in excellent agreement with experimental results.

In the **scanning tunneling electron microscope** developed in the 1980s, a thin space between a material specimen and a tiny probe acts as a barrier to electrons bound in the specimen. A small voltage applied between the probe and specimen causes the electrons to *tunnel* through the vacuum separating the two surfaces if the surfaces are close enough together. The tunneling current is extremely sensitive to the size of the gap between the probe and specimen. If a constant tunneling current is maintained as the probe scans the specimen, the surface of the specimen can be mapped out by the motions of the probe. In this way, the surface features of a specimen can be measured with a resolution of the order of the size of an atom.

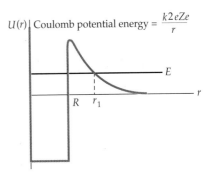

Figure 36-17 Model of a potential-energy function for an α particle in a radioactive nucleus. The strong attractive nuclear force when r is less than the nuclear radius R can be approximately described by the potential well shown. Outside the nucleus the nuclear force is negligible, and the potential is given by Coulomb's law, $U(r) = +k(2e)(Ze)/r$, where Ze is the nuclear charge and $2e$ is the charge of the α particle.

xploring

The Scanning Tunneling Microscope

Ellen Williams
University of Maryland

Although all scientists believe that matter consists of atoms, the evidence that atoms exist is mostly indirect. Three methods are known for actually imaging individual atoms: transmission electron microscopy, field ion microscopy, and scanning tunneling microscopy. Of these, the last is the most recently developed and the most versatile. Gert Binnig and Heinrich Rohrer were awarded the 1986 Nobel Prize in Physics (which they shared with a third scientist) for developing this technique.

To understand scanning tunneling microscopy (STM), we first need to look at the behavior of electrons in metals. Classically, the most weakly bound electrons—those at the Fermi energy—can never leave the metal unless they are given the energy necessary to go over the potential barrier represented by their work function ϕ. Quantum mechanically, however, electrons near the Fermi energy can *tunnel* through the potential barrier. By placing two pieces of metal close to one another, as shown in Figure 1, a finite square-well barrier can be created. The probability that electrons at the Fermi energy will tunnel through the barrier is proportional to $e^{-\alpha a}$, where a is the distance separating the two pieces of metal, and α depends on the barrier height—in this case, the work function. This exponential dependence of the transmission probability on separation is what makes STM possible.

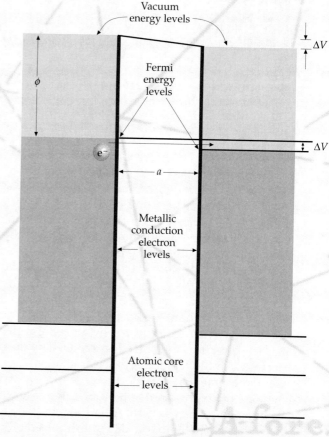

Figure 1 Energy-level diagram for the conduction electrons in two metals separated by a distance a. The electrons move as essentially free particles within the metal. The most weakly bound electrons have the highest energy (the Fermi energy E_F) and are held within the solid by a potential-energy barrier known as the work function ϕ. To induce a measurable electron tunneling current, a small voltage difference ΔV is applied between the two metals. This offsets their energy levels as shown, so that electrons can flow from the occupied states near E_F on the more negative metal (on the left in the figure) to the unoccupied states just above E_F on the more positive metal (on the right in the figure).

Ellen Williams earned her BS in chemistry at Michigan State University and her doctorate at the California Institute of Technology, studying the properties of thin layers of molecules on surfaces. Since then, she has been at the Department of Physics and Astronomy at the University of Maryland.

The mechanism of STM is illustrated in Figure 2. If a pointed metal probe is placed sufficiently close to a sample and a small voltage is applied between the probe and the sample, electron tunneling can occur. The net flow of electrons can be measured as a tunneling current proportional to the transmission probability. If we then scan the probe back and forth above the sample, any bumps on the sample surface will change the separation. Because of the exponential relationship between the separation and the transmission probability, changes as small as 0.01 nm result in measurable changes in the tunneling current. Measurement of the tunneling current while scanning thus generates a topographic map of the surface. In principle, it is possible to image individual atoms on the surface using STM.

Three formidable problems had to be overcome before individual atoms could be imaged. Vibration was an important challenge, because the separation between the sample and the probe must be very small—typically, only a few nanometers (comparable to the size of atoms). A minor perturbation can jam the probe into the sample, ruining the experiment. The most common problem is floor vibration, typically with an amplitude of about 1 μm—a thousand times larger than the allowable separation between tip and sample. Thus, very careful engineering is required.

The second problem is probe sharpness, which determines how small a structure can be imaged. Electrochemical etching can sharpen the end of a metal wire to a radius of about 1 μm (1000 nm). A probe with such a large surface area would allow tunneling to occur over too large a region of the sample surface. To resolve small features such as atoms, the probe must be comparable in size to the features. In trying to fabricate such a probe, we are rewarded for our technological shortcomings. Our most polished wires are rough on an atomic scale, having mini-tips much like the one illustrated in Figure 2. The end of such a mini-tip will present one or a few atoms close to the surface. The exponential dependence of transmission probability on separation then guarantees that tunneling can occur preferentially from the end of the mini-tip.

The third problem in STM is position control. How can the probe be moved around to an accuracy of less than 0.1 nm? The answer lies in a type of material known as piezoelectric ceramic, which expands and contracts when an external voltage is applied to electrodes on opposite faces. Typically, expansions are on the order of a few tenths of a nanometer per volt. A probe attached to a piece of piezoelectric ceramic can be moved with great precision.

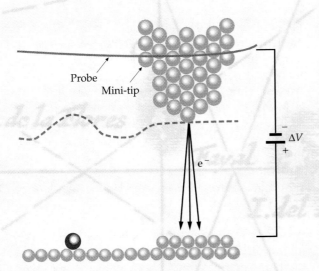

Probe
Mini-tip

ΔV

e^-

Figure 2 Schematic illustration of the path of a probe (dashed line) scanned across a surface while constant tunneling current is maintained. If the probe is very large (as illustrated by the solid line), tunneling occurs over a large area and atomic features cannot be resolved. However, if the probe has a mini-tip of atomic dimensions, then tunneling occurs into a small area, allowing very small features (even individual atoms) to be imaged.

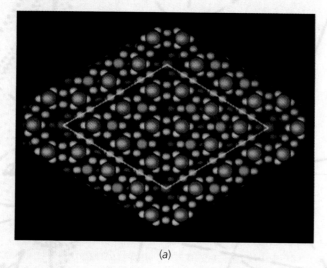

(a)

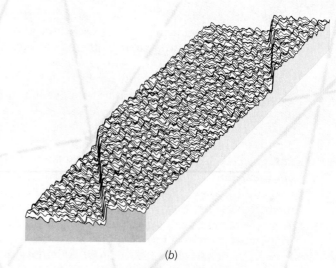

(b)

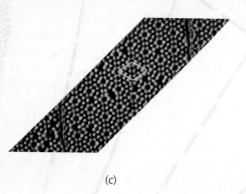

(c)

Figure 3 STM imaging of atoms and steps on a silicon surface. (*a*) Model of the atoms on a silicon surface. The red circles represent atoms that protrude highest above the surface, the blue circles are atoms in a lower layer, and the gray circles are atoms in a still lower layer. The white diamond shows the repeating unit of the structure. The length of each side of this unit cell is 2.7 nm. (*b*) Traces of height versus position (line scans) obtained by an STM over an area of approximately 10 by 35 nm. (*c*) The same data as presented in (*b*) in a gray-scale representation, allowing the imaging of the highest layer of atoms in the atomic model of (*a*) to be immediately recognized. The unit cell of the structure is indicated for comparison with (*a*).

The power of STM is illustrated in Figure 3, where an atomic model of a silicon surface is compared with STM images of the real surface. The data from an STM scan consist of values of surface height versus position and can be presented as a line scan (the dashed line in Figure 2). The line scan image of Figure 3*b* is easier to visualize if the data are represented by a gray scale (Figure 3*c*): Height is represented by intensity of color—ranging from white for the highest points to black for the lowest. Note the striking correspondence to the model in Figure 3*a*. The deep holes correspond to the positions of missing atoms in the model, and the bright spots are due to the atoms that protrude above the average surface plane. Two abrupt changes in height also appear in this image; these surface steps are important in practical processes such as crystal growth and microfabrication.

In addition to studies of atoms at surfaces, STM has a range of practical applications, in part because STM is quite insensitive to its microscopic environment. Tunneling microscopes operate in vacuum, air, liquid helium, oil, water, and even electrolytic solutions. Thus, STM is useful, for example, in imaging DNA in a biological environment and observing the surfaces of battery electrodes while they are operating. Variations of STM have also been developed that can image samples that are not conductors (atomic force microscopy) or image magnetic properties at surfaces. Most stunning is the ability of STM to write with atomic resolution. At the ultimate limit, the probe has been used to pull *individual atoms* of xenon around on a surface (Figure 4).

Scanning tunneling microscopy is a practical demonstration of quantum mechanics and an illustration that understanding basic concepts of physics can yield tremendous gains in advanced technology. It is also an object lesson in the long-term and often unforeseeable benefits that accrue from developing fundamental ideas.

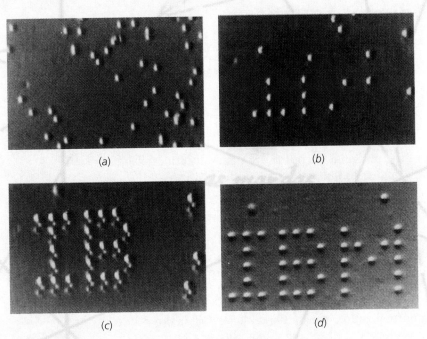

Figure 4 Sequence of STM images taken during the construction of a patterned array of xenon atoms on a nickel surface at a temperature of 4 K. Xenon atoms were allowed to stick randomly on the surface from the gas phase (upper left). The STM tip was then used to "nudge" the atoms one by one across the surface to spell out the name of the company that sponsored the development of STM.

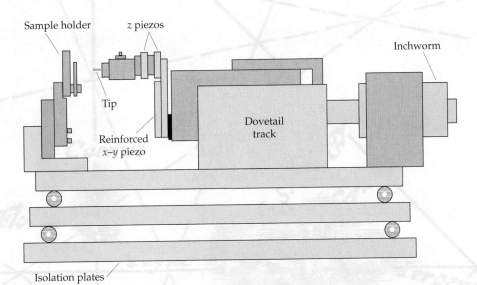

Figure 5 Schematic diagram of an STM. The sample holder is rigidly mounted to the top plate in a stack of isolation plates. The tip is fixed onto the x, y, z scanning piezos. To position the tip so tunneling can occur, these piezos are mounted on a heavy block that can slide on a dovetail track. The block is then pushed by an electronically controlled "walker" device called an inchworm, which can move forward and back in steps of 4 nm.

36-5   The Schrödinger Equation in Three Dimensions

The one-dimensional time-independent Schrödinger equation is easily extended to three dimensions. In rectangular coordinates, it is

$$-\frac{\hbar^2}{2m}\left(\frac{\partial^2 \psi}{\partial x^2} + \frac{\partial^2 \psi}{\partial y^2} + \frac{\partial^2 \psi}{\partial z^2}\right) + U\psi = E\psi \qquad \text{36-30}$$

where the wave function ψ and the potential energy U are generally functions of all three coordinates, x, y, and z. To illustrate some of the features of problems in three dimensions, we consider a particle in a three-dimensional infinite square well given by $U(x, y, z) = 0$ for $0 < x < L$, $0 < y < L$, and $0 < z < L$. Outside this cubical region $U(x, y, z) = \infty$. For this problem, the wave function must be zero at the edges of the well.

There are standard methods in partial differential equations for solving Equation 36-30. We can guess the form of the solution from our knowledge of probability. For a one-dimensional box along the x axis, we have found the probability that a particle is in the region dx at x to be $A_1^2 \sin^2 k_1 x \, dx$ (from Equation 36-10), where A_1 is a normalization constant and $k_1 = n\pi/L$ is the wave number. Similarly, for a box along the y axis, the probability of a particle being in a region dy at y is $A_2^2 \sin^2 k_2 y \, dy$. The probability of two independent events occurring is the product of the probabilities of each event occurring.[†] So the probability of a particle being in region dx at x *and* in region dy at y is $A_1^2 \sin^2 k_1 x \, dx \, A_2^2 \sin^2 k_2 y \, dy = A_1^2 \sin^2 k_1 x \, A_2^2 \sin^2 k_2 y \, dx \, dy$. The probability of a particle being in the region dx, dy, and dz is $\psi(x, y, z) \, dx \, dy \, dz$, where $\psi(x, y, z)$ is the solution of Equation 36-30. This solution is of the form

$$\psi(x, y, z) = A \sin k_1 x \sin k_2 y \sin k_3 z \qquad \text{36-31}$$

where the constant A is determined by normalization. Inserting this solution into Equation 36-30, we obtain for the energy

$$E = \frac{\hbar^2}{2m}(k_1^2 + k_2^2 + k_3^2)$$

which is equivalent to $E = (p_x^2 + p_y^2 + p_z^2)/2m$, with $p_x = \hbar k_1$, and so on. The wave function will be zero at $x = L$ if $k_1 = n_1\pi/L$, where n_1 is an integer. Similarly, the wave function will be zero at $y = L$ if $k_2 = n_2\pi/L$, and it will be zero at $z = L$ if $k_3 = n_3\pi/L$. The energy is thus quantized to the values

$$E_{n_1, n_2, n_3} = \frac{\hbar^2 \pi^2}{2mL^2}(n_1^2 + n_2^2 + n_3^2) = E_1(n_1^2 + n_2^2 + n_3^2) \qquad \text{36-32}$$

where n_1, n_2, and n_3 are integers and E_1 is the ground-state energy of the one-dimensional well. Note that the energy and wave function are characterized by three quantum numbers, each arising from a boundary condition for one of the coordinates.

The lowest energy state (the ground state) for the cubical well occurs when $n_1 = n_2 = n_3 = 1$ and has the value

$$E_{1,1,1} = \frac{3\hbar^2 \pi^2}{2mL^2} = 3E_1$$

[†] For example, if you throw two dice, the probability of the first die coming up 6 is 1/6 and the probability of the second die coming up an odd number is 1/2. The probability of the first die coming up 6 *and* the second die coming up an odd number is (1/6)(1/2) = 1/12.

The first excited energy level can be obtained in three different ways: $n_1 = 2$, $n_2 = n_3 = 1$; $n_2 = 2$, $n_1 = n_3 = 1$; or $n_3 = 2$, $n_1 = n_2 = 1$. Each has a different wave function. For example, the wave function for $n_1 = 2$ and $n_2 = n_3 = 1$ is

$$\psi_{2,1,1} = A \sin \frac{2\pi x}{L} \sin \frac{\pi y}{L} \sin \frac{\pi z}{L} \qquad\qquad 36\text{-}33$$

There are thus three different quantum states as described by the three different wave functions corresponding to the same energy level. An energy level with which more than one wave function is associated is said to be **degenerate**. In this case, there is threefold degeneracy. Degeneracy is related to the spatial symmetry of the problem. If, for example, we consider a noncubic well where $U = 0$ for $0 < x < L_1$, $0 < y < L_2$, and $0 < z < L_3$, the boundary conditions at the edges would lead to the quantum conditions $k_1 L_1 = n_1 \pi$, $k_2 L_2 = n_2 \pi$, and $k_3 L_3 = n_3 \pi$, and the total energy would be

$$E_{n_1,n_2,n_3} = \frac{\hbar^2 \pi^2}{2m}\left(\frac{n_1^2}{L_1^2} + \frac{n_2^2}{L_2^2} + \frac{n_3^2}{L_3^2}\right) \qquad\qquad 36\text{-}34$$

These energy levels are not degenerate if L_1, L_2, and L_3 are not equal. Figure 36-18 shows the energy levels for the ground state and first two excited states for an infinite cubic well in which the excited states are degenerate and for a noncubic infinite well in which L_1, L_2, and L_3 are slightly different so that the excited levels are slightly split apart and the degeneracy is removed.

$L_1 = L_2 = L_3$

$E_{1,2,2} = E_{2,1,2} = E_{2,2,1} = 9E_1$ ————

$E_{2,1,1} = E_{1,2,1} = E_{1,1,2} = 6E_1$ ————

$E_{1,1,1} = 3E_1$ ————

(a)

$L_1 < L_2 < L_3$

———— $E_{2,2,1}$
———— $E_{2,1,2}$
———— $E_{1,2,2}$

———— $E_{2,1,1}$
———— $E_{1,2,1}$
———— $E_{1,1,2}$

————

(b)

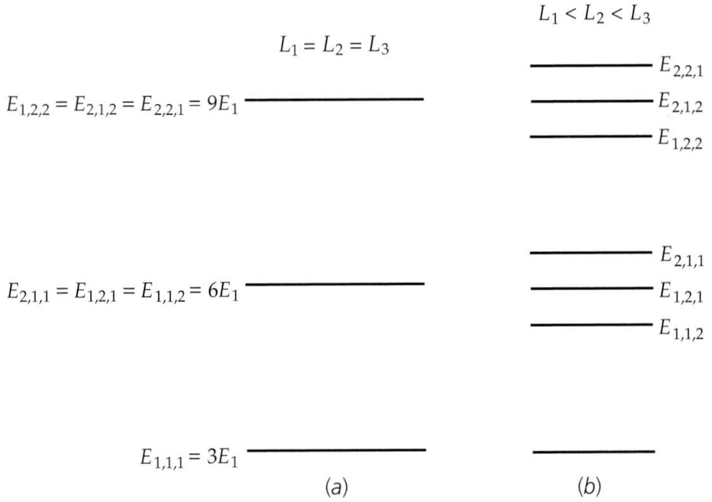

Figure 36-18 Energy-level diagrams for (a) a cubic infinite well and (b) a noncubic infinite well. In (a) the energy levels are degenerate; that is, there are two or more wave functions having the same energy. The degeneracy is removed when the symmetry of the potential is removed, as in (b).

Example 36-3

A particle is in a three-dimensional box with $L_3 = L_2 = 2L_1$. Give the quantum numbers n_1, n_2, n_3 that correspond to the lowest ten quantum states of this box.

Picture the Problem We can use Equation 36-34 to write the energies in terms of the ratios $L_2/L_1 = 2$ and $L_3/L_1 = 2$, then find by inspection the values of the quantum numbers that give the lowest energies.

1. The energy of a level is given by Equation 36-34:

$$E_{n_1,n_2,n_3} = \frac{\hbar^2 \pi^2}{2m}\left(\frac{n_1^2}{L_1^2} + \frac{n_2^2}{L_2^2} + \frac{n_3^2}{L_3^2}\right)$$

2. Factor out $1/L_1^2$:

$$E_{n_1,n_2,n_3} = \frac{\hbar^2 \pi^2}{2mL_1^2}\left(n_1^2 + n_2^2 \frac{L_1^2}{L_2^2} + n_3^2 \frac{L_1^2}{L_3^2}\right)$$

$$= E_1(n_1^2 + n_2^2/4 + n_3^2/4)$$

3. The lowest energy is $E_{1,1,1}$:

$$E_{1,1,1} = E_1 (1^2 + 1^2/4 + 1^2/4) = 1.5E_1 \quad (\text{1st})$$

4. The energy increases the least when we increase n_2 or n_3. Try various values of the quantum numbers:

$$E_{1,2,1} = E_{1,1,2} = E_1(1^2 + 2^2/4 + 1/4) = 2.25E_1 \quad \text{(2nd and 3rd)}$$

$$E_{1,2,2} \qquad\quad = E_1(1^2 + 2^2/4 + 2^2/4) = 3.0E_1 \quad \text{(4th)}$$

$$E_{1,3,1} = E_{1,1,3} = E_1(1^2 + 3^2/4 + 1^2/4) = 3.50E_1 \quad \text{(5th and 6th)}$$

$$E_{1,3,2} = E_{1,2,3} = E_1(1^2 + 3^2/4 + 2^2/4) = 4.25E_1 \quad \text{(7th and 8th)}$$

$$E_{2,1,1} \qquad\quad = E_1(2^2 + 1^2/4 + 1^2/4) = 4.5E_1 \quad \text{(9th)}$$

$$\left.\begin{array}{l} E_{2,2,1} = E_{2,1,2} = E_1(2^2 + 2^2/4 + 1^2/4) = 5.25E_1 \\ E_{1,4,1} = E_{1,1,4} = E_1(1^2 + 4^2/4 + 1^2/4) = 5.25E_1 \end{array}\right\} \quad (10^{\text{th}}, 11^{\text{th}}, 12^{\text{th}}, \text{and } 13^{\text{th}})$$

Remarks Note the degeneracy of the levels.

Exercise Find the quantum numbers and energies of the next four levels in step 4. (Answer $E_{1,3,3} = 5.5E_1$, $E_{1,4,2} = E_{1,2,4} = E_{2,2,2} = 6.0E_1$)

| Example 36-4 | *try it yourself* |

Write the degenerate wave functions for the 4th and 5th excited states (levels 5 and 6) of the results in step 4 of Example 36-3

Picture the Problem Use Equation 36-33 with $k_i = n_i\pi/L$.

Cover the column to the right and try these on your own before looking at the answers.

Steps   *Answers*

1. Write the wave functions corresponding to the energies $E_{1,3,1}$ and $E_{1,1,3}$.

$$\psi_{1,3,1} = A \sin\frac{\pi x}{L} \sin\frac{3\pi y}{L} \sin\frac{\pi z}{L}$$

$$\psi_{1,1,3} = A \sin\frac{\pi x}{L} \sin\frac{\pi y}{L} \sin\frac{3\pi z}{L}$$

36-6   The Schrödinger Equation for Two Identical Particles

Our discussion of quantum mechanics has thus far been limited to situations in which a single particle moves in some force field characterized by a potential-energy function U. The most important physical problem of this type is the hydrogen atom, in which a single electron moves in the Coulomb potential of the proton nucleus. This problem is actually a two-body problem since the proton also moves in the field of the electron. However, the motion of the much more massive proton requires only a very small correction to the energy of the atom that is easily made in both classical and quantum mechanics. When we consider more complicated problems, such as the helium atom, we must apply quantum mechanics to two or more electrons moving in an external field. Such problems are complicated by the interaction of the electrons with each other and also by the fact that the electrons are identical.

The interaction of two electrons with each other is electromagnetic and is essentially the same as the classical interaction of two charged particles. The Schrödinger equation for an atom with two or more electrons cannot be solved exactly, so approximation methods must be used. This is not very different from the situation in classical problems with three or more particles. How-ever, the complications arising from the identity of electrons are purely

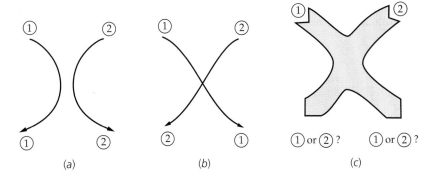

(a)   (b)   (c)

Figure 36-19 Two possible classical electron paths (*a* and *b*). If electrons were classical particles they could be distin-guished by the path followed. But because of the quantum-mechanical wave proper-ties of electrons, the paths are spread out, as indicated by the shaded region in (*c*). It is impossible to distinguish which electron is which after they separate.

quantum mechanical and have no classical counterpart. They are due to the fact that it is impossible to keep track of which electron is which. Classically, identical particles can be identified by their positions, which can be deter-mined with unlimited accuracy. This is impossible quantum mechanically because of the uncertainty principle. Figure 36-19 offers a schematic illustra-tion of the problem.

The indistinguishability of identical particles has important consequences. For instance, consider the very simple case of two identical, noninteracting particles in a one-dimensional infinite square well. The time-independent Schrödinger equation for two particles, each of mass m, is

$$-\frac{\hbar^2}{2m}\frac{\partial^2 \psi(x_1, x_2)}{\partial x_1^2} - \frac{\hbar^2}{2m}\frac{\partial^2 \psi(x_1, x_2)}{\partial x_2^2} + U\psi(x_1, x_2) = E\psi(x_1, x_2) \qquad 36\text{-}35$$

where x_1 and x_2 are the coordinates of the two particles. If the particles inter-act, the potential energy U contains terms with both x_1 and x_2 that cannot be separated into separate terms containing only x_1 or x_2. For example, the elec-trostatic repulsion of two electrons in one dimension is represented by the potential energy $ke^2/|x_2 - x_1|$. However, if the particles do not interact (as we are assuming here), we can write $U = U_1(x_1) + U_2(x_2)$. For the infinite square well, we need only solve the Schrödinger equation inside the well where $U = 0$, and require that the wave function be zero at the walls of the well. With $U = 0$, Equation 36-35 looks just like the expression for a particle in a two-dimensional well (Equation 36-30, with no z and with y replaced by x_2).

Solutions of this equation can be written in the form[†]

$$\psi_{n,m} = \psi_n(x_1)\psi_m(x_2) \qquad 36\text{-}36$$

where ψ_n and ψ_m are the single-particle wave functions for a particle in an in-finite well and n and m are the quantum numbers of particles 1 and 2, respec-tively. For example, for $n = 1$ and $m = 2$, the wave function is

$$\psi_{1,2} = A \sin \frac{\pi x_1}{L} \sin \frac{2\pi x_2}{L} \qquad 36\text{-}37$$

The probability of finding particle 1 in dx_1 *and* particle 2 in dx_2 is $\psi_{n,m}^2(x_1, x_2)\, dx_1\, dx_2$, which is just the product of the separate probabilities $\psi_n^2(x_1)\, dx_1$ and $\psi_m^2(x_2)\, dx_2$. However, even though we have labeled the par-ticles 1 and 2, we cannot distinguish which is in dx_1 and which is in dx_2 if they are identical. The mathematical descriptions of identical particles must be the same if we interchange the labels. The probability density $\psi^2(x_1, x_2)$ must therefore be the same as $\psi^2(x_2, x_1)$:

$$\psi^2(x_2, x_1) = \psi^2(x_1, x_2) \qquad 36\text{-}38$$

[†] Again, this result can be obtained by solving Equation 36-35, but it also can be understood in terms of our knowledge of probability. The probability of electron 1 being in region dx_1 *and* electron 2 being in region dx_2 is the product of the individual probabilities.

Equation 36-38 is satisfied if $\psi(x_2, x_1)$ is either **symmetric** or **antisymmetric** on the exchange of particles—that is, if either

$$\psi(x_2, x_1) = \psi(x_1, x_2) \qquad \text{symmetric} \qquad\qquad 36\text{-}39$$

or

$$\psi(x_2, x_1) = -\psi(x_1, x_2) \qquad \text{antisymmetric} \qquad\qquad 36\text{-}40$$

Note that the wave functions given by Equations 36-36 and 36-37 are neither symmetric nor antisymmetric. If we interchange x_1 and x_2 in these wave functions, we get a different wave function, which implies that the particles can be distinguished.

We can find symmetric and antisymmetric wave functions that are solutions of the Schrödinger equation by adding or subtracting $\psi_{n,m}$ and $\psi_{m,n}$. Adding them, we obtain

$$\psi_S = A'[\psi_n(x_1)\psi_m(x_2) + \psi_n(x_2)\psi_m(x_1)] \quad \text{symmetric} \qquad\qquad 36\text{-}41$$

and subtracting them, we obtain

$$\psi_A = A'[\psi_n(x_1)\psi_m(x_2) - \psi_n(x_2)\psi_m(x_1)] \quad \text{antisymmetric} \qquad\qquad 36\text{-}42$$

For example, the symmetric and antisymmetric wave functions for the first excited state of two identical particles in an infinite square well would be

$$\psi_S = A'\left(\sin\frac{\pi x_1}{L}\sin\frac{2\pi x_2}{L} + \sin\frac{\pi x_2}{L}\sin\frac{2\pi x_1}{L}\right) \qquad\qquad 36\text{-}43$$

and

$$\psi_A = A'\left(\sin\frac{\pi x_1}{L}\sin\frac{2\pi x_2}{L} - \sin\frac{\pi x_2}{L}\sin\frac{2\pi x_1}{L}\right) \qquad\qquad 36\text{-}44$$

There is an important difference between antisymmetric and symmetric wave functions. If $n = m$, the antisymmetric wave function is identically zero for all values of x_1 and x_2 whereas the symmetric wave function is not. Thus, if the wave function describing two identical particles is antisymmetric, the quantum numbers n and m of two particles cannot be the same. This is an example of the **Pauli exclusion principle,** which was first stated by Wolfgang Pauli for electrons in an atom:

> No two electrons in an atom can have the same quantum numbers.

Pauli exclusion principle

It is found that electrons, protons, neutrons, and some other particles have antisymmetric wave functions and obey the Pauli exclusion principle. These particles are called **fermions.** Other particles, such as α particles, deuterons, photons, and mesons, have symmetric wave functions and do not obey the Pauli exclusion principle. These particles are called **bosons.**

Summary

1. The Schrödinger equation is a differential equation that relates the second spatial derivative of a wave function to its first time derivative. Wave functions that describe physical situations are solutions of this differential equation.

2. Because a wave function must be normalizable, it must be well behaved; that is, it must approach zero as x approaches infinity. For bound systems such as a particle in a

box, a simple harmonic oscillator, or an electron in an atom, this requirement leads to energy quantization.

3. The well-behaved wave functions for bound systems describe standing waves.

Topic	Remarks and Relevant Equations

1. Time-Independent Schrödinger Equation

$$-\frac{\hbar^2}{2m}\frac{d^2\psi(x)}{dx^2} + U(x)\psi(x) = E\psi(x) \qquad \text{36-4}$$

Allowable solutions

In addition to satisfying the Schrödinger equation, a wave function $\psi(x)$ must be continuous and (if U is not infinite) must have a continuous first derivative $d\psi/dx$. Because the probability of finding an electron somewhere must be 1, the wave function must obey the normalization condition

$$\int_{-\infty}^{\infty} |\psi|^2\, dx = 1$$

This condition implies the boundary condition that ψ must approach 0 as x approaches $\pm\infty$. Such boundary conditions lead to the quantization of energy.

2. Confined Particles

When the total energy E is greater than the potential energy $U(x)$ in some region (the classically allowed region) and less than $U(x)$ outside that region, the wave function oscillates within the classically allowed region and increases or decreases exponentially outside that region. The wave function approaches zero as x approaches ∞ only for certain values of the total energy E. The energy is thus quantized.

Finite square well

In a finite well of height U_0, there are only a finite number of allowed energies, and these are slightly less than the corresponding energies in an infinite well.

Simple harmonic oscillator

In the oscillator potential energy function $U(x) = \frac{1}{2}m\omega_0^2 x^2$, the allowed energies are equally spaced and given by

$$E_n = (n + \tfrac{1}{2})\hbar\omega_0 \qquad \text{36-26}$$

The ground-state wave function is given by

$$\psi_0(x) = A_0 e^{-ax^2} \qquad \text{36-23}$$

where A_0 is the normalization constant and $a = m\omega_0/2\hbar$.

3. Reflection and Barrier Penetration

When the potential changes abruptly over a small distance, a particle may be reflected even though $E > U(x)$. A particle may penetrate a region in which $E < U(x)$. Reflection and penetration of electron waves are similar to those for other kinds of waves.

4. Schrödinger Equation in Three Dimensions

The wave function for a particle in a three-dimensional box can be written

$$\psi(x\,y, z) = \psi_1(x)\psi_2(y)\psi_3(z)$$

where ψ_1, ψ_2, and ψ_3 are wave function for a one-dimensional box.

Degeneracy

When more than one wave function is associated with the same energy level, the energy level is said to be degenerate. Degeneracy arises because of spatial symmetry.

5. Schrödinger Equation for Two Identical Particles

A wave function that describes two identical particles must be either symmetric or antisymmetric when the coordinates of the particles are exchanged. Fermions, which include electrons, protons, and neutrons, are described by antisymmetric wave functions and obey the Pauli exclusion principle, which states that no two particles can have the same values for their quantum number. Bosons, which include α particles, deuterons, photons, and mesons, have symmetric wave functions and do not obey the Pauli exclusion principle.

Problem-Solving Guide

Begin by drawing a neat diagram that includes the important features of the problem.

Summary of Worked Examples

Type of Calculation	Procedure and Relevant Examples
1. Harmonic Oscillator	
Verify that a given function satisfies the Schrödinger equation	Take the first and second derivative with respect to x and substitute into the Schrödinger equation. **Example 36-1**
2. Transmission and Reflection	
Calculate the probability of reflection or transmission when $E > U_0$.	Calculate the reflection coefficient given by Equation 36-27. For transmission use $R + T = 1$ **Example 36-2**
3. Schrödinger Equation in Three Dimensions	
Find the quantum numbers, energies, and wave functions of the lowest states in a three-dimensional square well.	Write the energies in terms of the quantum numbers and try the various combinations of the smallest quantum numbers. The wave functions are determined by the quantum numbers. **Examples 36-3, 36-4**

Problems

Conceptual Problems

Problems from Optional and Exploring sections

In a few problems, you are given more data than you actually need; in a few other problems, you are required to supply data from your general knowledge, outside sources, or informed estimates.

- • Single-concept, single-step, relatively easy
- •• Intermediate-level, may require synthesis of concepts
- ••• Challenging, for advanced students

A Particle in a Finite Square Well

1 • True or false: Boundary conditions on the wave function lead to energy quantization.

2 • Sketch (a) the wave function and (b) the probability distribution for the $n = 4$ state for the finite square-well potential.

3 • Sketch (a) the wave function and (b) the probability distribution for the $n = 5$ state for the finite square-well potential.

The Harmonic Oscillator

4 •• Show that the expectation value $<x> = \int x|\psi|^2\, dx$ is zero for both the ground and the first excited states of the harmonic oscillator.

5 •• Use the procedure of Example 36-1 to verify that the energy of the first excited state of the harmonic oscillator is $E_1 = \frac{3}{2}\hbar\omega_0$. (*Note*: Rather than solve for a again, use the result $a = m\omega_0/2\hbar$ obtained in Example 36-1.)

6 ••• Show that the normalization constant A_0 of Equation 36-23 is $A_0 = (2m\omega_0/h)^{1/4}$.

7 ••• Find the normalization constant A_1 for the wave function of the first excited state of the harmonic oscillator, Equation 36-25.

8 ••• Find the expectation value $<x^2> = \int x^2|\psi|^2\, dx$ for the ground state of the harmonic oscillator. Use it to show that the average potential energy equals half the total energy.

9 ••• Verify that $\psi_1(x) = A_1 x e^{-ax^2}$ is the wave function corresponding to the first excited state of a harmonic oscillator by substituting it into the time-independent Schrödinger equation and solving for a and E.

10 ••• Find the expectation value $<x^2> = \int x^2|\psi|^2\, dx$ for the first excited state of the harmonic oscillator.

11 ••• Classically, the average kinetic energy of the harmonic oscillator equals the average potential energy. We may assume that this is also true for the quantum mechanical harmonic oscillator. Use this condition to determine the expectation value of p^2 for the ground state of the harmonic oscillator.

12 ••• We know that for the classical harmonic oscillator, $p_{av} = 0$. It can be shown that for the quantum mechanical harmonic oscillator $<p> = 0$. Use the results of Problems 4, 6, and 11 to determine the uncertainty product $\Delta x \, \Delta p$ for the ground state of the harmonic oscillator.

Reflection and Transmission of Electron Waves: Barrier Penetration

13 •• A free particle of mass m with wave number k_1 is traveling to the right. At $x = 0$, the potential jumps from zero to U_0 and remains at this value for positive x. (a) If the total energy is $E = \hbar^2 k_1^2 / 2m = 2U_0$, what is the wave number k_2 in the region $x > 0$? Express your answer in terms of k_1 and in terms of U_0. (b) Calculate the reflection coefficient R at the potential step. (c) What is the transmission coefficient T? (d) If one million particles with wave number k_1 are incident upon the potential step, how many particles are expected to continue along in the positive x direction? How does this compare with the classical prediction?

14 •• Suppose that the potential jumps from zero to $-U_0$ at $x = 0$ so that the free particle speeds up instead of slowing down. The wave number for the incident particle is again k_1, and the total energy is $2U_0$. (a) What is the wave number for the particle in the region of positive x? (b) Calculate the reflection coefficient R at $x = 0$. (c) What is the transmission coefficient T? (d) If one million particles with wave number k_1 are incident upon the potential step, how many particles are expected to continue along in the positive x direction? How does this compare with the classical prediction?

15 •• Work Problem 13 for the case in which the energy of the incident particle is $1.01U_0$ instead of $2U_0$.

16 •• A particle of energy E approaches a step barrier of height U. What should be the ratio E/U so that the reflection coefficient is $\frac{1}{2}$?

17 •• Use Equation 36-29 to calculate the order of magnitude of the probability that a proton will tunnel out of a nucleus in one collision with the nuclear barrier if it has energy 6 MeV below the top of the potential barrier and the barrier thickness is 10^{-15} m.

18 •• A 10-eV electron is incident on a potential barrier of height 25 eV and width of 1 nm. (a) Use Equation 36-29 to calculate the order of magnitude of the probability that the electron will tunnel through the barrier. (b) Repeat your calculation for a width of 0.1 nm.

The Schrödinger Equation in Three Dimensions

19 • A particle is confined to a three-dimensional box that has sides L_1, $L_2 = 2L_1$, and $L_3 = 3L_1$. Give the quantum numbers n_1, n_2, n_3 that correspond to the lowest ten quantum states of this box.

20 • Give the wave functions for the lowest ten quantum states of the particle in Problem 19.

21 • (a) Repeat Problem 19 for the case $L_2 = 2L_1$ and $L_3 = 4L_1$. (b) What quantum numbers correspond to degenerate energy levels?

22 • Give the wave functions for the lowest ten quantum states of the particle in Problem 21.

23 • A particle moves in a potential well given by $U(x, y, z) = 0$ for $-L/2 < x < L/2$, $0 < y < L$, and $0 < z < L$, and $U = \infty$ outside these ranges. (a) Write an expression for the ground-state wave function for this particle. (b) How do the allowed energies compare with those for a box having $U = 0$ for $0 < x < L$, rather than for $-L/2 < x < L/2$?

24 •• A particle moves freely in the two-dimensional region defined by $0 \le x \le L$ and $0 \le y \le L$. (a) Find the wave function satisfying Schrödinger's equation. (b) Find the corresponding energies. (c) Find the lowest two states that are degenerate. Give the quantum numbers for this case. (d) Find the lowest three states that have the same energy. Give the quantum numbers for the three states having the same energy.

25 •• What is the next energy level above those found in Problem 24c for a particle in a two-dimensional square box for which the degeneracy is greater than 2?

Identical Particles

26 • Show that Equation 36-37 satisfies Equation 36-35 with $U = 0$, and find the energy of this state.

27 • What is the ground-state energy of ten noninteracting bosons in a one-dimensional box of length L?

28 • What is the ground-state energy of ten noninteracting fermions, such as neutrons, in a one-dimensional box of length L? (Because the quantum number associated with spin can have two values, each spatial state can hold two neutrons.)

Orthogonality of Wave Functions

The integral of two functions over some space interval is somewhat analogous to the dot product of two vectors. If this integral is zero, the functions are said to be orthogonal, which is analogous to two vectors being perpendicular. The following problems illustrate the general principle that any two wave functions corresponding to different energy levels in the same potential are orthogonal.

29 •• Show that the ground-state wave function and that of the first excited state of the harmonic oscillator are orthogonal; i.e., show that $\int \psi_0(x)\psi_1(x)\, dx = 0$.

30 •• The wave function for the state $n = 2$ of the harmonic oscillator is $\psi_2(x) = A_2(2ax^2 - \frac{1}{2})e^{-ax^2}$, where A_2 is the normalization constant for this wave function. Show that the wave functions for the states $n = 1$ and $n = 2$ of the harmonic oscillator are orthogonal.

31 ••• For the wave functions $\psi_n(x) = \sqrt{2/L} \sin(n\pi x/L)$ corresponding to a particle in an infinite square well potential from 0 to L, show that $\int \psi_n(x)\psi_m(x)\, dx = 0$, that is, ψ_n and ψ_m are orthogonal.

General Problems

32 •• Consider a particle in a one-dimensional box of length L that is centered at the origin. (a) What are the values of $\psi_1(0)$ and $\psi_2(0)$? (*b*) What are the values of $<x>$ for the states $n = 1$ and $n = 2$? (*c*) Evaluate $<x^2>$ for the states $n = 1$ and $n = 2$. (See Problem 59 in Chapter 17.)

33 •• Eight identical noninteracting fermions (such as neutrons) are confined to a two-dimensional square box of side length L. Determine the energies of the three lowest states. (See Problem 26.)

34 •• A particle is confined to a two-dimensional box defined by the following boundary conditions: $U(x, y) = 0$ for $-L/2 \leq x \leq L/2$ and $-3L/2 \leq y \leq 3L/2$; and $U(x, y) = \infty$ elsewhere. (*a*) Determine the energies of the lowest three bound states. Are any of these states degenerate? (*b*) Identify the lowest doubly degenerate bound state by appropriate quantum numbers and determine its energy.

35 •• A particle moves in a potential given by $U(x) = A|x|$. Without attempting to solve the Schrödinger equation, sketch the wave function for (*a*) the ground-state energy of a particle inside this potential and (*b*) the first excited state for this potential.

36 ••• The classical probability distribution function for a particle in a one-dimensional box of length L is $P = 1/L$. (See Example 17-5.) (*a*) Show that the classical expectation value of x^2 for a particle in a one-dimensional box of length L centered at the origin (Problem 32) is $L^2/12$. (*b*) Find the quantum expectation value of x^2 for the nth state of a particle in the one-dimensional box of Problem 32 and show that it approaches the classical limit $L^2/12$ for $n \gg 1$.

37 ••• Show that Equations 36-27 and 36-28 imply that the transmission coefficient for particles of energy E incident on a step barrier $U_0 < E$ is given by

$$T = \frac{4k_1k_2}{(k_1 + k_2)^2} = \frac{4r}{(1 + r)^2}$$

where $r = k_2/k_1$.

38 ••• (*a*) Show that for the case of a particle of energy E incident on a step barrier $U_0 < E$, the wave numbers k_1 and k_2 are related by

$$\frac{k_2}{k_1} = r = \sqrt{1 - \frac{U_0}{E}}$$

Use this and the results of Problem 37 to calculate the transmission coefficient T and the reflection coefficient R for the case (*b*) $E = 1.2U_0$, (*c*) $E = 2.0U_0$, and (*d*) $E = 10.0U_0$.

39 ••• Determine the normalization constant A_2 in Problem 30.

40 ••• Consider the time-independent one-dimensional Schrödinger equation when the potential function is symmetric about the origin, i.e., when $U(x) = U(-x)$. (*a*) Show that if $\psi(x)$ is a solution of the Schrödinger equation with energy E, then $\psi(-x)$ is also a solution with the same energy E, and that, therefore, $\psi(x)$ and $\psi(-x)$ can differ by only a multiplicative constant. (*b*) Write $\psi(x) = C\psi(-x)$, and show that $C = \pm 1$. Note that $C = +1$ means that $\psi(x)$ is an even function of x, and $C = -1$ means that $\psi(x)$ is an odd function of x.

41 ••• In this problem you will derive the ground-state energy of the harmonic oscillator using the precise form of the uncertainty principle, $\Delta x \, \Delta p \geq \hbar/2$, where Δx and Δp are defined to be the standard deviations $(\Delta x)^2 = [(x - x_{av})^2]_{av}$ and $(\Delta p)^2 = [(p - p_{av})^2]_{av}$ (see Equation 18-35). Proceed as follows:

1. Write the total classical energy in terms of the position x and momentum p using $U(x) = \frac{1}{2}m\omega^2x^2$ and $K = p^2/2m$.

2. Use the result of Equation 18-35 to write $(\Delta x)^2 = [(x - x_{av})^2]_{av} = (x^2)_{av} - x_{av}^2$ and $(\Delta p)^2 = [(p - p_{av})^2]_{av} = (p^2)_{av} - p_{av}^2$.

3. Use the symmetry of the potential energy function to argue that x_{av} and p_{av} must be zero, so that $(\Delta x)^2 = (x^2)_{av}$ and $(\Delta p)^2 = (p^2)_{av}$.

4. Assume that $\Delta p = \hbar/2\Delta x$ to eliminate $(p^2)_{av}$ from the average energy $E_{av} = (p^2)_{av}/2m + \frac{1}{2}m\omega^2(x^2)_{av}$ and write E_{av} as $E_{av} = \hbar^2/8mZ + \frac{1}{2}m\omega^2Z$, where $Z = (x^2)_{av}$.

5. Set $dE/dZ = 0$ to find the value of Z for which E is a minimum.

6. Show that the minimum energy is given by $(E_{av})_{min} = +\frac{1}{2}\hbar\omega$.

42 ••• A particle of mass m near the earth's surface at $z = 0$ can be described by the potential energy

$$U = mgz, \quad z > 0$$

$$U = \infty, \quad z < 0$$

For some positive value of total energy E, indicate the classically allowed region on a sketch of $U(z)$ versus z. Sketch also the classical kinetic energy versus z. The Schrödinger equation for this problem is quite difficult to solve. Using arguments similar to those in Section 36-2 about the curvature of the wave function as given by the Schrödinger equation, sketch your "guesses" for the shape of the wave function for the ground state and the first two excited states.

CHAPTER 37

Atoms

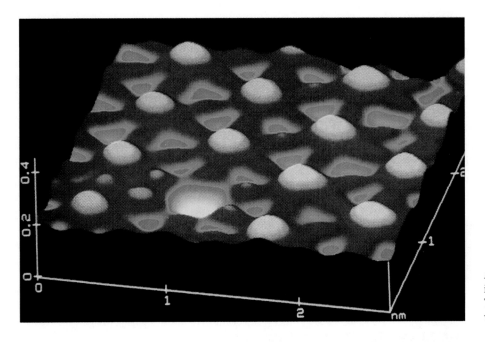

A scanning tunneling microscope image of iodine atoms (pink) adsorbed on platinum. The yellow pocket represents the gap where an iodine atom has been dislodged.

Slightly more than 100 different elements have been discovered. Each is characterized by an atom that contains a number of protons Z, an equal number of electrons, and a number of neutrons N. The number of protons Z is called the **atomic number.** The lightest atom, hydrogen (H), has $Z = 1$; the next lightest, helium (He), has $Z = 2$; the next lightest, lithium (Li), has $Z = 3$; and so forth. Nearly all the mass of the atom is concentrated in a tiny nucleus, which contains the protons and neutrons. The nuclear radius is typically about 1 to 10 fm (1 fm $= 10^{-15}$ m). The distance between the nucleus and the electrons is about 0.1 nm $= 100,000$ fm. This distance determines the "size" of the atom.

The chemical and physical properties of an element are determined by the number and arrangement of the electrons in the atom. Because each proton has a positive charge $+e$, the nucleus has a total positive charge $+Ze$. The electrons are negatively charged ($-e$), so they are attracted to the nucleus and repelled by each other. Since electrons and protons have equal but opposite charges and there are an equal number of electrons and protons in an atom, atoms are electrically neutral. Atoms that lose or gain one or more electrons are then electrically charged and are called *ions*.

We will begin our study of atoms by discussing the Bohr model, a semi-classical model developed by Niels Bohr in 1913 to explain the spectra emitted by hydrogen atoms. Although this "pre-quantum mechanics" model has many shortcomings, it provides a useful framework for the discussion of atomic phenomena. For example, even though we now know that the electron does not circle the nucleus in well-defined orbits as in the Bohr model but instead is described by a wave function that satisfies the Schrödinger equation, the probability distributions that follow from the full quantum theory do in fact have maxima at the positions of the Bohr orbits. After discussing the Bohr model, we will apply our knowledge of quantum mechanics from Chapter 36 to give a qualitative description of the hydrogen atom. We will then discuss the structure of other atoms and the periodic table of the elements.

37-1 The Nuclear Atom

Atomic Spectra

By the beginning of the twentieth century, a large body of data had been collected on the emission of light by atoms in a gas when they are excited by an electric discharge. Viewed through a spectroscope with a narrow-slit aperture, this light appears as a discrete set of lines of different colors or wavelengths; the spacing and intensities of the lines are characteristic of the element. The wavelengths of these spectral lines could be accurately determined, and much effort went into finding regularities in the spectra. Figure 37-1 shows line spectra for hydrogen and mercury.

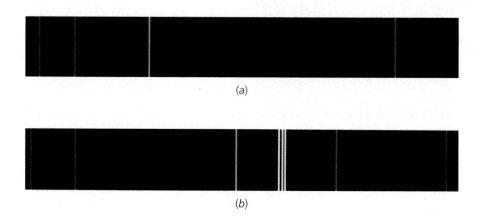

(a)

(b)

Figure 37-1 (*a*) Line spectrum of hydrogen; (*b*) line spectrum of mercury.

In 1884 a Swiss school teacher, Johann Balmer, found that the wavelengths of the lines in the visible spectrum of hydrogen can be represented by the formula

$$\lambda = (364.6 \text{ nm})\frac{m^2}{m^2 - 4}, \qquad m = 3, 4, 5, \ldots \qquad \text{37-1}$$

Balmer suggested that this might be a special case of a more general expression that would be applicable to the spectra of other elements. Such an expression, found by Johannes R. Rydberg and Walter Ritz and known as the **Rydberg–Ritz formula,** gives the reciprocal wavelength as

$$\frac{1}{\lambda} = R\left(\frac{1}{n_2^2} - \frac{1}{n_1^2}\right) \qquad \text{37-2}$$

where n_1 and n_2 are integers with $n_1 > n_2$ and R is the **Rydberg constant,** which is the same for all spectral series of the same element and varies only slightly in a regular way from element to element. For hydrogen, R has the value

$$R_{\mathrm{H}} = 1.096776 \times 10^{7}\ \mathrm{m}^{-1}$$

The Rydberg–Ritz formula gives the wavelengths for all the lines in the spectra of hydrogen as well as alkali elements such as lithium and sodium.

Many attempts were made to construct a model of the atom that would yield these formulas for its radiation spectrum. The most popular model, due to J. J. Thomson, considered various arrangements of electrons embedded in some kind of fluid that contained most of the mass of the atom and had enough positive charge to make the atom electrically neutral. Thomson's model, called the "plum pudding" model, is illustrated in Figure 37-2. Since classical electromagnetic theory predicted that a charge oscillating with frequency f would radiate electromagnetic energy of that frequency, Thomson searched for configurations that were stable and had normal modes of vibration of frequencies equal to those of the spectrum of the atom. A difficulty of this model and all others was that according to classical physics, electric forces alone cannot produce stable equilibrium. Thomson was unsuccessful in finding a model that predicted the observed frequencies for any atom.

Figure 37-2 J. J. Thomson's plum pudding model of the atom. In this model, the negative electrons are embedded in a fluid of positive charge. For a given configuration in such a system, the resonance frequencies of oscillations of the electrons can be calculated. According to classical theory, the atom should radiate light of frequency equal to the frequency of oscillation of the electrons. Thomson could not find any configuration that would give frequencies in agreement with the measured frequencies of the spectrum of any atom.

The Thomson model was essentially ruled out by a set of experiments by H. W. Geiger and E. Marsden under the supervision of E. Rutherford about 1911, in which alpha particles from radioactive radium were scattered by atoms in a gold foil. Rutherford showed that the number of alpha particles scattered at large angles could not be accounted for by an atom in which the positive charge was distributed throughout the atomic size (known to be about 0.1 nm in diameter) but required that the positive charge and most of the mass of the atom be concentrated in a very small region, now called the nucleus, of diameter of the order of 10^{-6} nm = 1 fm.

37-2 The Bohr Model of the Hydrogen Atom

Niels Bohr, working in the Rutherford laboratory in 1913, proposed a model of the hydrogen atom that extended the work of Planck, Einstein, and Rutherford and successfully predicted the observed spectra. According to Bohr's model, the electron of the hydrogen atom moves under the influence of the Coulomb attraction to the positive nucleus according to classical mechanics, which predicts circular or elliptical orbits with the force center at one focus, as in the motion of the planets around the sun. For simplicity he chose a circular orbit as shown in Figure 37-3.

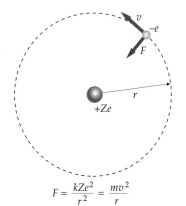

Figure 37-3 Electron of charge $-e$ traveling in a circular orbit of radius r around the nuclear charge $+Ze$. The attractive electrical force kZe^2/r^2 provides the centripetal force holding the electron in its orbit.

$$F = \frac{kZe^2}{r^2} = \frac{mv^2}{r}$$

Energy in a Circular Orbit

Consider an electron of charge $-e$ moving in a circular orbit of radius r about a positive charge Ze such as the nucleus of a hydrogen atom ($Z = 1$) or of a singly ionized helium atom ($Z = 2$). The total energy of the electron can be related to the radius of the orbit. The potential energy of the electron of charge $-e$ at a distance r from a positive charge Ze is

$$U = \frac{kq_1q_2}{r} = \frac{k(Ze)(-e)}{r} = -\frac{kZe^2}{r} \qquad \text{37-3}$$

where k is the Coulomb constant. The kinetic energy K can be obtained as a function of r by using Newton's second law, $\Sigma\vec{F} = m\vec{a}$. Setting the Coulomb attractive force equal to the mass times the centripetal acceleration gives

$$\frac{kZe^2}{r^2} = m\frac{v^2}{r} \qquad \text{37-4}a$$

Then

$$K = \frac{1}{2}mv^2 = \frac{1}{2}\frac{kZe^2}{r} \qquad \text{37-4}b$$

The kinetic energy thus varies inversely with r like the potential energy. Note that the magnitude of the potential energy is twice that of the kinetic energy,

$$U = -2K \qquad \text{37-5}$$

This is a general result in $1/r^2$ force fields. It also holds for circular orbits in a gravitational field (see Example 11-6 in Section 11-3). The total energy is the sum of the kinetic energy and potential energy

$$E = K + U = \frac{1}{2}\frac{kZe^2}{r} - \frac{kZe^2}{r}$$

or

$$E = -\frac{1}{2}\frac{kZe^2}{r} \qquad \text{37-6}$$

Energy in a circular orbit for a $1/r^2$ force

Although mechanical stability is achieved because the Coulomb attractive force provides the centripetal force necessary for the electron to remain in orbit, classical *electromagnetic* theory says that such an atom would be unstable electrically because the electron must accelerate when moving in a circle and therefore radiate electromagnetic energy of frequency equal to that of its motion. According to the classical theory, such an atom would quickly collapse, the electron spiraling into the nucleus as it radiates away its energy.

Bohr's Postulates

Bohr "solved" the difficulty of the collapsing atom by *postulating* that only certain orbits, called stationary states, are allowed, and in these orbits the electron does not radiate. An atom radiates only when the electron makes a transition from one allowed orbit (stationary state) to another.

> The electron in the hydrogen atom can move only in certain nonradiating, circular orbits called stationary states.

Bohr's first postulate: nonradiating orbits

The second postulate relates the frequency of radiation to the energies of the stationary states. If E_i and E_f are the initial and final energies of the atom, the frequency of the emitted radiation during a transition is given by

$$f = \frac{E_i - E_f}{h} \qquad\qquad \text{37-7}$$

Bohr's second postulate: photon frequency from energy conservation

where h is Planck's constant. This postulate is equivalent to the assumption of conservation of energy with the emission of a photon of energy hf. Combining Equations 37-6 and 37-7, we obtain for the frequency

$$f = \frac{E_1 - E_2}{h} = \frac{1}{2}\frac{kZe^2}{h}\left(\frac{1}{r_2} - \frac{1}{r_1}\right) \qquad\qquad \text{37-8}$$

where r_1 and r_2 are the radii of the initial and final orbits.

To obtain the frequencies implied by the Rydberg–Ritz formula, $f = c/\lambda = cR(1/n_2^2 - 1/n_1^2)$, it is evident that the radii of stable orbits must be proportional to the squares of integers. Bohr searched for a quantum condition for the radii of the stable orbits that would yield this result. After much trial and error, he found that he could obtain it if he postulated that the angular momentum of the electron in a stable orbit equals an integer times $\hbar$ ("h bar," Planck's constant divided by 2π). Since the angular momentum of a circular orbit is just mvr, this postulate is

$$mvr = \frac{nh}{2\pi} = n\hbar, \qquad n = 1, 2, \dots \qquad\qquad \text{37-9}$$

Bohr's third postulate: quantized angular momentum

where $\hbar = h/2\pi = 1.055 \times 10^{-34}\,\text{J·s} = 6.582 \times 10^{-16}\,\text{eV·s}$.

Equation 37-9 relates the speed v to the radius r. Equation 37-4a from Newton's second law gives us another equation relating the speed to the radius:

$$\frac{kZe^2}{r^2} = m\frac{v^2}{r}$$

or

$$v^2 = \frac{kZe^2}{mr} \qquad\qquad \text{37-10}$$

We can determine r by eliminating v between Equations 37-9 and 37-10. Solving Equation 37-9 for v and squaring gives

$$v^2 = n^2\frac{\hbar^2}{m^2r^2}$$

Comparing this result with Equation 37-10, we get

$$n^2\frac{\hbar^2}{m^2r^2} = \frac{kZe^2}{mr}$$

Solving for r, we get

$$r = n^2\frac{\hbar^2}{mkZe^2} = n^2\frac{a_0}{Z} \qquad\qquad \text{37-11}$$

Radius of the Bohr orbits

where a_0 is called the **first Bohr radius**.

$$a_0 = \frac{\hbar^2}{mke^2} \approx 0.0529\,\text{nm} \qquad\qquad \text{37-12}$$

First Bohr radius

Substituting the expressions for r in Equation 37-11 into Equation 37-8 for the frequency gives

$$f = \frac{1}{2}\frac{kZe^2}{h}\left(\frac{1}{r_2} - \frac{1}{r_1}\right) = Z^2\frac{mk^2e^4}{4\pi\hbar^3}\left(\frac{1}{n_2^2} - \frac{1}{n_1^2}\right)$$ 37-13

If we compare this expression with $Z = 1$ for $f = c/\lambda$ with the empirical Rydberg–Ritz formula (Equation 37-2), we obtain for the Rydberg constant

$$R = \frac{mk^2e^4}{4\pi c\hbar^3}$$ 37-14

Using the values of m, e, and $\hbar$ known in 1913, Bohr calculated R and found his result to agree (within the limits of the uncertainties of the constants) with the value obtained from spectroscopy.

Example 37-1

For waves in a circle, the standing wave condition is $n\lambda = 2\pi r$. Show that this condition for electron waves implies quantization of angular momentum.

1. Write the standing-wave condition: $n\lambda = 2\pi r$

2. Use the de Broglie relation to relate the momentum to λ: $p = \dfrac{h}{\lambda} = \dfrac{h}{2\pi r/n} = n\dfrac{h}{2\pi r} = n\dfrac{\hbar}{r}$

3. The angular momentum of an electron in a circular orbit is mvr: $L = mvr = pr = n\hbar$

Energy Levels

The total energy of the electron in the hydrogen atom is related to the radius of the circular orbit by Equation 37-6. If we substitute the quantized values of r as given by Equation 37-11, we obtain

$$E_n = -\frac{1}{2}\frac{kZe^2}{r} = -\frac{1}{2}\frac{kZe^2}{n^2a_0/Z} = -\frac{1}{2}\frac{kZe^2}{n^2\hbar^2/mkZe^2}$$

or

$$E_n = -\frac{mk^2e^4}{2\hbar^2}\frac{Z^2}{n^2} = -Z^2\frac{E_0}{n^2}, \qquad n = 1, 2, \ldots$$ 37-15

Energy levels

where

$$E_0 = \frac{mk^2e^4}{2\hbar^2} = \frac{1}{2}\frac{ke^2}{a_0} \approx 13.6 \text{ eV}$$ 37-16

The energies E_n with $Z = 1$ are the quantized allowed energies for the hydrogen atom.

Transitions between these allowed energies result in the emission or absorption of a photon whose frequency is given by $f = (E_i - F_f)/h$, and whose wavelength is

$$\lambda = \frac{c}{f} = \frac{hc}{E_i - E_f}$$ 37-17

As we found in Chapter 17, it is convenient to have the value of hc in electron-volt nanometers:

$$hc = 1240 \text{ eV·nm} \qquad\qquad 37\text{-}18$$

Since the energies are quantized, the frequencies and wavelengths of the radiation emitted by the hydrogen atom are quantized in agreement with the observed line spectrum.

Figure 37-4 shows the energy-level diagram for hydrogen. The energy of the hydrogen atom in the ground state is $E_1 = -13.6$ eV. As n approaches infinity the energy approaches zero, the highest energy state. The process of removing an electron from an atom is called ionization, and the energy required to remove the electron is the **ionization energy**. The ionization energy of the hydrogen atom, which is also the binding energy of the atom, is 13.6 eV. A few transitions from a higher to a lower state are indicated in Figure 37-4. When Bohr published his model of the hydrogen atom, the Balmer series, corresponding to $n_2 = 2$ and $n_1 = 3, 4, 5, \ldots$, and the Paschen series, corresponding to $n_2 = 3$ and $n_1 = 4, 5, 6, \ldots$, were known. In 1916, T. Lyman found the series corresponding to $n_2 = 1$, and in 1922 and 1924, F. Brackett and H. A. Pfund, respectively, found series corresponding to $n_2 = 4$ and $n_2 = 5$. Only the Balmer series lies in the visible portion of the electromagnetic spectrum.

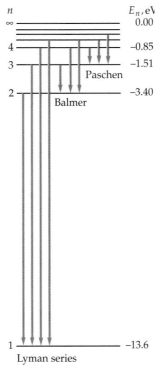

Figure 37-4 Energy-level diagram for hydrogen showing the first few transitions in each of the Lyman, Balmer, and Paschen series. The energies of the levels are given by Equation 37-15.

Example 37-2

Find (a) the energy and (b) wavelength of the line with the longest wavelength in the Lyman series.

Picture the Problem From Figure 37-4, we can see that the Lyman series corresponds to transitions ending at the ground-state energy, $E_f = E_1 = -13.6$ eV. Since λ varies inversely with energy, the transition with the longest wavelength is the transition with the lowest energy, which is that from the first excited state $n = 2$ to the ground state $n = 1$.

(a) The energy of the photon is the difference in the energies of the initial and final atomic state:

$$E = E_i - E_f$$
$$= E_2 - E_1 = \frac{-13.6 \text{ eV}}{2^2} - \frac{-13.6 \text{ eV}}{1^2}$$
$$= -3.40 \text{ eV} + 13.6 \text{ eV} = 10.2 \text{ eV}$$

(b) The wavelength of the photon is:

$$\lambda = \frac{hc}{E_2 - E_1} = \frac{1240 \text{ eV·nm}}{10.2 \text{ eV}} = 121.6 \text{ nm}$$

Remark This photon is outside the visible spectrum, in the ultraviolet region. Since all the other lines in the Lyman series have even greater energies and shorter wavelengths, the Lyman series is completely in the ultraviolet region.

Exercise Find the shortest wavelength for a line in the Lyman series. (Answer 91.2 nm)

Despite its spectacular successes, the Bohr model of the hydrogen atom had many shortcomings. There was no justification for the postulates of stationary states, or for the quantization of angular momentum other than the fact that these postulates led to energy levels that agreed with spectroscopic data. Furthermore, attempts to apply the model to more complicated atoms had little success. The quantum-mechanical theory resolves these difficulties. The stationary states of the Bohr model correspond to the standing-wave solutions of the Schrödinger equation analogous to the standing electron waves for a particle in a box discussed in Chapters 17 and 36. Energy quantization is a direct consequence of the standing-wave solutions of the Schrödinger equation. For hydrogen these quantized energies agree with those obtained from the Bohr model and with experiment. The quantization of angular momentum that had to be postulated in the Bohr model is predicted by the quantum theory.

37-3   Quantum Theory of Atoms

The Schrödinger Equation in Spherical Coordinates

In quantum theory, the electron is described by its wave function ψ. The absolute square of the electron wave function $|\psi|^2$ gives the probability of finding the electron in some region of space. Boundary conditions on the wave function lead to the quantization of the wavelengths and frequencies and thereby to the quantization of the electron energy.

Consider a single electron of mass m moving in three dimensions in a region in which the potential energy is U. The time-independent Schrödinger equation for such a particle is given by Equation 36-30:

$$-\frac{\hbar^2}{2m}\left(\frac{\partial^2\psi}{\partial x^2} + \frac{\partial^2\psi}{\partial y^2} + \frac{\partial^2\psi}{\partial z^2}\right) + U\psi = E\psi \qquad 37\text{-}19$$

For an isolated atom, the potential energy U depends only on the radial distance $r = \sqrt{x^2 + y^2 + z^2}$. The problem is then most conveniently treated using the spherical coordinates r, θ, and ϕ, which are related to the rectangular coordinates x, y, and z by

$$z = r\cos\theta$$
$$x = r\sin\theta\cos\phi \qquad 37\text{-}20$$
$$y = r\sin\theta\sin\phi$$

These relations are shown in Figure 37-5. The transformation of the three-dimensional Schrödinger equation into spherical coordinates is straightforward but involves much tedious calculation, which we will omit. The result is

$$-\frac{\hbar^2}{2mr^2}\frac{\partial}{\partial r}\left(r^2\frac{\partial\psi}{\partial r}\right) - \frac{\hbar^2}{2mr^2}\left[\frac{1}{\sin\theta}\frac{\partial}{\partial\theta}\left(\sin\theta\frac{\partial\psi}{\partial\theta}\right) + \frac{1}{\sin^2\theta}\frac{\partial^2\psi}{\partial\phi^2}\right] + U(r)\psi = E\psi$$

$$37\text{-}21$$

Despite the formidable appearance of this equation, it was not difficult for Schrödinger to solve because it is similar to other partial differential equations in classical physics that had been thoroughly studied. We will not solve this equation but merely discuss qualitatively some of the interesting features of the wave functions that satisfy it.

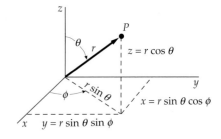

Figure 37-5 Geometric relations between spherical and rectangular coordinates.

The first step in the solution of a partial differential equation such as Equation 37-21 is to separate the variables by writing the wave function $\psi(r, \theta, \phi)$ as a product of functions of each single variable:

$$\psi(r, \theta, \phi) = R(r)f(\theta)g(\phi) \qquad \text{37-22}$$

where R depends only on the radial coordinate r, f depends only on θ, and g depends only on ϕ. When this form of $\psi(r, \theta, \phi)$ is substituted into Equation 37-21, the partial differential equation can be transformed into three ordinary differential equations, one for $R(r)$, one for $f(\theta)$, and one for $g(\phi)$. The potential energy $U(r)$ appears only in the equation for $R(r)$, which is called the **radial equation.** The particular form of $U(r)$ given in Equation 37-19 therefore has no effect on the solutions of the equations for $f(\theta)$ and $g(\phi)$, and therefore has no effect on the angular dependence of the wave function $\psi(r, \theta, \phi)$. These solutions are applicable to any problem in which the potential energy depends only on r.

Quantum Numbers in Spherical Coordinates

In three dimensions, the requirement that the wave function be continuous and normalizable introduces three quantum numbers, one associated with each dimension. In spherical coordinates the quantum number associated with r is labeled n, that associated with θ is labeled ℓ, and that associated with ϕ is labeled m. The quantum numbers n_1, n_2, and n_3 that we found in Chapter 36 for a particle in a three-dimensional square well in rectangular coordinates x, y, and z were independent of one another, but the quantum numbers associated with wave functions in spherical coordinates are interdependent. The possible values of these quantum numbers are

$$n = 1, 2, 3, \ldots$$

$$\ell = 0, 1, 2, \ldots, n - 1 \qquad \text{37-23}$$

$$m = -\ell, (-\ell + 1), \ldots 0, 1, 2, \ldots \ell$$

Quantum numbers in spherical coordinates

That is, n can be any positive integer; ℓ can be 0 or any positive integer up to $n - 1$; and m can have $2\ell + 1$ possible values, ranging from $-\ell$ to $+\ell$ in integral steps.

The number n is called the **principal quantum number.** It is associated with the dependence of the wave function on the distance r and therefore with the probability of finding the electron at various distances from the nucleus. The quantum numbers ℓ and m are associated with the angular momentum of the electron and with the angular dependence of the electron wave function. The quantum number ℓ is called the **orbital quantum number.** The magnitude of the orbital angular momentum L of the electron is related to ℓ by

$$L = \sqrt{\ell(\ell + 1)}\,\hbar \qquad \text{37-24}$$

The quantum number m is called the **magnetic quantum number.** It is related to the component of the angular momentum in some direction. Ordinarily, all directions are equivalent, but one particular direction can be specified by placing the atom in a magnetic field. If the z direction is chosen for the magnetic field, the z component of the angular momentum of the electron is given by the quantum condition

$$L_z = m\hbar \qquad \text{37-25}$$

This quantum condition arises from the boundary condition on the coordinate ϕ that the probability of finding the electron at some angle ϕ_1 must be the same as that of finding it at angle $\phi_1 + 2\pi$ because these are the same points in space.

If we measure the angular momentum of the electron in units of $\hbar$, we see that the angular momentum is quantized to the value $\sqrt{\ell(\ell + 1)}$ units and that its component along any direction can have only the $2\ell + 1$ values ranging from $-\ell$ to $+\ell$ units. Figure 37-6 shows a vector-model diagram illustrating the possible orientations of the angular-momentum vector for $\ell = 2$. Note that only specific values of θ are allowed; that is, the directions in space are quantized.

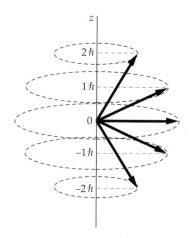

Figure 37-6 Vector-model diagram illustrating the possible values of the z component of the angular-momentum vector for the case $\ell = 2$. The magnitude of the angular momentum is $L = \hbar\sqrt{\ell(\ell + 1)} = \hbar\sqrt{2(2 + 1)} = \hbar\sqrt{6}$.

Example 37-3

If the angular momentum is characterized by the quantum number $\ell = 2$, what are the possible values of L_z, and what is the smallest possible angle between $\vec{L}$ and the z axis?

Picture the Problem The possible orientations of $\vec{L}$ and an arbitrary z axis are shown in Figure 37-6.

1. Write the possible values of L_z:
$$L_z = m\hbar, \quad \text{where } m = -2, -1, 0, +1, \text{ or } +2$$

2. Express the angle θ between $\vec{L}$ and the z axis in terms of L and L_z:
$$\cos\theta = \frac{L_z}{L} = \frac{m\hbar}{\sqrt{\ell(\ell + 1)}\,\hbar} = \frac{m}{\sqrt{\ell(\ell + 1)}}$$

3. The smallest angle occurs when $m = +\ell$ or $-\ell$. Calculate this angle for $\ell = 2$:
$$\cos\theta = \frac{2}{\sqrt{2(2 + 1)}} = \frac{2}{\sqrt{6}} = 0.816$$
$$\theta = 35.3°$$

Remark We note the somewhat strange result that the angular-momentum vector cannot lie along the z axis.

Exercise An atom has an angular momentum characterized by the quantum number $\ell = 4$. What are the possible values of m? (*Answer* $-4, -3, -2, -1, 0, 1, 2, 3, 4$)

37-4 Quantum Theory of the Hydrogen Atom

We can treat the simplest atom, the hydrogen atom, as a stationary nucleus, the proton, that has a single particle, an electron, moving with kinetic energy $p^2/2m$ and potential energy $U(r)$ due to the electrostatic attraction of the proton*:

$$U(r) = -\frac{kZe^2}{r}$$
37-26

---

* We include the factor Z, which is 1 for hydrogen, so that we can apply our results to other one-electron atoms, such as ionized helium He^+, for which $Z = 2$.

For this potential energy, the Schrödinger equation can be solved exactly. In the lowest energy state, the ground state, the principal quantum number n has the value 1, ℓ is 0, and m is 0.

Energy Levels

The allowed energies of the hydrogen atom that result from the solution of the Schrödinger equation are

$$E_n = -\frac{mk^2e^4}{2\hbar^2}\frac{Z^2}{n^2} = -Z^2\frac{E_0}{n^2}, \qquad n = 1, 2, 3, \ldots \qquad \text{37-27}$$

Energy levels for hydrogen

where

$$E_0 = \frac{mk^2e^4}{2\hbar^2} \approx 13.6 \text{ eV} \qquad \text{37-28}$$

These energies are the same as in the Bohr model. Note that the energy is negative, indicating that the electron is bound to the nucleus (thus the term *bound state*), and that the energy depends only on the principal quantum number n. The fact that the energy does not depend on the orbital quantum number ℓ is a peculiarity of the inverse-square force and holds only for an inverse r potential such as Equation 37-26. For more complicated atoms having several electrons, the interaction of the electrons leads to a dependence of the energy on ℓ. In general, the lower the value of ℓ, the lower the energy for such atoms. Since there is usually no preferred direction in space, the energy for any atom does not ordinarily depend on the magnetic quantum number m, which is related to the z component of the angular momentum. The energy does depend on m if the atom is in a magnetic field.

Figure 37-7 shows an energy-level diagram for hydrogen. This diagram is similar to Figure 37-4, except that the states with the same value of n but different values of ℓ are shown separately. These states (called *terms*) are referred to by giving the value of n along with a code letter: S for $\ell = 0$, P for $\ell = 1$, D for $\ell = 2$, and F for $\ell = 3$.* When an atom makes a transition from one allowed energy state to another, electromagnetic radiation in the form of a photon is emitted or absorbed. Such transitions result in spectral lines that are characteristic of the atom. The transitions obey the **selection rules**

$$\Delta m = 0 \quad \text{or} \quad \pm 1$$
$$\Delta \ell = \pm 1 \qquad \qquad \text{37-29}$$

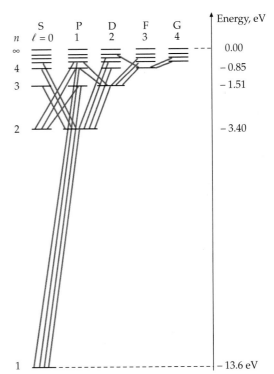

Figure 37-7 Energy-level diagram for hydrogen. The diagonal lines show transitions that involve emission or absorption of radiation and obey the selection rule $\Delta\ell = \pm 1$. States with the same value of n but different values of ℓ have the same energy $-E_0/n^2$, where $E_0 = 13.6$ eV as in the Bohr model.

These selection rules are related to the conservation of angular momentum and to the fact that the photon itself has an intrinsic angular momentum that has a maximum component along any axis of $1\hbar$. The wavelengths of the spectral lines emitted by hydrogen (and by other atoms) are related to the energy levels by

$$hf = \frac{hc}{\lambda} = E_i - E_f \qquad \text{37-30}$$

where E_i and E_f are the energies of the initial and final states.

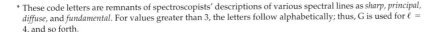

* These code letters are remnants of spectroscopists' descriptions of various spectral lines as *sharp, principal, diffuse,* and *fundamental.* For values greater than 3, the letters follow alphabetically; thus, G is used for $\ell = 4$, and so forth.

Wave Functions and Probability Densities

The solutions of the Schrödinger equation in spherical coordinates are characterized by the quantum numbers n, ℓ, and m, and are written $\psi_{n\ell m}$. For any given value of n, there are n possible values of ℓ ($\ell = 0, 1, \ldots, n - 1$), and for each value of ℓ, there are $2\ell + 1$ possible values of m. For hydrogen, the energy depends only on n, so there are generally many different wave functions that correspond to the same energy (except at the lowest energy level, for which $n = 1$ and therefore ℓ and m must be 0). These energy levels are therefore degenerate (see Section 36-5). The origins of this degeneracy are the $1/r$ dependence of the potential energy and the fact that, in the absence of any external fields, there is no preferred direction in space.

The Ground State   In the lowest energy state, the ground state, the principal quantum number n has the value 1, ℓ is 0, and m is 0. The energy is -13.6 eV, and the angular momentum is zero. Note that this differs from the Bohr model in which the angular momentum in the ground state is $1\hbar$. The wave function for the ground state is

$$\psi_{1,0,0} = C_{1,0,0}e^{-Zr/a_0} \qquad\qquad 37\text{-}31$$

where

$$a_0 = \frac{\hbar^2}{mke^2} = 0.0529 \text{ nm}$$

is the first Bohr radius and $C_{1,0,0}$ is a constant that is determined by normalization. In three dimensions, the normalization condition is

$$\int |\psi|^2 \, dV = 1$$

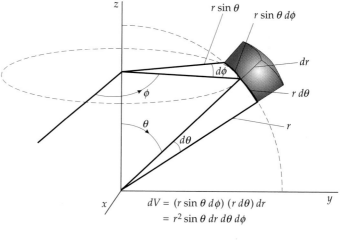

Figure 37-8 Volume element in spherical coordinates.

where dV is a volume element and the integration is performed over all space. In spherical coordinates, the volume element (Figure 37-8) is

$$dV = (r \sin\theta \, d\phi)(r \, d\theta) \, dr = r^2 \sin\theta \, dr \, d\theta \, d\phi$$

We integrate over all space by integrating over ϕ from $\phi = 0$ to $\phi = 2\pi$, over θ from $\theta = 0$ to $\theta = \pi$, and over r from $r = 0$ to $r = \infty$. The normalization condition is thus

$$\int |\psi|^2 \, dV = \int_0^\infty \int_0^\pi \int_0^{2\pi} |\psi|^2 r^2 \sin\theta \, dr \, d\theta \, d\phi$$

$$= \int_0^{2\pi} d\phi \int_0^\pi \sin\theta \, d\theta \int_0^\infty |\psi|^2 r^2 \, dr$$

$$= \int_0^{2\pi} d\phi \int_0^\pi \sin\theta \, d\theta \int_0^\infty C_{1,0,0}^2 e^{-2Zr/a_0} r^2 \, dr = 1$$

Since there is no θ or ϕ dependence in $\psi_{1,0,0}$, the integration over the angles gives 4π. From a table of integrals, we obtain

$$\int_0^\theta e^{-2Zr/a_0} r^2 \, dr = \frac{a_0^3}{4Z^3}$$

Then

$$4\pi C_{1,0,0}^2 \left(\frac{a_0^3}{4Z^3}\right) = 1$$

and

$$C_{1,0,0} = \frac{1}{\sqrt{\pi}}\left(\frac{Z}{a_0}\right)^{3/2} \qquad \text{37-32}$$

The normalized ground-state wave function is thus

$$\psi_{1,0,0} = \frac{1}{\sqrt{\pi}}\left(\frac{Z}{a_0}\right)^{3/2} e^{-Zr/a_0} \qquad \text{37-33}$$

The probability of finding the electron in a volume dV is $|\psi|^2\, dV$. The probability density $|\psi|^2$ is illustrated in Figure 37-9. Note that this probability density is spherically symmetric; that is, it depends only on r and not on θ or ϕ. The probability density is maximum at the origin.

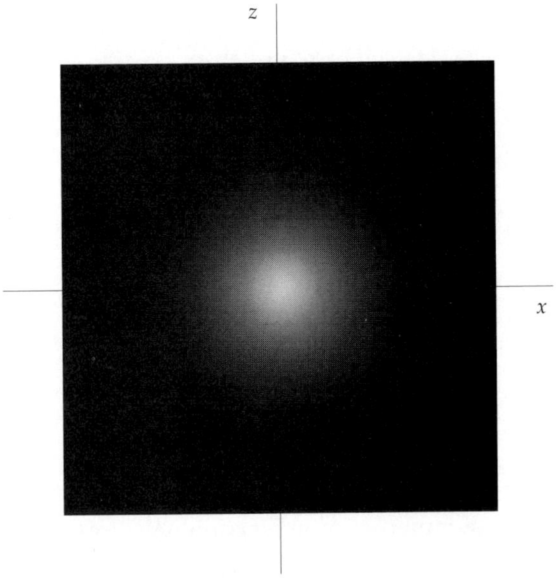

We are more often interested in the probability of finding the electron at some radial distance r between r and $r + dr$. This radial probability $P(r)\, dr$ is the probability density $|\psi|^2$ times the volume of the spherical shell of thickness dr, which is $dV = 4\pi r^2\, dr$. The probability of finding the electron in the range from r to $r + dr$ is thus $P(r)\, dr = |\psi|^2 4\pi r^2\, dr$, and the **radial probability density** is

$$P(r) = 4\pi r^2|\psi|^2 \qquad \text{37-34}$$

Radial probability density

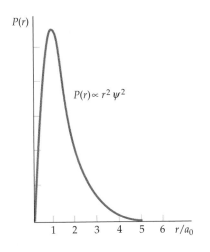

Figure 37-9 Computer-generated picture of the probability density $|\psi|^2$ for the ground state of hydrogen. The quantity $e|\psi|^2$ can be thought of as the electron charge density in the atom. The density is spherically symmetric, is greatest at the origin, and decreases exponentially with r.

For the hydrogen atom in the ground state, the radial probability density is

$$P(r) = 4\pi r^2|\psi|^2 = 4\pi C_{100}^2 r^2 e^{-2Zr/a_0} = 4\left(\frac{Z}{a_0}\right)^3 r^2 e^{-2Zr/a_0} \qquad \text{37-35}$$

Figure 37-10 shows the radial probability density $P(r)$ as a function of r. The maximum value of $P(r)$ occurs at $r = a_0/Z$, which for $Z = 1$ is the first Bohr radius. In contrast to the Bohr model, in which the electron stays in a well-defined orbit at $r = a_0$, we see that it is possible for the electron to be found at any distance from the nucleus. However, the most probable distance is a_0 (assuming $Z = 1$), and the chance of finding the electron at a much different distance is small. It is often useful to think of the electron in an atom as a charged cloud of charge density $\rho = e|\psi|^2$, but we should remember that when it interacts with matter, an electron is always observed as a single charge.

Figure 37-10 Radial probability density $P(r)$ versus r/a_0 for the ground state of the hydrogen atom. $P(r)$ is proportional to $r^2\psi^2$. The value of r for which $P(r)$ is maximum is the most probable distance $r = a_0$.

$P(r)$

$P(r) \propto r^2\, \psi^2$

1 2 3 4 5 6 r/a_0

Example 37-4

Find the probability of finding the electron in the range $\Delta r = 0.06a_0$ at (a) $r = a_0$ and (b) $r = 2a_0$ for the ground state of the hydrogen atom.

Picture the Problem Because the range Δr is so small, the variation in the radial probability density $P(r)$ can be neglected. The probability of finding the electron in some small range Δr is then $P(r)\,\Delta r$:

(a) Use Equation 37-35 with
$Z = 1$ and $r = a_0$:

$$P(r)\,\Delta r = \left[4\left(\frac{1}{a_0}\right)^3 r^2 e^{-2r/a_0}\right]\Delta r = \left[4\left(\frac{1}{a_0}\right)^3 a_0^2 e^{-2}\right](0.06a_0) = 0.0325$$

(b) Use Equation 37-35 with
$Z = 1$ and $r = 2a_0$:

$$P(r)\,\Delta r = \left[4\left(\frac{1}{a_0}\right)^3 r^2 e^{-2r/a_0}\right]\Delta r = \left[4\left(\frac{1}{a_0}\right)^3 4a_0^2 e^{-4}\right](0.06a_0) = 0.0176$$

Remark There is about a 3% chance of finding the electron in this range at $r = a_0$, but at $r = 2a_0$ the chance is only about 0.2%.

The First Excited State

In the first excited state, $n = 2$ and ℓ can be either 0 or 1. For $\ell = 0, m = 0$, and we again have a spherically symmetric wave function, this time given by

$$\psi_{2,0,0} = C_{2,0,0}\left(2 - \frac{Zr}{a_0}\right)e^{-Zr/2a_0} \tag{37-36}$$

For $\ell = 1$, m can be $+1, 0$, or -1. The corresponding wave functions are

$$\psi_{2,1,0} = C_{2,1,0}\frac{Zr}{a_0}e^{-Zr/2a_0}\cos\theta \tag{37-37}$$

$$\psi_{2,1\pm1} = C_{2,1,1}\frac{Zr}{a_0}e^{-Zr/2a_0}\sin\theta\, e^{\pm i\phi} \tag{37-38}$$

where $C_{2,0,0}$, $C_{2,1,0}$, and $C_{2,1,1}$ are normalization constants. The probability densities are given by

$$\psi_{2,0,0}^2 = C_{2,0,0}^2\left(2 - \frac{Zr}{a_0}\right)^2 e^{-Zr/a_0} \tag{37-39}$$

$$\psi_{2,1,0}^2 = C_{2,1,0}^2\left(\frac{Zr}{a_0}\right)^2 e^{-Zr/a_0}\cos^2\theta \tag{37-40}$$

$$|\psi_{2,1\pm1}|^2 = C_{2,1,1}^2\left(\frac{Zr}{a_0}\right)^2 e^{-Zr/a_0}\sin^2\theta \tag{37-41}$$

The wave functions and probability densities for $\ell \neq 0$ are not spherically symmetric, but instead depend on the angle θ. The probability densities do not depend on ϕ. Figure 37-11 shows the probability density $|\psi|^2$ for $n = 2$, $\ell = 0, m = 0$ (Figure 37-11a); for $n = 2, \ell = 1, m = 0$ (Figure 37-11b); and for $n = 2, \ell = 1, m = \pm1$ (Figure 37-11c). An important feature of these plots is that the electron cloud is spherically symmetric for $\ell = 0$ and is not spherically symmetric for $\ell \neq 0$. These angular distributions of the electron charge density depend only on the values of ℓ and m and not on the radial part of the wave function. Similar charge distributions for the valence electrons of more complicated atoms play an important role in the chemistry of molecular bonding.

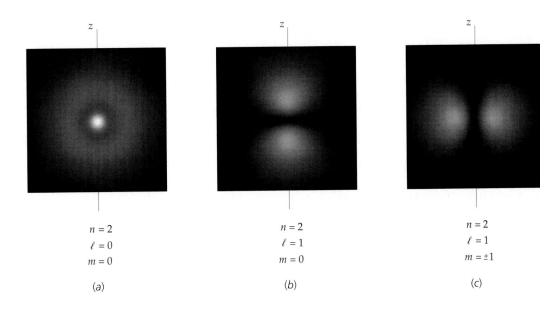

Figure 37-11 Computer-generated picture of the probability densities $|\psi|^2$ for the electron in the $n = 2$ states of hydrogen. (a) For $\ell = 0$, $|\psi|^2$ is spherically symmetric. (b) For $\ell = 1$ and $m = 0$, $|\psi|^2$ is proportional to $\cos^2 \theta$. (c) For $\ell = 1$ and $m = +1$ or -1, $|\psi|^2$ is proportional to $\sin^2 \theta$.

$n = 2$
$\ell = 0$
$m = 0$

(a)

$n = 2$
$\ell = 1$
$m = 0$

(b)

$n = 2$
$\ell = 1$
$m = \pm 1$

(c)

Figure 37-12 shows the probability of finding the electron at a distance r as a function of r for $n = 2$ when $\ell = 1$ and when $\ell = 0$. We can see from the figure that the probability distribution depends on ℓ as well as on n.

For $n = 1$, we found that the most likely distance between the electron and the nucleus is a_0, the first Bohr radius, whereas for $n = 2$, $\ell = 1$, it is $4a_0$. These are the orbital radii for the first and second Bohr orbits (Equation 37-11). For $n = 3$ (and $\ell = 2$),* the most likely distance between the electron and nucleus is $9a_0$, the radius of the third Bohr orbit.

---

* The correspondence with the Bohr model is closest for the maximum value of ℓ, which is $n - 1$.

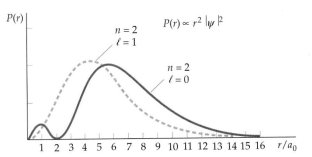

Figure 37-12 Radial probability density $P(r)$ versus r/a_0 for the $n = 2$ states of hydrogen. For $\ell = 1$, $P(r)$ is maximum at the Bohr value $r = 2^2 a_0$. For $\ell = 0$, there is a maximum near this value and a smaller submaximum near the origin.

37-5 # The Spin–Orbit Effect and Fine Structure

In general, an electron in an atom has both orbital angular momentum characterized by the quantum number ℓ and spin angular momentum characterized by the quantum number s. Analogous classical systems that have two kinds of angular momentum are the earth, which is spinning about its axis of rotation in addition to revolving about the sun, and a precessing gyroscope that has angular momentum of precession in addition to its spin. The total angular momentum $\vec{J}$ is the sum of the orbital angular momentum $\vec{L}$ and the spin angular momentum $\vec{S}$

$$\vec{J} = \vec{L} + \vec{S} \qquad\qquad 37\text{-}42$$

Classically $\vec{J}$ is an important quantity because the resultant torque on a system equals the rate of change of the total angular momentum, and in the case of central forces, the total angular momentum is conserved. For a classical system, the magnitude of the total angular momentum J can have any value between $L + S$ and $L - S$. In quantum mechanics, angular momentum is more complicated. Both L and S are quantized, and their directions are restricted.

Quantum mechanics also limits the possible values of the total angular momentum J. For an electron with orbital angular momentum characterized by the quantum number ℓ and spin $s = \frac{1}{2}$, the total angular momentum J has the magnitude $\sqrt{j(j + 1)}\,\hbar$, where the quantum number j can be either

$$j = \ell + \tfrac{1}{2} \quad \text{or} \quad j = \ell - \tfrac{1}{2}, \quad \ell \neq 0 \qquad \text{37-43}$$

(For $\ell = 0$, the total angular momentum is simply the spin and $j = \frac{1}{2}$.) Figure 37-13 is a vector model illustrating the two possible combinations $j = \frac{3}{2}$ and $j = \frac{1}{2}$ for the case of $\ell = 1$. The lengths of the vectors are proportional to $\sqrt{\ell(\ell + 1)}\,\hbar$, $\sqrt{s(s + 1)}\,\hbar$, and $\sqrt{j(j + 1)}\,\hbar$. The spin and orbital angular momentum are said to be "parallel" when $j = \ell + s$ and "antiparallel" when $j = \ell - s$.

Atomic states with the same n and ℓ values but different j values have slightly different energies because of the interaction of the spin of the electron with its orbital motion. This effect is called the **spin–orbit effect**. The resulting splitting of spectral lines is called **fine-structure splitting**.

In spectroscopic notation, the total angular-momentum quantum number of an atomic state is written as a subscript after the code letter describing the orbital angular momentum. For example, the ground state of hydrogen is written $1S_{1/2}$, where the 1 indicates the value of n. The $n = 2$ states can have either $\ell = 0$ or $\ell = 1$, and the $\ell = 1$ state can have either $j = \frac{3}{2}$ or $j = \frac{1}{2}$. These states are thus denoted by $2S_{1/2}$, $2P_{3/2}$, and $2P_{1/2}$. Because of the spin–orbit effect, the $2P_{3/2}$ and $2P_{1/2}$ states have slightly different energies resulting in the fine-structure splitting of the transitions $2P_{3/2} \to 2S_{1/2}$ and $2P_{1/2} \to 2S_{1/2}$.

We can understand the spin–orbit effect qualitatively from a simple Bohr-model picture as shown in Figure 37-14. In this picture, the electron moves in a circular orbit around a fixed proton. In Figure 37-14a, the orbital angular momentum $\vec{L}$ is up. In the reference frame of the electron (Figure 37-14b), the proton is moving in a circle around it, thus constituting a circular loop of current that produces a magnetic field $\vec{B}$ at the position of the electron. The direction of $\vec{B}$ is up, parallel to $\vec{L}$. The energy of the electron depends on its spin because of the magnetic moment $\vec{\mu}_s$ associated with its spin. The energy is lowest when $\vec{\mu}_s$ is parallel to $\vec{B}$ and highest when it is antiparallel. This energy is given by (Equation 28-16)

$$U = -\vec{\mu}_s \cdot \vec{B} \qquad \text{37-44}$$

Since $\vec{\mu}_s$ is directed opposite to its spin (because the electron has a negative charge), the energy is lowest when the spin is antiparallel to $\vec{B}$ and thus to $\vec{L}$. The energy of the $2P_{1/2}$ state in hydrogen, in which $\vec{L}$ and $\vec{S}$ are antiparallel (Figure 37-15),

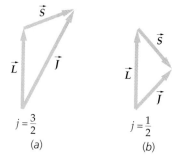

Figure 37-13 Vector diagrams illustrating the addition of orbital and spin angular momentum for the case $\ell = 1$ and $s = \frac{1}{2}$. There are two possible values of the quantum number for the total angular momentum: $j = \ell + s = \frac{3}{2}$ and $j = \ell - s = \frac{1}{2}$.

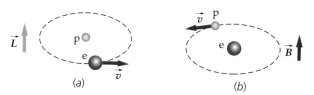

Figure 37-14 (a) An electron moving about a proton in a circular orbit in the horizontal plane with angular momentum $\vec{L}$ up. (b) The magnetic field $\vec{B}$ seen by the electron due to the apparent (relative) motion of the proton is also up. When the electron spin is parallel to $\vec{L}$, the magnetic moment is antiparallel to $\vec{L}$ and $\vec{B}$, so the spin–orbit energy is at its greatest.

Figure 37-15 Fine-structure energy-level diagram. On the left, the levels in the absence of a magnetic field are shown. The effect of an applied field is shown on the right. Because of the spin–orbit interaction, the magnetic field splits the 2P level into two energy levels, with the $j = \frac{3}{2}$ level having slightly greater energy than the $j = \frac{1}{2}$ level. The spectral line due to the transition $2P \to 1S$ is therefore split into two lines of slightly different wavelengths.

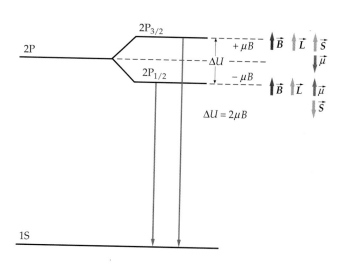

is therefore slightly lower than that of the $2P_{3/2}$ state, in which $\vec{L}$ and $\vec{S}$ are parallel.

Example 37-5

As a consequence of fine-structure splitting, the energies of the $2P_{3/2}$ and $2P_{1/2}$ levels in hydrogen differ by 4.5×10^{-5} eV. If the 2p electron sees an internal magnetic field B, the spin–orbit energy splitting will be of the order of $\Delta E = 2\mu_B B$, where μ_B is the Bohr magneton.* From this, estimate the magnetic field that the 2p electron in hydrogen experiences.

*The Bohr magneton unit is discussed in Section 29-5.

1. Write the energy splitting in terms of the magnetic moment:

$$\Delta E = 2\mu_B B = 4.5 \times 10^{-5}\,\text{eV}$$

2. Solve for the magnetic field B:

$$B = \frac{4.5 \times 10^{-5}\,\text{eV}}{2\mu_B} = \frac{4.5 \times 10^{-5}\,\text{eV}}{2(5.79 \times 10^{-5}\,\text{eV/T})} = 0.389\,\text{T}$$

37-6   The Periodic Table

For atoms with more than one electron, the Schrödinger equation cannot be solved exactly. However, powerful approximation methods allow us to determine the energy levels of the atoms and wave functions of the electrons to a high degree of accuracy. As a first approximation, the Z electrons in an atom are assumed to be noninteracting. The Schrödinger equation can then be solved, and the resulting wave functions used to calculate the interaction of the electrons, which in turn can be used to better approximate the wave functions.[†] Because the spin of an electron can have two possible components along an axis, there is an additional quantum number m_s, which can have the possible values $+\frac{1}{2}$ or $-\frac{1}{2}$. The state of each electron is thus described by the four quantum numbers n, ℓ, m, and m_s. The energy of the electron is determined mainly by the principal quantum number n (which is related to the radial dependence of the wave function) and by the orbital angular-momentum quantum number ℓ. Generally, the lower the values of n and ℓ, the lower the energy. The dependence of the energy on ℓ is due to the interaction of the electrons in the atom with each other. In hydrogen, of course, there is only one electron, and the energy is independent of ℓ. The specification of n and ℓ for each electron in an atom is called the **electron configuration.** Customarily, ℓ is specified according to the same code used to label the states of the hydrogen atom rather than by its numerical value. The code is

	s	p	d	f	g	h
ℓ value	0	1	2	3	4	5

The n values are sometimes referred to as shells, which are identified by another letter code: $n = 1$ denotes the K shell; $n = 2$, the L shell; and so on.

The electron configuration of atoms is governed by the Pauli exclusion principle, which states that no two electrons in an atom can be in the same quantum state; that is, no two electrons can have the same set of values for

---

[†] This approximation method, called perturbation theory, is similar to that used to find the orbits of planets around the sun. The orbits are first found neglecting any interaction between them, then these orbits are used to calculate the perturbation in the orbit due to the forces exerted by one planet on another.

the quantum numbers n, ℓ, m, and m_s. Using the exclusion principle and the restrictions on the quantum numbers discussed in the previous sections (n is an integer, ℓ is an integer that ranges from 0 to $n - 1$, m can have $2\ell + 1$ values from $-\ell$ to ℓ in integral steps, and m_s can be either $+\frac{1}{2}$ or $-\frac{1}{2}$), we can understand much of the structure of the periodic table.

We have already discussed the lightest element, hydrogen, which has just one electron. In the ground (lowest energy) state, the electron has $n = 1$ and $\ell = 0$, with $m = 0$ and $m_s = +\frac{1}{2}$ or $-\frac{1}{2}$. We call this a 1s electron. The 1 signifies that $n = 1$, and the s signifies that $\ell = 0$.

As electrons are added to make the heavier atoms, the electrons go into those states that will give the lowest total energy consistent with the Pauli exclusion principle.

Helium (Z = 2)

The next element after hydrogen is helium ($Z = 2$), which has two electrons. In the ground state, both electrons are in the K shell with $n = 1$, $\ell = 0$, and $m = 0$; one electron has $m_s = +\frac{1}{2}$ and the other has $m_s = -\frac{1}{2}$. This configuration is lower in energy than any other two-electron configuration. The resultant spin of the two electrons is zero. Since the orbital angular momentum is also zero, the total angular momentum is zero. The electron configuration for helium is written $1s^2$. The 1 signifies that $n = 1$, the s signifies that $\ell = 0$, and the superscript 2 signifies that there are two electrons in this state. Since ℓ can be only 0 for $n = 1$, these two electrons fill the K ($n = 1$) shell. The energy required to remove the most loosely bound electron from an atom in the ground state is called the **ionization energy**. This energy is the binding energy of the last electron placed in the atom. For helium, the ionization energy is 24.6 eV, which is relatively large. Helium is therefore basically inert.

| Example 37-6 |

(*a*) **Use the measured ionization energy to calculate the energy of interaction of the two electrons in the ground state of the helium atom.** (*b*) **Use your result to estimate the average separation of the two electrons.**

Picture the Problem   The energy of one electron in the ground state of helium is E_1 (which is negative) given by Equation 37-27 with $n = 1$ and $Z = 2$. If the electrons did not interact, the energy of the second electron would also be E_1, the same as that of the first electron, and the ground–state energy would be $E_{non} = 2E_1$. This is represented by the lowest level in Figure 37-16. Because of the interaction energy, the ground–state energy is greater than this as represented by the higher level labeled E_g in the figure. When we add $E_{ion} = 24.6$ eV to ionize He, we obtain ionized helium, written He$^+$, which has just one electron and therefore energy E_1.

Figure 37-16

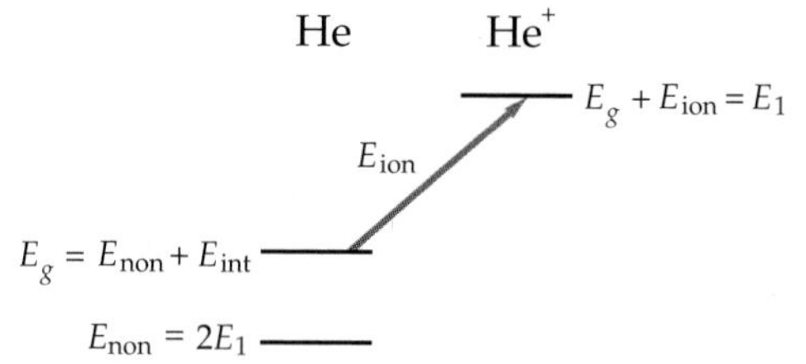

(*a*)1. The energy of interaction plus the energy of two noninteracting electrons equals the ground–state energy of helium:

$$E_{int} + E_{non} = E_g$$

2. Solve for E_{int} and substitute $E_{non} = 2E_1$:

$$E_{int} = E_g - E_{non} = E_g - 2E_1$$

3. Use Equation 37-27 to calculate the energy E_1 of one electron in the ground state:

$$E_1 = -Z^2\frac{E_0}{n^2} = -(2)^2\frac{13.6\text{ eV}}{1^2} = -54.4\text{ eV}$$

4. Substitute this value for E_1:

$$E_{int} = E_g - 2E_1 = E_g - (2)(-54.4 \text{ eV})$$
$$= E_g + 108.8 \text{ eV}$$

5. The ground-state energy of He, E_g, plus the ionization energy equals the energy of He$^+$, which is E_1:

$$E_g + E_{ion} = E_1 = -54.4 \text{ eV}$$

6. Substitute $E_{ion} = 24.6$ eV to calculate E_g:

$$E_g = -54.4 \text{ eV} - E_{ion} = -54.4 \text{ eV} - 24.6 \text{ eV}$$
$$= -79 \text{ eV}$$

7. Substitute this result for E_g to obtain E_{int}:

$$E_{int} = E_g + 108.8 \text{ eV} = -79 \text{ eV} + 108.8 \text{ eV}$$
$$= 29.8 \text{ eV}$$

(b)1. The energy of interaction of two electrons a distance r apart is the potential energy:

$$U = +\frac{ke^2}{r}$$

2. Set U equal to 29.8 eV, and solve for r. It is convenient to express r in terms of a_0, the radius of the first Bohr orbit in hydrogen:

$$r = \frac{ke^2}{U} = \frac{ke^2/a_0}{U} a_0 = \frac{13.6 \text{ eV}}{29.8 \text{ eV}} a_0 = 0.456 a_0$$

Check the Result This separation is approximately equal to the radius of the first Bohr orbit for an electron in helium, which is $r_1 = a_0/Z = 0.50 a_0$.

Lithium (Z = 3)

The next element, lithium, has three electrons. Since the K shell is completely filled with two electrons, the third electron must go into a higher energy shell. The next lowest energy shell after $n = 1$ is the $n = 2$ or L shell. The outer electron is much farther from the nucleus than are the two inner, $n = 1$ electrons. It is most likely to be found at the radius of the second Bohr orbit, which is four times the radius of the first Bohr orbit.

The nuclear charge is partially screened from the outer electron by the two inner electrons. Recall that the electric field outside a spherically symmetric charge density is the same as if all the charge were at the center of the sphere. If the outer electron were completely outside of the charge cloud of the two inner electrons, the electric field it would see would be that of a single charge $+e$ at the center due to the nuclear charge of $+3e$ and the charge $-2e$ of the inner electron cloud. However, the outer electron does not have a well-defined orbit; instead, it is itself a charge cloud that penetrates the charge cloud of the inner electrons to some extent. Because of this penetration, the effective nuclear charge $Z'e$ is somewhat greater than $+1e$. The energy of the outer electron at a distance r from a point charge $+Z'e$ is given by Equation 37-6 with the nuclear charge $+Ze$ replaced by $+Z'e$:

$$E = -\frac{1}{2}\frac{kZ'e^2}{r} \qquad\qquad 37\text{-}45$$

The greater the penetration of the inner electron cloud, the greater the effective nuclear charge $Z'e$ and the lower the energy. Because the penetration is greater for lower ℓ values (see Figure 37-12), the energy of the outer electron in lithium is lower for the s state ($\ell = 0$) than for the p state ($\ell = 1$). The electron configuration of lithium in the ground state is therefore $1s^2 2s$. The ionization energy of lithium is only 5.39 eV. Because its outer electron is so loosely bound to the atom, lithium is very active chemically. It behaves like a "one-electron atom," similar to hydrogen.

Example 37-7

If the outer electron in lithium moved in the $n = 2$ Bohr orbit, the nuclear charge would be shielded by the two inner electrons and the effective nuclear charge would be $Z'e = 1e$. Then the energy of the electron would be $-13.6 \text{ eV}/2^2 = -3.4 \text{ eV}$. However, the ionization energy of lithium is 5.39 eV, not 3.4 eV. Use this fact to calculate the effective nuclear charge Z' seen by the outer electron in lithium.

Picture the Problem Because the outer electron moves in the $n = 2$ shell, we will take $r = 4a_0$ for its average distance from the nucleus. We can then calculate Z' from Equation 37-45. Since r is given in terms of a_0, it will be convenient to use the fact that $E_0 = ke^2/2a_0 = 13.6 \text{ eV}$ (Equation 37-16).

1. Equation 37-45 relates the energy of the outer electron to its average distance r and the effective nuclear charge Z':

$$E = \frac{1}{2}\frac{kZ'e^2}{r}$$

2. Substitute the given values $r = 4a_0$ and $E = -5.39 \text{ eV}$:

$$-5.39 \text{ eV} = -\frac{1}{2}\frac{kZ'e^2}{4a_0} = -\frac{Z'}{4}\left(\frac{ke^2}{2a_0}\right)$$

3. Use $ke^2/2a_0 = 13.6 \text{ eV}$ and solve for Z':

$$-5.39 \text{ eV} = -\frac{Z'}{4}\left(\frac{ke^2}{2a_0}\right) = -\frac{Z'}{4}(13.6 \text{ eV})$$

$$Z' = 4\frac{5.39 \text{ eV}}{13.6 \text{ eV}} = 1.59$$

Remark This calculation is interesting but not very rigorous. We essentially used the circular orbit from the semiclassical Bohr model and the measured ionization energy to calculate the effective inner charge seen by the outer electron. We know, of course, that this outer electron does not move in a circular orbit of constant radius, but is better represented by a stationary charged cloud of charge density $|\psi|^2$ that penetrates the charged clouds of the inner electrons.

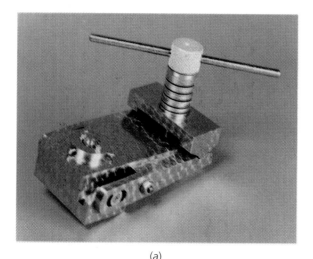

(a)

(b)

(a) A diamond anvil cell, in which the facets of two diamonds (about 1 mm² each) are used to compress a sample substance, subjecting it to very high pressure. (b) Samarium monosulfide (SmS) is normally a black, dull-looking semiconductor. When it is subjected to pressure above 7000 atm, an electron from the 4f state is dislocated into the 5d state. The resulting compound glitters like gold and behaves like a metal.

Beryllium (Z = 4)

The fourth electron has the least energy in the 2s state. There can be two electrons with $n = 2$, $\ell = 0$, and $m = 0$ because of the two possible values for the spin quantum number m_s. The configuration of beryllium is thus $1s^2 2s^2$.

Hydrogen

Boron to Neon (Z = 5 to Z = 10)

Since the 2s subshell is filled, the fifth electron must go into the next available (lowest energy) subshell, which is the 2p subshell, with $n = 2$ and $\ell = 1$. Since there are three possible values of m ($+1$, 0, and -1) and two values of m_s for each value of m, there can be six electrons in this subshell. The electron configuration for boron is $1s^2 2s^2 2p$. The electron configurations for the elements carbon ($Z = 6$) to neon ($Z = 10$) differ from that for boron only in the number of electrons in the 2p subshell. The ionization energy increases with Z for these elements, reaching the value of 21.6 eV for the last element in the group, neon. Neon has the maximum number of electrons allowed in the $n = 2$ shell. Its electron configuration is $1s^2 2s^2 2p^6$. Because of its very high ionization energy, neon, like helium, is basically chemically inert. The element just before neon, fluorine, has a "hole" in the 2p subshell; that is, it has room for one more electron. It readily combines with elements such as lithium that have one outer electron. Lithium, for example, will donate its single outer electron to the fluorine atom to make an F^- ion and an Li^+ ion. These ions then bond together to form a molecule of lithium fluoride.

Carbon

Silicon

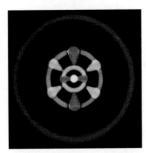

Iron

Sodium to Argon (Z = 11 to Z = 18)

The eleventh electron must go into the $n = 3$ shell. Since this electron is very far from the nucleus and from the inner electrons, it is weakly bound in the sodium ($Z = 11$) atom. The ionization energy of sodium is only 5.14 eV. Sodium therefore combines readily with atoms such as fluorine. With $n = 3$, the value of ℓ can be 0, 1, or 2. Because of the lowering of the energy due to penetration of the electron shield formed by the other ten electrons (similar to that discussed for lithium) the 3s state is lower than the 3p or 3d states. This energy difference between subshells of the same n value becomes greater as the number of electrons increases. The electron configuration of sodium is $1s^2 2s^2 2p^6 3s^1$. As we move to elements with higher values of Z, the 3s subshell and then the 3p subshell fill. These two subshells can accommodate $2 + 6 = 8$ electrons. The configuration of argon ($Z = 18$) is $1s^2 2s^2 2p^6 3s^2 3p^6$. One might expect the nineteenth electron to go into the third subshell (the d subshell with $\ell = 2$), but the penetration effect is now so strong that the energy of the next electron is lower in the 4s subshell than in the 3d subshell. There is thus another large energy difference between the eighteenth and nineteenth electrons, and so argon, with its full 3p subshell, is basically stable and inert.

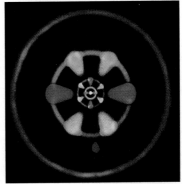

Silver

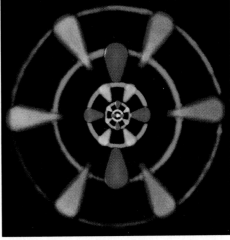

Europium

A schematic depiction of the electron configurations in atoms. The spherically symmetric s states can contain 2 electrons and are colored white and blue. The dumbbell-shaped p states can contain up to 6 electrons and are colored orange. The d states can contain up to 10 electrons and are colored yellow-green. The f states can contain up to 14 electrons and are colored purple.

Elements with Z > 18

The nineteenth electron in potassium ($Z = 19$) and the twentieth electron in calcium ($Z = 20$) go into the 4s rather than the 3d subshell. The electron configurations of the next ten elements, scandium ($Z = 21$) through zinc ($Z = 30$), differ only in the number of electrons in the 3d shell, except for chromium ($Z = 24$) and copper ($Z = 29$), each of which has only one 4s electron. These ten elements are called **transition elements.** Since their chemical properties are mainly due to their 4s electrons, they are quite similar chemically.

Figure 37-17 shows a plot of the ionization energy versus Z for $Z = 1$ to $Z = 60$. The peaks in ionization energy at $Z = 2, 10, 18, 36,$ and 54 mark the closing of a shell or subshell. Table 37-1 gives the electron configurations of all the elements.

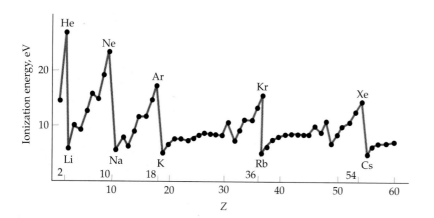

Figure 37-17 Ionization energy versus Z for $Z = 1$ to $Z = 60$. This energy is the binding energy of the last electron in the atom. The binding energy increases with Z until a shell is closed at $Z = 2, 10, 18, 36,$ and 54. Elements with a closed shell plus one outer electron, such as sodium ($Z = 11$), have very low binding energies because the outer electron is very far from the nucleus and is shielded by the inner core electrons.

Table 37-1

Electron Configurations of the Atoms in Their Ground States
For some of the rare-earth elements (Z = 57 to 71) and the heavy elements (Z > 89) the configurations are not firmly established.

			Shell (*n*): K (1)	L (2)		M (3)			N (4)				O (5)				P (6)			Q (7)
				s	p	s	p	d	s	p	d	f	s	p	d	f	s	p	d	s
Z		Element	Subshell (ℓ): (0)	(0)	(1)	(0)	(1)	(2)	(0)	(1)	(2)	(3)	(0)	(1)	(2)	(3)	(0)	(1)	(2)	(1)
1	H	hydrogen	1																	
2	He	helium	2																	
3	Li	lithium	2	1																
4	Be	beryllium	2	2																
5	B	boron	2	2	1															
6	C	carbon	2	2	2															
7	N	nitrogen	2	2	3															
8	O	oxygen	2	2	4															
9	F	fluorine	2	2	5															
10	Ne	neon	2	2	6															
11	Na	sodium	2	2	6	1														
12	Mg	magnesium	2	2	6	2														
13	Al	aluminum	2	2	6	2	1													
14	Si	silicon	2	2	6	2	2													
15	P	phosphorus	2	2	6	2	3													
16	S	sulfur	2	2	6	2	4													
17	Cl	chlorine	2	2	6	2	5													
18	Ar	argon	2	2	6	2	6													

Table 37-1 (*Continued*)

Electron Configurations of the Atoms in Their Ground States

For some of the rare-earth elements (Z = 57 to 71) and the heavy elements (Z > 89) the configurations are not firmly established.

Z	Element	K (1) s(0)	L (2) s(0)	L p(1)	M (3) s(0)	M p(1)	M d(2)	N (4) s(0)	N p(1)	N d(2)	N f(3)	O (5) s(0)	O p(1)	O d(2)	O f(3)	P (6) s(0)	P p(1)	P d(2)	Q (7) s(1)
19	K potassium	2	2	6	2	6	.	1											
20	Ca calcium	2	2	6	2	6	.	2											
21	Sc scandium	2	2	6	2	6	1	2											
22	Ti titanium	2	2	6	2	6	2	2											
23	V vanadium	2	2	6	2	6	3	2											
24	Cr chromium	2	2	6	2	6	5	1											
25	Mn manganese	2	2	6	2	6	5	2											
26	Fe iron	2	2	6	2	6	6	2											
27	Co cobalt	2	2	6	2	6	7	2											
28	Ni nickel	2	2	6	2	6	8	2											
29	Cu copper	2	2	6	2	6	10	1											
30	Zn zinc	2	2	6	2	6	10	2											
31	Ga gallium	2	2	6	2	6	10	2	1										
32	Ge germanium	2	2	6	2	6	10	2	2										
33	As arsenic	2	2	6	2	6	10	2	3										
34	Se selenium	2	2	6	2	6	10	2	4										
35	Br bromine	2	2	6	2	6	10	2	5										
36	Kr krypton	2	2	6	2	6	10	2	6										
37	Rb rubidium	2	2	6	2	6	10	2	6	.	.	1							
38	Sr strontium	2	2	6	2	6	10	2	6	.	.	2							
39	Y yttrium	2	2	6	2	6	10	2	6	1	.	2							
40	Zr zirconium	2	2	6	2	6	10	2	6	2	.	2							
41	Nb niobium	2	2	6	2	6	10	2	6	4	.	1							
42	Mo molybdenum	2	2	6	2	6	10	2	6	5	.	1							
43	Tc technetium	2	2	6	2	6	10	2	6	6	.	1							
44	Ru ruthenium	2	2	6	2	6	10	2	6	7	.	1							
45	Rh rhodium	2	2	6	2	6	10	2	6	8	.	1							
46	Pd palladium	2	2	6	2	6	10	2	6	10	.	.							
47	Ag silver	2	2	6	2	6	10	2	6	10	.	1							
48	Cd cadmium	2	2	6	2	6	10	2	6	10	.	2							
49	In indium	2	2	6	2	6	10	2	6	10	.	2	1						
50	Sn tin	2	2	6	2	6	10	2	6	10	.	2	2						
51	Sb antimony	2	2	6	2	6	10	2	6	10	.	2	3						
52	Te tellurium	2	2	6	2	6	10	2	6	10	.	2	4						
53	I iodine	2	2	6	2	6	10	2	6	10	.	2	5						
54	Xe xenon	2	2	6	2	6	10	2	6	10	.	2	6						
55	Cs cesium	2	2	6	2	6	10	2	6	10	.	2	6	.	.	1			
56	Ba barium	2	2	6	2	6	10	2	6	10	.	2	6	.	.	2			
57	La lanthanum	2	2	6	2	6	10	2	6	10	.	2	6	1	.	2			
58	Ce cerium	2	2	6	2	6	10	2	6	10	1	2	6	1	.	2			
59	Pr praseodymium	2	2	6	2	6	10	2	6	10	3	2	6	.	.	2			
60	Nd neodymium	2	2	6	2	6	10	2	6	10	4	2	6	.	.	2			
61	Pm promethium	2	2	6	2	6	10	2	6	10	5	2	6	.	.	2			
62	Sm samarium	2	2	6	2	6	10	2	6	10	6	2	6	.	.	2			
63	Eu europium	2	2	6	2	6	10	2	6	10	7	2	6	.	.	2			
64	Gd gadolinium	2	2	6	2	6	10	2	6	10	7	2	6	1	.	2			
65	Tb terbium	2	2	6	2	6	10	2	6	10	9	2	6	.	.	2			
66	Dy dysprosium	2	2	6	2	6	10	2	6	10	10	2	6	.	.	2			
67	Ho holmium	2	2	6	2	6	10	2	6	10	11	2	6	.	.	2			
68	Er erbium	2	2	6	2	6	10	2	6	10	12	2	6	.	.	2			
69	Tm thulium	2	2	6	2	6	10	2	6	10	13	2	6	.	.	2			
70	Yb ytterbium	2	2	6	2	6	10	2	6	10	14	2	6	.	.	2			
71	Lu lutetium	2	2	6	2	6	10	2	6	10	14	2	6	1	.	2			
72	Hf hafnium	2	2	6	2	6	10	2	6	10	14	2	6	2	.	2			
73	Ta tantalum	2	2	6	2	6	10	2	6	10	14	2	6	3	.	2			
74	W tungsten (wolfram)	2	2	6	2	6	10	2	6	10	14	2	6	4	.	2			

Continued on page 1192

Table 37-1 (*Continued*)

Electron Configurations of the Atoms in Their Ground States
For some of the rare-earth elements (Z = 57 to 71) and the heavy elements (Z > 89) the configurations are not firmly established.

Z	Element	K (1) s (0)	L (2) s (0)	L (2) p (1)	M (3) s (0)	M (3) p (1)	M (3) d (2)	N (4) s (0)	N (4) p (1)	N (4) d (2)	N (4) f (3)	O (5) s (0)	O (5) p (1)	O (5) d (2)	O (5) f (3)	P (6) s (0)	P (6) p (1)	P (6) d (2)	Q (7) s (1)
75	Re rhenium	2	2	6	2	6	10	2	6	10	14	2	6	5	.	2			
76	Os osmium	2	2	6	2	6	10	2	6	10	14	2	6	6	.	2			
77	Ir iridium	2	2	6	2	6	10	2	6	10	14	2	6	7	.	2			
78	Pt platinum	2	2	6	2	6	10	2	6	10	14	2	6	9	.	1			
79	Au gold	2	2	6	2	6	10	2	6	10	14	2	6	10	.	1			
80	Hg mercury	2	2	6	2	6	10	2	6	10	14	2	6	10	.	2			
81	Tl thallium	2	2	6	2	6	10	2	6	10	14	2	6	10	.	2	1		
82	Pb lead	2	2	6	2	6	10	2	6	10	14	2	6	10	.	2	2		
83	Bi bismuth	2	2	6	2	6	10	2	6	10	14	2	6	10	.	2	3		
84	Po polonium	2	2	6	2	6	10	2	6	10	14	2	6	10	.	2	4		
85	At astatine	2	2	6	2	6	10	2	6	10	14	2	6	10	.	2	5		
86	Rn radon	2	2	6	2	6	10	2	6	10	14	2	6	10	.	2	6		
87	Fr francium	2	2	6	2	6	10	2	6	10	14	2	6	10	.	2	6	.	1
88	Ra radium	2	2	6	2	6	10	2	6	10	14	2	6	10	.	2	6	.	2
89	Ac actinium	2	2	6	2	6	10	2	6	10	14	2	6	10	.	2	6	1	2
90	Th thorium	2	2	6	2	6	10	2	6	10	14	2	6	10	.	2	6	2	2
91	Pa protactinium	2	2	6	2	6	10	2	6	10	14	2	6	10	1	2	6	2	2
92	U uranium	2	2	6	2	6	10	2	6	10	14	2	6	10	3	2	6	1	2
93	Np neptunium	2	2	6	2	6	10	2	6	10	14	2	6	10	4	2	6	1	2
94	Pu plutonium	2	2	6	2	6	10	2	6	10	14	2	6	10	6	2	6	.	2
95	Am americium	2	2	6	2	6	10	2	6	10	14	2	6	10	7	2	6	.	2
96	Cm curium	2	2	6	2	6	10	2	6	10	14	2	6	10	7	2	6	1	2
97	Bk berkelium	2	2	6	2	6	10	2	6	10	14	2	6	10	8	2	6	1	2
98	Cf californium	2	2	6	2	6	10	2	6	10	14	2	6	10	10	2	6	.	2
99	Es einsteinium	2	2	6	2	6	10	2	6	10	14	2	6	10	11	2	6	.	2
100	Fm fermium	2	2	6	2	6	10	2	6	10	14	2	6	10	12	2	6	.	2
101	Md mendelevium	2	2	6	2	6	10	2	6	10	14	2	6	10	13	2	6	.	2
102	No nobelium	2	2	6	2	6	10	2	6	10	14	2	6	10	14	2	6	.	2
103	Lw lawrencium	2	2	6	2	6	10	2	6	10	14	2	6	10	14	2	6	1	2

37-7   Optical and X-Ray Spectra

When an atom is in an excited state (that is, when it is in an energy state above the ground state), it makes transitions to lower energy states, and in doing so emits electromagnetic radiation. The wavelength of the electromagnetic radiation emitted is related to the initial and final states by the Bohr formula (Equation 37-17), $\lambda = hc/(E_i - E_f)$, where E_i and E_f are the initial and final energies and h is Planck's constant. The atom can be excited to a higher energy state by bombarding it with a beam of electrons as in a spectral tube with a high voltage across it. Since the excited energy states of an atom form a discrete (rather than continuous) set, only certain wavelengths are emitted. These wavelengths of the emitted radiation constitute the emission spectrum of the atom.

Optical Spectra

To understand atomic spectra we thus need to understand the excited states of the atom. The situation for an atom with many electrons is, in general, much more complicated than that of hydrogen with just one electron. An excited state of the atom may involve a change in the state of any one of the electrons, or even two or more electrons. Fortunately, in most cases, an excited state of an atom involves the excitation of just one of the electrons in the atom. The energies of excitation of the outer, valence electrons of an atom are of the order of a few electron volts. Transitions involving these electrons result in photons in or near the visible or **optical spectrum**. (Recall that the energies of visible photons range from about 1.5 to 3 eV.) The excitation energies can often be calculated from a simple model in which the atom is pictured as a single electron plus a stable core consisting of the nucleus plus the other, inner electrons. This model works particularly well for the alkali metals: Li, Na, K, Rb, and Cs. These elements are in the first column of the periodic table. The optical spectra of these elements are similar to that of hydrogen.

Figure 37-18 shows an energy-level diagram for the optical transitions in sodium, whose electrons form a neon core plus one outer electron. Since the spin angular momentum of the core adds up to zero, the spin of each state of sodium is $\frac{1}{2}$. Because of the spin–orbit effect, the states with $j = \ell - \frac{1}{2}$ have a slightly lower energy than those with $j = \ell + \frac{1}{2}$. Each state (except for the S states) is therefore split into two states, called a doublet. The doublet splitting is very small and not evident on the energy scale of this diagram. The states are labeled by the usual spectroscopic notation with the superscript 2 before the letter indicating that the state is a doublet. Thus $^2P_{3/2}$, read as

A neon sign outside a Chinatown restaurant in Paris. Neon atoms in the tube are excited by an electron current passing through the tube. The excited neon atoms emit light in the visible range as they decay toward their ground states. The colors of neon signs result from the characteristic red-orange spectrum of neon plus the color of the glass tube itself.

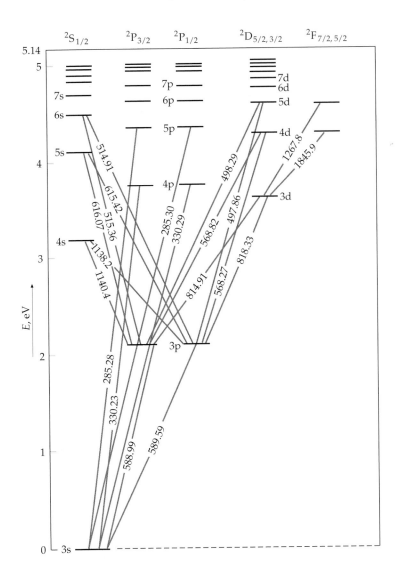

Figure 37-18 Energy-level diagram for sodium. The diagonal lines show observed optical transitions, with wavelengths given in nanometers. The energy of the ground state has been chosen as the zero point for the scale on the left.

"doublet P three halves," denotes a state in which $\ell = 1$ and $j = \frac{3}{2}$. (The S states are customarily labeled as if they were doublets even though they are not.) In the first excited state, the outer electron is excited from the 3s level to the 3p level, which is about 2.1 eV above the ground state. The energy difference between the $P_{3/2}$ and $P_{1/2}$ states due to the spin–orbit effect is about 0.002 eV. Transitions from these states to the ground state give the familiar sodium yellow doublet:

$$3p(^2P_{1/2}) \rightarrow 3s(^2S_{1/2}), \qquad \lambda = 589.6 \text{ nm}$$
$$3p(^2P_{3/2}) \rightarrow 3s(^2S_{1/2}), \qquad \lambda = 589.0 \text{ nm}$$

The energy levels and spectra of other alkali metal atoms are similar to those for sodium. The optical spectrum for atoms such as helium, beryllium, and magnesium that have two outer electrons is considerably more complex because of the interaction of the two outer electrons.

X-Ray Spectra

X rays are usually produced by bombarding a target element with a high-energy beam of electrons in an X-ray tube. The result (Figure 37-19) consists of a continuous spectrum that depends only on the energy of the bombarding electrons, and a line spectrum that is characteristic of the target element. The characteristic spectrum results from excitation of the inner core electrons in the target element.

The energy needed to excite an inner core electron—for example, an electron in the $n = 1$ state (K shell)—is much greater than that required to excite an outer, valence electron. An inner electron cannot be excited to any of the filled states (for example, the $n = 2$ states in sodium) because of the exclusion principle. The energy required to excite an inner core electron to an unoccupied state is typically of the order of several keV. If an electron is knocked out of the $n = 1$ state (K shell), there is a vacancy left in this shell. This vacancy can be filled if an electron in the L shell (or in a higher shell) makes a transition into the K shell. The photons emitted by electrons making such transitions also have energies of the order of keV and produce the sharp peaks in the X-ray spectrum, as shown in Figure 37-18. The K_α line arises from transitions from the $n = 2$ (L) shell to the $n = 1$ (K) shell. The K_β line arises from transitions from the $n = 3$ shell to the $n = 1$ shell. These and other lines arising from transitions ending at the $n = 1$ shell make up the K series of the characteristic X-ray spectrum of the target element. Similarly, a second series, the L series, is produced by transitions from higher energy states to a vacated place in the $n = 2$ (L) shell.

We can use the Bohr theory to calculate approximately the frequencies of the characteristic X-ray spectra. According to the Bohr model, the energy of a single electron in a state n is given by

$$E_n = -Z^2 \frac{13.6 \text{ eV}}{n^2}$$

Since for any atom other than hydrogen, there are two electrons in the innermost shell, the K shell, the effective charge seen by one of the electrons is less than Ze because of the shielding due to the other electron. Assuming that the effective charge is $(Z - 1)e$, the energy of an electron in the K shell is given by this equation with $n = 1$ and Z replaced by $Z - 1$:

$$E_1 = -(Z - 1)^2 (13.6 \text{ eV})$$

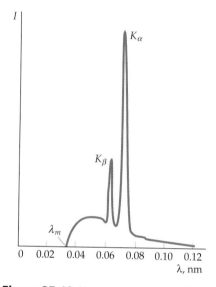

Figure 37-19 X-ray spectrum of molybdenum. The sharp peaks labeled K_α and K_β are characteristic of the element. The cutoff wavelength λ_m is independent of the target element and is related to the voltage V of the X-ray tube by $\lambda_m = hc/eV$.

The energy of an electron in the state n (assuming the same effective charge) is given by

$$E_n = -(Z-1)^2 \frac{13.6 \text{ eV}}{n^2}$$

When an electron from state n drops into the vacated state in the $n = 1$ shell, a photon of energy $E_n - E_1$ is emitted. The wavelength of this photon is

$$\lambda = \frac{hc}{E_n - E_1} = \frac{hc}{(Z-1)^2(13.6 \text{ eV})(1 - 1/n^2)} \qquad \text{37-46}$$

In 1913 the English physicist H. Moseley measured the wavelengths of the characteristic X-ray spectra for about 40 elements. From his data he was able to accurately determine the atomic number Z for each element.

Example 37-8

The wavelength of the K_α X-ray line for a certain element is $\lambda = 0.0721$ nm. What is the element?

Picture the Problem The K_α line corresponds to a transition from $n = 2$ to $n = 1$. The wavelength is related to the atomic number Z by Equation 37-46 with $n = 2$.

1. Solve Equation 37-46 for $(Z-1)^2$:

$$\lambda = \frac{hc}{(Z-1)^2(13.6 \text{ eV})(1 - 1/n^2)}$$

$$(Z-1)^2 = \frac{hc}{\lambda(13.6 \text{ eV})(1 - 1/n^2)}$$

2. Substitute the given data and solve for Z:

$$(Z-1)^2 = \frac{1240 \text{ eV} \cdot \text{nm}}{(0.0721 \text{ nm})(13.6 \text{ eV})(1 - 1/4)} = 1686$$

$$Z - 1 = \sqrt{1686} = 41.06$$

$$Z = 42$$

Remark Since Z is an integer, we round to the nearest integer. The element is molybdenum.

Summary

1. The Bohr model is important historically because it was the first successful model to explain the discrete optical spectrum of atoms in terms of the quantization of energy.

2. The quantum theory of atoms results from the application of the Schrödinger equation to a bound system consisting of nucleus of charge $+Ze$ and Z electrons of charge $-e$.

3. For the simplest atom, hydrogen, consisting of one proton and one electron, the Schrödinger equation can be solved exactly to obtain the wave functions ψ, which depend on the quantum numbers n, ℓ, m, and m_s.

4. The electron configuration of atoms is governed by the Pauli exclusion principle, which states that no two electrons in an atom can have the same set of values for the quantum numbers n, ℓ, m, and m_s. Using the exclusion principle and the restrictions on the quantum numbers, we can understand much of the structure of the periodic table.

Topic	Remarks and Relevant Equations	
1. Bohr Model		
Postulates for the hydrogen atom		
Nonradiating orbits	The electron moves in a circular nonradiating orbit around the proton.	
Frequency of radiation related to energy states	$f = \dfrac{E_i - E_f}{h}$	37-7
Quantization of angular momentum	$mvr = \dfrac{nh}{2\pi} = n\hbar$	37-9
Radius of allowed orbits	$r = n^2 \dfrac{\hbar^2}{mkZe^2} = n^2 \dfrac{a_0}{Z}$	37-11
Bohr radius	$a_0 = \dfrac{\hbar^2}{mke^2} \approx 0.0529 \text{ nm}$	37-12
Energy levels in the H atom	$E_n = -\dfrac{mk^2e^4}{2\hbar^2}\dfrac{Z^2}{n^2} = -Z^2 \dfrac{E_0}{n^2}, \quad n = 1, 2, \ldots$	37-15
	$E_0 = \dfrac{mk^2e^4}{2\hbar^2} = \dfrac{1}{2}\dfrac{ke^2}{a_0} \approx 13.6 \text{ eV}$	37-16
Wavelengths emitted by the H atom	$\lambda = \dfrac{c}{f} = \dfrac{hc}{E_i - E_f} = \dfrac{1240 \text{ eV·nm}}{E_i - E_f}$	37-17, 37-18
2. Quantum Theory of Atoms	The electron is described by a wave function ψ that obeys the Schrödinger equation. Energy quantization arises from standing wave conditions. ψ is described by the quantum numbers n, ℓ, m, and the spin quantum number $m_s = \pm\frac{1}{2}$.	
Schrödinger equation in spherical coordinates	$-\dfrac{\hbar^2}{2mr^2}\dfrac{\partial}{\partial r}\left(r^2 \dfrac{\partial \psi}{\partial r}\right) - \dfrac{\hbar^2}{2mr^2}\left[\dfrac{1}{\sin\theta}\dfrac{\partial}{\partial\theta}\left(\sin\theta\dfrac{\partial\psi}{\partial\theta}\right) + \dfrac{1}{\sin^2\theta}\dfrac{\partial^2\psi}{\partial\phi^2}\right] + U(r)\psi = E\psi$	37-21
The solutions can be written as products of functions of r, θ, and ϕ separately.	$\psi(r, \theta, \phi) = R(r)f(\theta)g(\phi)$	37-22
Quantum numbers		
Principal quantum number	$n = 1, 2, 3, \ldots$	
Orbital quantum number	$\ell = 0, 1, 2, 3, \ldots, n - 1$	
Magnetic quantum number	$m = -\ell, (-\ell + 1), \ldots, 0, 1, 2, \ldots, \ell$	37-23
Orbital angular momentum	$L = \sqrt{\ell(\ell + 1)}\,\hbar$	37-24

z component of angular momentumn	$L_z = m\hbar$	37-25

3. Quantum Theory of the Hydrogen Atom

Energy levels (same as for Bohr model)	$E_n = -\dfrac{mk^2e^4}{2\hbar^2}\dfrac{Z^2}{n^2} = -Z^2\dfrac{E_0}{n^2},\qquad n = 1, 2, 3, \ldots$	37-27
	where $E_0 = mk^2e^4/2\hbar^2 \approx 13.6\text{ eV}$	37-28
Wavelengths emitted by H atom (same as for Bohr model)	$\lambda = \dfrac{c}{f} = \dfrac{hc}{E_i - E_f} = \dfrac{1240\text{ eV·nm}}{E_i - E_f}$	37-17, 37-18

Wave functions

Ground state	$\psi_{1,0,0} = C_{1,0,0}e^{-Zr/a_0} = \dfrac{1}{\sqrt{\pi}}\left(\dfrac{Z}{a_0}\right)^{3/2}e^{-Zr/a_0}$	37-31, 37-33
First excited state	$\psi_{2,0,0} = C_{2,0,0}\left(2 - \dfrac{Zr}{a_0}\right)e^{-Zr/2a_0}$	37-36
	$\psi_{2,1,0} = C_{2,1,0}\dfrac{Zr}{a_0}e^{-Zr/2a_0}\cos\theta$	37-37
	$\psi_{2,1\pm1} = C_{2,1,1}\dfrac{Zr}{a_0}e^{-Zr/2a_0}\sin\theta\,e^{\pm i\phi}$	37-38

Probability densities	For $\ell = 0$, $|\psi|^2$ is spherically symmetric. For $\ell \neq 0$, $|\psi|^2$ depends on the angle θ.	
Radial probability density	$P(r) = 4\pi r^2|\psi|^2$	37-34

The radial probability density is maximum at the distances corresponding roughly to the Bohr orbits.

4. Spin–Orbit Effect and Fine Structure

The total angular momentum of an electron in an atom is a combination of the orbital angular momentum and spin angular momentum. It is characterized by the quantum number j, which can be either $\ell - \frac{1}{2}$ or $\ell + \frac{1}{2}$. Because of the interaction of the orbital and spin magnetic moments, the state $j = \ell - \frac{1}{2}$ has lower energy than the state $j = \ell + \frac{1}{2}$. This small splitting of the energy states gives rise to a small splitting of the spectral lines called fine structure.

5. Periodic Table

Beginning with hydrogen, each larger neutral atom adds one electron. The electrons go into those states that will give the lowest energy consistent with the Pauli exclusion principle.

The state of an atom is described by its electron configuration, which gives the values of n and ℓ for each electron. The ℓ values are specified by a code:

	s	p	d	f	g	h
ℓ values	0	1	2	3	4	5

Pauli exclusion principle	No two electrons in an atom can have the same set of values for the quantum numbers n, ℓ, m, and m_s.

6. Atomic Spectra

Atomic spectra include optical spectra and X-ray spectra. Optical spectra result from transitions between energy levels of a single outer electron moving in the field of the nucleus and core electrons of the atom. Characteristic X-ray spectra result from the excitation of an inner core electron and the subsequent filling of the vacancy by other electrons in the atom.

Selection rules

Transitions between energy states with the emission of a photon are governed by the following selection rules:

$$\Delta m = 0 \quad \text{or} \quad \pm 1$$

$$\Delta \ell = \pm 1$$

37-29

Problem-Solving Guide

Begin by drawing a neat diagram that includes the important features of the problem.

Summary of Worked Examples

Type of Calculation	Procedure and Relevant Examples
1. Calculate Wavelengths of Emitted Radiation	
Find the wavelength of radiation emitted by an atom.	Use $\lambda = hc/(E_i - E_f)$, where $hc = 1240$ eV·nm and E_i and E_f are the energies of the initial and final states. **Example 37-2**
2. Angular Momentum	
Find the possible values of L_z given ℓ.	The possible values are $m\hbar$, where m ranges in integral steps from $-\ell$ to $+\ell$. **Example 37-3**
Find the angle between $\vec{L}$ and the z axis.	Use $\cos\theta = L_z/L$, where $L_z = m\hbar$ and $L = \sqrt{\ell(\ell+1)}\,\hbar$. **Example 37-3**
3. Hydrogen Atom Wave Functions	
Determine the probability of finding the electron in some range Δr in the hydrogen atom.	For small Δr, the probability is given by $P(r)\,\Delta r = 4\pi r^2 |\psi|^2\,\Delta r$, where ψ is the appropriate wave function. **Example 37-4**
4. Spin–Orbit Effect	
Estimate the magnetic field seen by an electron given the fine-structure splitting of the energy levels.	Use $\Delta E = 2\mu_B B$. **Example 37-5**
5. Periodic Table	
Estimate the energy of interaction of two electrons in an atom.	Compare the ionization energy with the total energy of the electrons, neglecting their interaction. **Example 37-6**
Find the effective nuclear charge seen by an outer electron.	Choose the orbit radius from the Bohr model and calculate the energy from $E = -kZ'e^2/2r$ using the value of E determined from the ionization energy. **Example 37-7**
6. X-Ray Spectra	
Find the element given the wavelength of a line in the K series.	Use $$\lambda = \frac{hc}{E_n - E_1} = \frac{hc}{(Z-1)^2(13.6\text{ eV})(1 - 1/n^2)}$$ **Example 37-8**

Problems

In a few problems, you are given more data than you actually need; in a few other problems, you are required to supply data from your general knowledge, outside sources, or informed estimates.

•   Single-concept, single-step, relatively easy
•• Intermediate-level, may require synthesis of concepts
••• Challenging, for advanced students

The Bohr Model

1 • As n increases, does the spacing of adjacent energy levels increase or decrease?

2 • The energy of the ground state of doubly ionized lithium ($Z = 3$) is _____ , where $E_0 = 13.6$ eV.

(a) $-9E_0$
(b) $-3E_0$
(c) $-E_0/3$
(d) $-E_0/9$

3 • Bohr's quantum condition on electron orbits requires

(a) that the angular momentum of the electron about the hydrogen nucleus equal $n\hbar$.
(b) that no more than one electron occupy a given stationary state.
(c) that the electrons spiral into the nucleus while radiating electromagnetic waves.
(d) that the energies of an electron in a hydrogen atom be equal to nE_0, where E_0 is a constant energy and n is an integer.
(e) none of the above.

4 •• If an electron moves to a larger orbit, does its total energy increase or decrease? Does its kinetic energy increase or decrease?

5 •• The kinetic energy of the electron in the ground state of hydrogen is 13.6 eV = E_0. The kinetic energy of the electron in the state $n = 2$ is _____ .

(a) $4E_0$
(b) $2E_0$
(c) $E_0/2$
(d) $E_0/4$

6 • The radius of the $n = 1$ orbit in the hydrogen atom is $a_0 = 0.053$ nm. What is the radius of the $n = 5$ orbit?

(a) $5a_0$
(b) $25a_0$
(c) a_0
(d) $\frac{1}{5}a_0$
(e) $\frac{1}{25}a_0$

7 • Use the known values of the constants in Equation 37-11 to show that a_0 is approximately 0.0529 nm.

8 • The longest wavelength of the Lyman series was calculated in Example 37-2. Find the wavelengths for the transitions (a) $n_1 = 3$ to $n_2 = 1$ and (b) $n_1 = 4$ to $n_2 = 1$.

9 • Find the photon energy for the three longest wavelengths in the Balmer series and calculate the wavelengths.

10 • (a) Find the photon energy and wavelength for the series limit (shortest wavelength) in the Paschen series ($n_2 = 3$). (b) Calculate the wavelengths for the three longest wavelengths in this series and indicate their positions on a horizontal linear scale.

11 • Repeat Problem 10 for the Brackett series ($n_2 = 4$).

12 • A hydrogen atom is in its tenth excited state according to the Bohr model ($n = 11$). (a) What is the radius of the Bohr orbit? (b) What is the angular momentum of the electron? (c) What is the electron's kinetic energy? (d) What is the electron's potential energy? (e) What is the electron's total energy?

13 •• The binding energy of an electron is the minimum energy required to remove the electron from its ground state to a large distance from the nucleus. (a) What is the binding energy for the hydrogen atom? (b) What is the binding energy for He$^+$? (c) What is the binding energy for Li^{2+}? [Singly ionized helium (He$^+$) and doubly ionized lithium (Li^{2+}) are "hydrogen-like" in that the system consists of a positively charged nucleus and a single bound electron.]

14 •• The electron of a hydrogen atom is in the $n = 2$ state. The electron makes a transition to the ground state. (a) What is the energy of the photon according to the Bohr model? (b) The linear momentum of the emitted photon is related to its energy by $p = E/c$. If we assume conservation of linear momentum, what is the recoil velocity of the atom? (c) Find the recoil kinetic energy of the atom in electron volts. By what percentage must the energy of the photon calculated in part (a) be corrected to account for this recoil energy?

15 •• Show that the speed of an electron in the nth Bohr orbit of hydrogen is given by $v_n = e^2/2\epsilon_0 hn$.

16 •• In this problem you will estimate the radius and the energy of the lowest stationary state of the hydrogen atom using the uncertainty principle. The total energy of the electron of momentum p and mass m a distance r from the proton in the hydrogen atom is given by $E = p^2/2m - ke^2/r$, where k is the Coulomb constant. Assume that the minimum value of p^2 is $p^2 \approx (\Delta p)^2 = \hbar^2/r^2$, where Δp is the uncertainty in p and we have taken $\Delta r \sim r$ for the order of magnitude of the uncertainty in position; the energy is then $E = \hbar^2/2mr^2 - ke^2/r$. Find the radius r_m for which this energy is a minimum, and calculate the minimum value of E in electron volts.

17 •• In a reference frame with the origin at the center of mass of an electron and the nucleus of an atom, the electron and nucleus have equal and opposite momenta of magnitude p. (a) Show that the total kinetic energy of the electron and nucleus can be written $K = p^2/2m_r$, where $m_r = m_e M/(M + m_e)$ is called the reduced mass, m_e is the mass of the electron, and M is the mass of the nucleus. It can be shown that the motion of the nucleus can be accounted for by replacing the mass of the electron by the reduced mass. In general, the reduced mass for a two-body problem with masses m_1 and m_2 is given by

$$m_r = \frac{m_1 m_2}{m_1 + m_2} \qquad \text{37-47}$$

(b) Use Equation 37-14 with m replaced by m_r to calculate the Rydberg constant for hydrogen ($M = m_p$) and for a very massive nucleus ($M = \infty$). (c) Find the percentage correction for the ground-state energy of the hydrogen atom due to the motion of the proton.

Quantum Numbers in Spherical Coordinates

18 • For the principal quantum number $n = 4$, how many different values can the orbital quantum number ℓ have?

(a) 4
(b) 3
(c) 7
(d) 16
(e) 6

19 • For the principal quantum number $n = 4$, how many different values can the magnetic quantum number m have?

(a) 4
(b) 3
(c) 7
(d) 16
(e) 6

20 • For $\ell = 1$, find (a) the magnitude of the angular momentum L and (b) the possible values of m. (c) Draw to scale a vector diagram showing the possible orientations of $\vec{L}$ with the z axis.

21 • Work Problem 20 for $\ell = 3$.

22 • A compact disk has a moment of inertia of about 2.3×10^{-5} kg·m^2. (a) Find its angular momentum L when it is rotating at 500 rev/min. (b) Find the approximate value of the quantum number ℓ for this angular momentum.

23 • If $n = 3$, (a) what are the possible values of ℓ? (b) For each value of ℓ in (a), list the possible values of m. (c) Using the fact that there are two quantum states for each value of ℓ and m because of electron spin, find the total number of electron states with $n = 3$.

24 • Find the total number of electron states with (a) $n = 2$ and (b) $n = 4$. (See Problem 23.)

25 • Find the minimum value of the angle θ between $\vec{L}$ and the z axis for (a) $\ell = 1$, (b) $\ell = 4$, and (c) $\ell = 50$.

26 • What are the possible values of n and m if (a) $\ell = 3$, (b) $\ell = 4$, and (c) $\ell = 0$?

27 • What are the possible values of n and ℓ if (a) $m = 0$, (b) $m = -1$, and (c) $m = 2$?

Quantum Theory of the Hydrogen Atom

28 • For the ground state of the hydrogen atom, find the values of (a) ψ, (b) ψ^2, and (c) the radial probability density $P(r)$ at $r = a_0$. Give your answers in terms of a_0.

29 • (a) If spin is not included, how many different wave functions are there corresponding to the first excited energy level $n = 2$ for hydrogen? (b) List these functions by giving the quantum numbers for each state.

30 • For the ground state of the hydrogen atom, calculate the probability of finding the electron in the range $\Delta r = 0.03a_0$ at (a) $r = a_0$ and (b) $r = 2a_0$.

31 • The value of the constant $C_{2,0,0}$ in Equation 37-36 is

$$C_{2,0,0} = \frac{1}{4\sqrt{2\pi}}\left(\frac{Z}{a_0}\right)^{3/2}$$

Find the values of (a) ψ, (b) ψ^2, and (c) the radial probability density $P(r)$ at $r = a_0$ for the state $n = 2$, $\ell = 0$, $m = 0$ in hydrogen. Give your answers in terms of a_0.

32 •• Show that the radial probability density for the $n = 2$, $\ell = 1$, $m = 0$ state of a one-electron atom can be written as $P(r) = A \cos^2 \theta r^4 e^{-Zr/a_0}$, where A is a constant.

33 •• Calculate the probability of finding the electron in the range $\Delta r = 0.02a_0$ at (a) $r = a_0$ and (b) $r = 2a_0$ for the state $n = 2$, $\ell = 0$, $m = 0$ in hydrogen. (See Problem 31 for the value of $C_{2,0,0}$.)

34 •• The radial probability distribution function for a one-electron atom in its ground state can be written $P(r) = Cr^2 e^{-2Zr/a_0}$, where C is a constant. Show that $P(r)$ has its maximum value at $r = a_0/Z$.

35 •• Find the expectation value of r, $\langle r \rangle = \int_0^\infty rP(r)\, dr$ for hydrogen in its ground state.

36 ••• Show that the number of states in the hydrogen atom for a given n is $2n^2$.

37 ••• Calculate the probability that the electron in the ground state of a hydrogen atom is in the region $0 < r < a_0$.

The Spin–Orbit Effect

38 • The potential energy of a magnetic moment in an external magnetic field is given by $U = -\vec{u} \cdot \vec{B}$. (a) Calculate the difference in energy between the two possible orientations of an electron in a magnetic field $\vec{B} = 0.600$ T $\hat{k}$. (b) If these electrons are bombarded with photons of energy equal to this energy difference, "spin flip" transitions can be induced. Find the wavelength of the photons needed for such transitions. This phenomenon is called *electron spin resonance*.

39 • The total angular momentum of a hydrogen atom in a certain excited state has the quantum number $j = \frac{1}{2}$. What can you say about the orbital angular-momentum quantum number ℓ?

40 • The total angular momentum of a hydrogen atom in a certain excited state has the quantum number $j = 1\frac{1}{2}$. What can you say about the orbital angular-momentum quantum number ℓ?

41 • A hydrogen atom is in the state $n = 3$, $\ell = 2$. What are the possible values of j?

The Periodic Table

42 • The p state of an electronic configuration corresponds to

(a) $n = 2$.
(b) $\ell = 2$.
(c) $\ell = 1$.
(d) $n = 0$.
(e) $\ell = 0$.

43 •• Why is the energy of the 3s state considerably lower than that of the 3p state for sodium, whereas in hydrogen these states have essentially the same energy?

44 •• Discuss the evidence from the periodic table of the need for a fourth quantum number. How would the properties of He differ if there were only three quantum numbers, n, ℓ, and m?

45 •• The properties of iron ($Z = 26$) and cobalt ($Z = 27$), which have adjacent atomic numbers, are similar, whereas the properties of neon ($Z = 10$) and sodium ($Z = 11$), which also have adjacent atomic numbers, are very different. Explain why.

46 •• Separate the following six elements—potassium, calcium, titanium, chromium, manganese, and copper—into two groups of three each such that those in a group have similar properties.

47 • What element has the electron configuration (a) $1s^2 2s^2 2p^6 3s^2 3p^2$ and (b) $1s^2 2s^2 2p^6 3s^2 3p^6 4s^2$?

48 • The total number of quantum states of hydrogen with quantum number $n = 4$ is _____ .

(a) 4
(b) 16
(c) 32
(d) 36

49 • How many of oxygen's eight electrons are found in the p state?

(a) 0
(b) 2
(c) 4
(d) 6
(e) 8

50 • Write the electron configuration of (a) carbon and (b) oxygen.

51 • Write the electron configuration of (a) aluminum and (b) chromium.

52 • Give the possible values of the z component of the orbital angular momentum of (a) a d electron and (b) an f electron.

53 •• If the outer electron in sodium moves in the $n = 3$ Bohr orbit, the effective nuclear charge would be $Z'e = 1e$, and the energy of the electron would be $-13.6 \text{ eV}/3^2 = -1.51 \text{ eV}$. However, the ionization energy of sodium is 5.14 eV, not 1.51 eV. Use this fact and Equation 37-45 to calcu-

late the effective nuclear charge Z' seen by the outer electron in sodium. Assume that $r = 9a_0$ for the outer electron.

Optical and X-Ray Spectra

54 • The optical spectra of atoms with two electrons in the same outer shell are similar, but they are quite different from the spectra of atoms with just one outer electron because of the interaction of the two electrons. Separate the following elements into two groups such that those in each group have similar spectra: lithium, beryllium, sodium, magnesium, potassium, calcium, chromium, nickel, cesium, barium.

55 • Write down the possible electron configurations for the first excited state of (a) hydrogen, (b) sodium, and (c) helium.

56 • Indicate which of the following elements should have optical spectra similar to hydrogen and which should be similar to helium: Li, Ca, Ti, Rb, Hg, Ag, Cd, Ba, Fr, Ra.

57 • (a) Calculate the next two longest wavelengths in the K series (after the K_α line) of molybdenum. (b) What is the wavelength of the shortest wavelength in this series?

58 • The wavelength of the K_α line for a certain element is 0.3368 nm. What is the element?

59 • The wavelength of the K_α line for a certain element is 0.0794 nm. What is the element?

60 • Calculate the wavelength of the K_α line of rhodium.

61 • Calculate the wavelength of the K_α line in (a) magnesium ($Z = 12$) and (b) copper ($Z = 29$).

General Problems

62 • For the principal quantum number $n = 3$, what are the possible values of the quantum numbers ℓ and m?

63 • An electron in the L shell means that

(a) $\ell = 0$.
(b) $\ell = 1$.
(b) $n = 1$.
(c) $n = 2$.
(d) $m = 2$.
(e) none of the above are true.

64 •• The Bohr theory and the Schrödinger theory of the hydrogen atom give the same results for the energy levels. Discuss the advantages and disadvantages of each model.

65 •• In Figure 37-17, there are small dips in the ionization-energy curve at $Z = 31$ (gallium) and $Z = 49$ (indium) that are not labeled. Explain these dips using the· electron configurations of these atoms given in Table 37-1.

66 • What is the energy of the shortest wavelength photon emitted by the hydrogen atom?

67 • The wavelength of a spectral line of hydrogen is 97.254 nm. Identify the transition that results in this line.

68 • The wavelength of a spectral line of hydrogen is 1093.8 nm. Identify the transition that results in this line.

69 • Spectral lines of the following wavelengths are emitted by singly ionized helium: 164 nm, 230.6 nm, and 541 nm. Identify the transitions that result in these spectral lines.

70 •• We are often interested in finding the quantity ke^2/r in electron volts when r is given in nanometers. Show that $ke^2 = 1.44$ eV·nm.

71 •• The wavelengths of the photons emitted by potassium corresponding to transitions from the $4P_{3/2}$ and $4P_{1/2}$ states to the ground state are 766.41 and 769.90 nm. (a) Calculate the energies of these photons in electron volts. (b) The difference in the energies of these photons equals the difference in energy ΔE between the $4P_{3/2}$ and $4P_{1/2}$ states in potassium. Calculate ΔE. (c) Estimate the magnetic field that the 4p electron in potassium experiences.

72 •• To observe the characteristic K lines of the X-ray spectrum, one of the $n = 1$ electrons must be ejected from the atom. This is generally accomplished by bombarding the target material with electrons of sufficient energy to eject this tightly bound electron. What is the minimum energy required to observe the K lines of (a) tungsten, (b) molybdenum, and (c) copper?

73 •• The combination of physical constants $\alpha = e^2 k/\hbar c$, where k is the Coulomb constant, is known as the *fine-structure constant*. It appears in numerous relations in atomic physics. (a) Show that α is dimensionless. (b) Show that in the Bohr model of hydrogen $v_n = c\alpha/n$, where v_n is the speed of the electron in the stationary state of quantum number n.

74 •• The *positron* is a particle identical to the electron except that it carries a positive charge of e. *Positronium* is the bound state of an electron and positron. (a) Calculate the energies of the five lowest energy states of positronium using the reduced mass as given by Equation 37-47 in Problem 17. (b) Do transitions between any of the levels found in (a) fall in the visible range of wavelengths? If so, which transitions are these?

75 •• The deuteron, the nucleus of deuterium ("heavy hydrogen"), was first recognized from the spectrum of hydrogen. The deuteron has a mass twice that of the proton. (a) Calculate the Rydberg constant for hydrogen and for deuterium using the reduced mass as given by Equation 37-47 in Problem 17. (b) Using the result obtained in (a), determine the wavelength difference between the longest wavelength Balmer lines of hydrogen and deuterium.

76 •• The *muonium* atom is a hydrogen atom with the electron replaced by a μ^- particle. The μ^- is identical to an electron but has a mass 207 times as great as the electron. (a) Calculate the energies of the five lowest energy levels of muonium using the reduced mass as given by Equation 37-47 in Problem 17. (b) Do transitions between any of the levels found in (a) fall in the visible range of wavelengths, i.e., between $\lambda = 700$ nm and 400 nm? If so, which transitions are these?

77 •• The triton, a nucleus consisting of a proton and two neutrons, is unstable with a fairly long half-life of about 12 years. *Tritium* is the bound state of an electron and a triton. (a) Calculate the Rydberg constant of tritium using the reduced mass as given by Equation 37-47 in Problem 17. (b) Using the result obtained in (a) and in part (a) of Problem 75 determine the wavelength difference between the longest wavelength Balmer lines of tritium and deuterium and between tritium and hydrogen.

78 ••• Suppose that the interaction between an electron and proton were of the form $F = -Kr$, where K is a constant, rather than $1/r^2$. If the stationary state orbits are again limited by the angular momentum condition $L = n\hbar$, what are then the radii of these orbits? Show that for this case the total energies of the stationary states are given by $E = n\hbar\omega$, where ω is the angular frequency of the electron about the proton.

79 ••• The frequency of revolution of an electron in a circular orbit of radius r is $f_{rev} = v/2\pi r$, where v is the speed. (a) Show that in the nth stationary state

$$f_{rev} = \frac{k^2 Z^2 e^4 m}{2\pi h^3} \frac{1}{n^3}$$

(b) Show that when $n_1 = n$, $n_2 = n - 1$, and n is much greater than 1,

$$\frac{1}{n_2^2} - \frac{1}{n_1^2} \approx \frac{2}{n^3}$$

(c) Use your result in part (b) and Equation 37-13 to show that in this case the frequency of radiation emitted equals the frequency of motion. This result is an example of Bohr's correspondence principle: When n is large, so that the energy difference between adjacent states is a small fraction of the total energy, classical and quantum physics must give the same results.

CHAPTER 38

Molecules and Solids

Molten tin solidifies in a pattern of tree-shaped crystals called dendrites as it cools under controlled circumstances.

Most atoms bond together to form molecules or solids. Molecules may exist as separate entities as in gaseous O_2 or N_2 or they, too, may bond together to form liquids or solids. A molecule is the smallest constituent of a substance that retains its chemical properties.

In this chapter we use our understanding of quantum mechanics to discuss molecular bonding, the energy levels and spectra of diatomic molecules, the structure of solids, and solid-state semiconducting devices. Much of our discussion will be qualitative because, as in atomic physics, the quantum-mechanical calculations are very difficult. Before studying semiconductor devices you should review the material in Chapter 27 on the microscopic theory of electrical conduction.

38-1 Molecular Bonding

There are two extreme views that we can take of a molecule. Consider, for example, H_2. We can think of it either as two H atoms joined together, or as a quantum-mechanical system of two protons and two electrons. The latter picture is more fruitful in this case because neither of the electrons in the H_2 molecule can be identified as belonging to either proton. Instead, the wave function for each electron is spread out in space throughout the whole molecule. For more complicated molecules, however, an intermediate picture is useful. For example, the nitrogen molecule N_2 consists of 14 protons and 14 electrons, but only two of the electrons take part in the bonding. We therefore can consider this molecule as two N^+ ions and two electrons that belong to the molecule as a whole. The molecular wave functions for these bonding electrons are called **molecular orbitals.** In many cases these molecular wave functions can be constructed from combinations of the atomic wave functions with which we are familiar.

The two principal types of bonds responsible for the formation of molecules are the ionic bond and the covalent bond. Other types of bonds that are important in the bonding of liquids and solids are van der Waals bonds, metallic bonds, and hydrogen bonds. In many cases, bonding is a mixture of these mechanisms.

The Ionic Bond

The simplest type of bond is the **ionic bond,** which is found in salts such as sodium chloride (NaCl). The sodium atom has one 3s electron outside a stable core. The energy needed to remove this electron, the ionization energy, is just 5.14 eV (see Figure 37-15). The removal of this electron leaves a positive ion with a spherically symmetric, closed-shell electron core. Chlorine, on the other hand, is one electron short of having a closed shell. The energy released by an atom's acquisition of one electron is called its **electron affinity,** which in the case of chlorine is 3.62 eV. The acquisition of one electron by chlorine results in a negative ion with a spherically symmetric, closed-shell electron core. Thus, the formation of an Na^+ ion and a Cl^- ion by the donation of one electron of sodium to chlorine requires only 5.14 eV $-$ 3.62 eV $=$ 1.52 eV at infinite separation. The electrostatic potential energy of the two ions when they are a distance r apart is $-ke^2/r$. When the separation of the ions is less than about 0.95 nm, the negative potential energy of attraction is of greater magnitude than the 1.52 eV of energy needed to create the ions. Thus, at separation distances less than 0.95 nm it is energetically favorable (that is, the total energy of the system is reduced) for the sodium atom to donate an electron to the chlorine atom to form NaCl.

Since the electrostatic attraction increases as the ions get closer together, it might seem that equilibrium could not exist. However, when the separation of the ions is very small, there is a strong repulsion that is quantum mechanical in nature and is related to the exclusion principle. This **exclusion-principle repulsion** is responsible for the repulsion of the atoms in all molecules (except H_2)* for all bonding mechanisms. We can understand it qualitatively as follows. When the ions are very far apart, the wave function for a core electron in one of the ions does not overlap that of any electron in the other ion. We can distinguish the electrons by the ion to which they belong. This means that electrons in the two ions can have the same quantum numbers because they occupy different regions of space. However, as the distance between the ions decreases, the wave functions of the core electrons begin to overlap; that is, the electrons in the two ions begin to occupy the same region of space. Because of the exclusion principle, some of these electrons must go into higher energy quantum states.[†] But energy is required to shift the elec-

---

* In H_2, the repulsion is simply that of the two positively charged protons.

[†] Recall from our discussion in Chapter 36 that the exclusion principle is related to the fact that the wave function for two identical electrons is antisymmetric on the exchange of the electrons and that an antisymmetric wave function for two electrons with the same quantum numbers is zero if the space coordinates of the electrons are the same.

trons into higher energy quantum states. This increase in energy when the ions are pushed closer together is equivalent to a repulsion of the ions. It is not a sudden process. The energy states of the electrons change gradually as the ions are brought together. A sketch of the potential energy of the Na^+ and Cl^- ions versus separation is shown in Figure 38-1. The energy is lowest at an equilibrium separation of about 0.236 nm. At smaller separations, the energy rises steeply as a result of the exclusion principle. The energy required to separate the ions and form neutral sodium and chlorine atoms is called the **dissociation energy,** which is about 4.27 eV for NaCl.

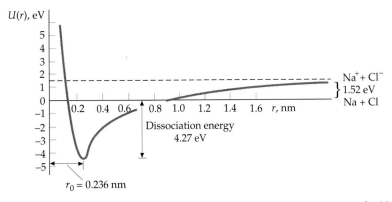

The equilibrium separation distance of 0.236 nm is for gaseous diatomic NaCl, which can be obtained by evaporating solid NaCl. Normally, NaCl exists in a cubic crystal structure, with the Na^+ and Cl^- ions at the alternate corners of a cube. The separation of the ions in a crystal is somewhat larger, about 0.28 nm. Because of the presence of neighboring ions of opposite charge, the electrostatic energy per ion pair is lower when the ions are in a crystal.

Figure 38-1 Potential energy for Na^+ and Cl^- ions as a function of separation distance r. The energy at infinite separation was chosen to be 1.52 eV, corresponding to the energy needed to form the ions from neutral atoms. The minimum energy is at the equilibrium separation $r_0 = 0.236$ nm for the ions in the molecule.

Example 38-1

The electron affinity of fluorine is 3.40 eV, and the equilibrium separation of sodium fluoride (NaF) is 0.193 nm. (a) How much energy is needed to form Na^+ and F^- ions from neutral sodium and fluorine atoms? (b) What is the electrostatic potential energy of the Na^+ and F^- ions at their equilibrium separation? (c) The dissociation energy of NaF is 5.38 eV. What is the energy due to repulsion of the ions at the equilibrium separation?

Picture the Problem (a) The energy ΔE needed to form Na^+ and F^- ions from the neutral sodium and fluorine atoms is the difference between the ionization energy of sodium (5.14 eV) and the electron affinity of fluorine. (b) The electrostatic potential energy with $U = 0$ at infinity is $U_e = \dfrac{-ke^2}{r}$. (c) If we choose the potential energy at infinity to be ΔE, the total potential energy is $U_{tot} = U_e + \Delta E + U_{rep}$, where U_{rep} is the energy of repulsion, which is found by setting the dissociation energy equal to $-U_{tot}$.

(a) Calculate the energy needed to form Na^+ and F^- ions from the neutral sodium and fluorine atoms:	$\Delta E = 5.14\ \text{eV} - 3.40\ \text{eV} = 1.74\ \text{eV}$
(b)1. Calculate the electrostatic potential energy at the equilibrium separation of $r = 0.193$ nm:	$U_e = -\dfrac{ke^2}{r} = -\dfrac{(8.99 \times 10^9\ \text{N·m}^2/\text{C}^2)(1.60 \times 10^{-19}\ \text{C})^2}{1.93 \times 10^{-10}\ \text{m}}$ $= -1.19 \times 10^{-18}\ \text{J}$
2. Convert from joules to electron volts:	$U_e = -1.19 \times 10^{-18}\ \text{J}\ \dfrac{1\ \text{eV}}{1.60 \times 10^{-19}\ \text{J}} = -7.45\ \text{eV}$
(c)1. The dissociation energy equals the negative of the total potential energy:	$E_d = -U_{tot} = -(U_e + \Delta E + U_{rep})$ $= -(-7.45\ \text{eV} + 1.74\ \text{eV} + U_{rep}) = 5.38\ \text{eV}$
2. Solve for U_{rep}:	$U_{rep} = 7.45\ \text{eV} - 1.74\ \text{eV} - 5.38\ \text{eV} = 0.33\ \text{eV}$

The Covalent Bond

A completely different mechanism, the **covalent bond,** is responsible for the bonding of identical or similar atoms to form such molecules as gaseous hydrogen (H_2), nitrogen (N_2), and carbon monoxide (CO). If we calculate the energy needed to form H^+ and H^- ions by the transfer of an electron from one atom to the other and then add this energy to the electrostatic potential energy, we find that there is no separation distance for which the total energy is negative. The bond thus cannot be ionic. Instead, the attraction of two hydrogen atoms is an entirely quantum-mechanical effect. The decrease in energy when two hydrogen atoms approach each other is due to the sharing of the two electrons by both atoms. It is intimately connected with the symmetry properties of the wave functions of electrons.

We can gain some insight into covalent bonding by considering a simple, one-dimensional quantum-mechanics problem of two finite square wells. We first consider a single electron that is equally likely to be in either well. Since the wells are identical, the probability distribution, which is proportional to $|\psi^2|$, must be symmetric about the midpoint between the wells. Then ψ must be either symmetric or antisymmetric with respect to the two wells. The two possibilities for the ground state are shown in Figure 38-2a for the case in which the wells are far apart and in Figure 38-2b for the case in which the wells are close together. An important feature of Figure 38-2b is that in the region between the wells the symmetric wave function is large and the antisymmetric wave function is small.

Now consider adding a second electron to the two wells. We saw in Chapter 36 that the wave functions for particles that obey the exclusion principle are antisymmetric on exchange of the particles. Thus the total wave function for the two electrons must be antisymmetric on exchange of the electrons. Note that exchanging the electrons in the wells here is the same as exchanging the wells. The total wave function for two electrons can be written as a product of a space part and a spin part. So an antisymmetric wave function can be the product of a symmetric space part and an antisymmetric spin part or of a symmetric spin part and an antisymmetric space part.

To understand the symmetry of the total wave function, we must therefore understand the symmetry of the spin part of the wave function. The spin of a single electron can have two possible values for its quantum number m_s: $m_s = +\frac{1}{2}$, which we call spin up, or $m_s = -\frac{1}{2}$, which we call spin down. We will use arrows to designate the spin wave function for a single electron: $\uparrow_1$ or $\uparrow_2$ for electron 1 or electron 2 with spin up, and $\downarrow_1$ or $\downarrow_2$ for electron 1 or electron 2 with spin down. The total spin quantum number for two electrons can be $S = 1$, with $m_S = +1$, 0, or -1, or $S = 0$, with $m_S = 0$. We use ϕ_{S,m_S} to denote the spin wave function for two electrons. The spin state $\phi_{1,+1}$, corresponding to $S = 1$ and $m_S = +1$, can be written

$$\phi_{1,+1} = \uparrow_1\uparrow_2, \qquad S = 1, m_S = +1 \qquad \text{38-1}$$

Similarly, the spin state for $S = 1$, $m_S = -1$ is

$$\phi_{1,-1} = \downarrow_1\downarrow_2, \qquad S = 1, m_S = -1 \qquad \text{38-2}$$

Note that both of these states are symmetric upon exchange of the electrons. The spin state corresponding to $S = 1$ and $m_S = 0$ is not quite so obvious. It turns out to be proportional to

$$\phi_{1,0} = \uparrow_1\downarrow_2 + \uparrow_2\downarrow_1, \qquad S = 1, m_S = 0 \qquad \text{38-3}$$

This spin state is also symmetric upon exchange of the electrons. The spin state for two electrons with antiparallel spins ($S = 0$) is

$$\phi_{0,0} = \uparrow_1\downarrow_2 - \uparrow_2\downarrow_1, \qquad S = 0, m_S = 0 \qquad \text{38-4}$$

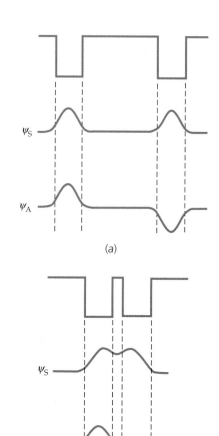

(a)

(b)

Figure 38-2 (a) Two square wells far apart. The electron wave function can be either symmetric (ψ_S) or antisymmetric (ψ_A) in space. The probability distributions and energies are the same for the two wave functions when the wells are far apart. (b) Two square wells that are close together. Between the wells the symmetric space wave function is larger than the antisymmetric space wave function.

This spin state is antisymmetric upon exchange of electrons.

We thus have the important result that the *spin* part of the wave function is symmetric for parallel spins ($S = 1$) and antisymmetric for antiparallel spins ($S = 0$). Because the total wave function is the product of the space function and spin function, we have the following important result:

> For the total wave function of two electrons to be antisymmetric, the space part of the wave function must be antisymmetric for parallel spins ($S = 1$) and symmetric for antiparallel spins ($S = 0$).

We can now consider the problem of two hydrogen atoms. Figure 38-3*a* shows a spatially symmetric wave function ψ_S and a spatially antisymmetric wave function ψ_A for two hydrogen atoms that are far apart, and Figure 38-3*b* shows the same two wave functions for two hydrogen atoms that are close together. The squares of these two wave functions are shown in Figure 38-3*c*. Note that the probability distribution $|\psi|^2$ in the region between the protons is large for the symmetric wave function and small for the antisymmetric wave function. Thus, when the space part of the wave function is symmetric ($S = 0$), the electrons are often found in the region between the protons. The negatively charged electron cloud representing these electrons is concentrated in the space between the protons, as shown in the upper part of Figure 38-3*c*, and the protons are bound together by this negatively charged cloud. Conversely, when the space part of the wave function is antisymmetric ($S = 1$), the electrons spend little time between the protons, and the atoms do not bind together to form a molecule. In this case, the electron cloud is not concentrated in the space between the protons, as shown in the lower part of Figure 38-3*c*.

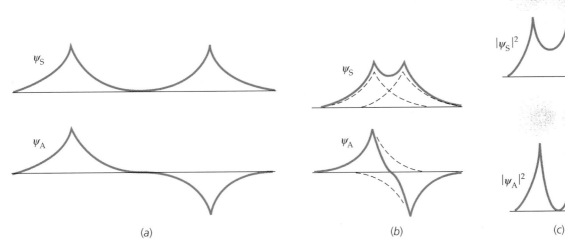

(a)   (b)   (c)

Figure 38-3 One-dimensional symmetric and antisymmetric wave functions for two hydrogen atoms (*a*) far apart and (*b*) close together. (*c*) Electron probability distribu- tions ($|\psi|^2$) for the wave functions in (*b*). For the symmetric wave function, the electron charge density is large between the protons. This negative charge density holds the protons together in the hydrogen molecule H_2. For the antisymmetric wave function, the electron charge density is not large between the protons.

The total electrostatic potential energy for the H_2 molecule consists of the positive energy of repulsion of the two electrons and the negative potential

energy of attraction of each electron for each proton. Figure 38-4 shows the electrostatic potential
energy for two hydrogen atoms versus separation
for the case in which the space part of the electron
wave function is symmetric (U_S) and for the case
in which it is antisymmetric (U_A). We can see that
the potential energy for the symmetric state is the
lower of the two and that the shape of this potential-energy curve is similar to that for ionic bonding. The equilibrium separation for H_2 is r_0 =
0.074 nm, and the binding energy is 4.52 eV. For
the antisymmetric state, the potential energy is
never negative and there is no bonding.

We can now see why three hydrogen atoms do
not bond to form H_3. If a third hydrogen atom is
brought near an H_2 molecule, the third electron
cannot be in a 1s state and have its spin antiparallel to the spin of both of the other electrons. If this
electron is in an antisymmetric space state with
respect to exchange with one of the electrons, the
repulsion of this atom is greater than the attrac-

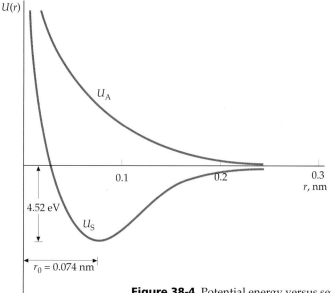

Figure 38-4 Potential energy versus separation for two hydrogen atoms. The curve
labeled U_S is for a wave function with a
symmetric space part, and the curve labeled U_A is for a wave function with an
antisymmetric space part.

tion of the other. As the three atoms are pushed together, the third electron is,
in effect, forced into a higher quantum-energy state by the exclusion principle. The bond between two hydrogen atoms is called a **saturated bond** because there is no room for another electron. The two shared electrons essentially fill the 1s states of both atoms.

We can also see why two helium atoms do not normally bond together to
form the He_2 molecule. There are no valence electrons that can be shared.
The electrons in the closed shells are forced into higher energy states when
the two atoms are brought together. At low temperatures or high pressures,
helium atoms do bond together due to van der Waals forces, which we will
discuss next. This bonding is so weak that at atmospheric pressure helium
boils at 4 K, and it does not form a solid at any temperature unless the pressure is greater than about 20 atm.

When two identical atoms bond, as in O_2 or N_2, the bonding is purely covalent. However, the bonding of two dissimilar atoms is often a mixture of
covalent and ionic bonding. Even in NaCl, the electron donated by sodium to
chlorine has some probability of being at the sodium atom because its wave
function does not suddenly fall to zero. Thus, this electron is partially shared
in a covalent bond, although this bonding is only a small part of the total
bond, which is mainly ionic.

A measure of the degree to which a bond is ionic or covalent can be obtained from the electric dipole moment of the molecule. For example, if the
bonding in NaCl were purely ionic, the center of positive charge would be at
the Na^+ ion and the center of negative charge would be at the Cl^- ion. The
electric dipole moment would have the magnitude

$$p_{ionic} = er_0 \qquad\qquad 38\text{-}5$$

where r_0 is the equilibrium separation of the ions. Thus, the dipole moment
of NaCl would be (from Figure 38-1)

$$p_{ionic} = er_0$$

$$= (1.60 \times 10^{-19}\,\text{C})(2.36 \times 10^{-10}\,\text{m}) = 3.78 \times 10^{-29}\,\text{C·m}$$

The actual measured electric dipole moment of NaCl is

$$p_{measured} = 3.00 \times 10^{-29}\,\text{C·m}$$

We can define the ratio of $p_{measured}$ to p_{ionic} as the fractional amount of ionic bonding. For NaCl, this ratio is $3.00/3.78 = 0.79$. Thus, the bonding in NaCl is about 79% ionic.

Exercise The equilibrium separation of HCl is 0.128 nm and its measured electric dipole moment is 3.60×10^{-30} C·m. What is the percentage of ionic bonding in HCl? (*Answer* 18%)

Other Bonding Types

The van der Waals Bond Any two separated molecules will be attracted to one another by electrostatic forces called van der Waals forces. So will any two atoms that do not form ionic or covalent bonds. The **van der Waals bonds** due to these forces are much weaker than the bonds already discussed. At high enough temperatures, these forces are not strong enough to overcome the ordinary thermal agitation of the atoms or molecules, but at sufficiently low temperatures, thermal agitation becomes negligible, and the van der Waals forces will cause virtually all substances to condense into a liquid and then a solid form.* The van der Waals forces arise from the interaction of the instantaneous electric dipole moments of the molecules.

Figure 38-5 shows how two polar molecules—molecules with *permanent* electric dipole moments, such as H_2O—can bond. The electric field due to the dipole moment of one molecule orients the other molecule such that the two dipole moments attract. Nonpolar molecules also attract other nonpolar molecules via the van der Waals forces. Although nonpolar molecules have zero electric dipole moments on the average, they have instantaneous dipole moments that are generally not zero because of fluctuations in the positions of the charges. When two nonpolar molecules are near each other, the fluctuations in the instantaneous dipole moments tend to become correlated so as to produce attraction. This is illustrated in Figure 38-6.

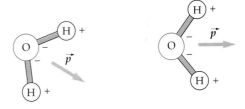

Figure 38-5 Bonding of H_2O molecules because of the attraction of the electric dipoles. The dipole moment of each molecule is indicated by $\vec{p}$. The field of one dipole orients the other dipole so the moments tend to be parallel. When the dipole moments are approximately parallel, the center of negative charge of one molecule is close to the center of positive charge of the other molecule, and the molecules attract.

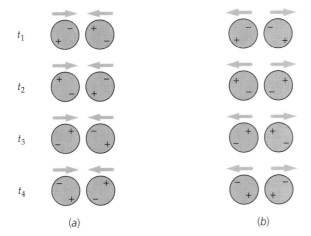

(a)   (b)

Figure 38-6 van der Waals attraction of molecules with zero average dipole moments. (*a*) Possible orientations of instantaneous dipole moments at different times leading to attraction. (*b*) Possible orientations leading to repulsion. The electric field of the instantaneous dipole moment of one molecule tends to polarize the other molecule; thus the orientations leading to attraction (*a*) are much more likely than those leading to repulsion (*b*).

The Hydrogen Bond Another bonding mechanism of great importance is the hydrogen bond, which is formed by the sharing of a proton (the nucleus of the hydrogen atom) between two atoms, frequently two oxygen atoms. This sharing of a proton is similar to the sharing of electrons responsible for the covalent bond already discussed. It is facilitated by the small mass of the proton and by the absence of inner-core electrons in hydrogen. The hydrogen bond often holds groups of molecules together and is responsible for the

---

* Helium is the only element that does not solidify at any temperature at atmospheric pressure.

cross-linking that allows giant biological molecules and polymers to hold their fixed shapes. The well-known helical structure of DNA is due to hydrogen-bond linkages across turns of the helix (Figure 38-7).

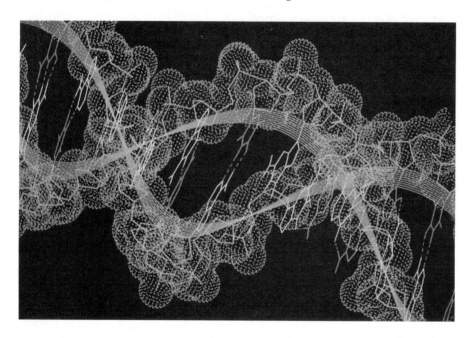

Figure 38-7 The DNA molecule.

The Metallic Bond   In a metal, two atoms do not bond together by exchanging or sharing an electron to form a molecule. Instead, each valence electron is shared by many atoms. The bonding is thus distributed throughout the entire metal. A metal can be thought of as a lattice of positive ions held together by a "gas" of essentially free electrons that roam throughout the solid. In the quantum-mechanical picture, these free electrons form a cloud of negative charge density between the positively charged lattice ions that holds the ions together. In this respect, the metallic bond is somewhat similar to the covalent bond. However, with the metallic bond, there are far more than just two atoms involved, and the negative charge is distributed uniformly throughout the volume of the metal. The number of free electrons varies from metal to metal but is of the order of one per atom.

38-2   Polyatomic Molecules

Molecules with more than two atoms range from such relatively simple molecules as water, which has a molecular mass number of 18, to such giants as proteins and DNA, which can have molecular masses of hundreds of thousands up to many millions. As with diatomic molecules, the structure of polyatomic molecules can be understood by applying basic quantum mechanics to the bonding of individual atoms. The bonding mechanisms for most polyatomic molecules are the covalent bond and the hydrogen bond. We will discuss only some of the simplest polyatomic molecules—H_2O, NH_3, and CH_4—to illustrate both the simplicity and complexity of the application of quantum mechanics to molecular bonding.

The basic requirement for the sharing of electrons in a covalent bond is that the wave functions of the valence electrons in the individual atoms must overlap as much as possible. As our first example, we will consider the water molecule. The ground-state configuration of the oxygen atom is $1s^2 2s^2 2p^4$. The 1s and 2s electrons are in closed-shell states and do not contribute to the bonding. The 2p shell has room for six electrons, two in each of the three space states corresponding to $\ell = 1$. In an isolated atom, we describe these space states by the hydrogen-like wave functions corresponding to $\ell = 1$ and $m = +1, 0,$ and -1. Since the energy is the same for these three space states, we could equally well use any linear combination of these wave functions. When an atom participates in molecular bonding, certain combinations of these atomic wave functions are important. These combinations are called the p_x, p_y, and p_z **atomic orbitals**. The angular dependence of these orbitals is

$$p_x \propto \sin\theta\cos\phi \qquad\qquad 38\text{-}6$$

$$p_y \propto \sin\theta\sin\phi \qquad\qquad 38\text{-}7$$

$$p_z \propto \cos\theta \qquad\qquad 38\text{-}8$$

The electron charge distribution is maximum along the x, y, or z axis, respectively, for these orbitals as shown in Figure 38-8.

Figure 38-8 Computer-generated dot plot illustrating the spatial dependence of the electron charge distribution in the p_x, p_y, and p_z atomic orbitals.

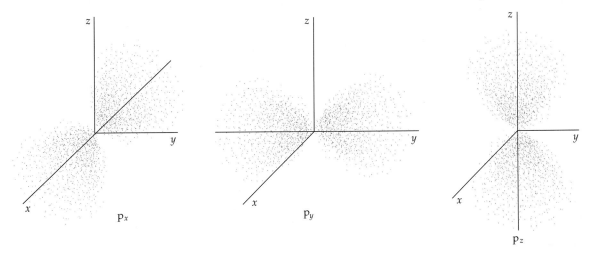

For the oxygen in an H$_2$O molecule, maximum overlap of the electron wave functions occurs when two of the four 2p electrons are paired with their spins antiparallel in one of the atomic orbitals (for this example, assume the p_z orbital), one of the other electrons is in a second orbital (the p_x orbital), and the other electron is in the third orbital (the p_y orbital). Each of the unpaired electrons (in the p_x and p_y orbitals, in this illustration) forms a bond with the electron of a hydrogen atom as shown in Figure 38-9. Because of the repulsion of the two hydrogen atoms, the angle between the O–H bonds is actually greater than 90°. The effect of this repulsion can be calculated, and the result is in agreement with the measured angle of 104.5°.

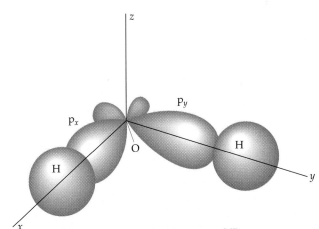

Figure 38-9 Electron charge distribution in the H$_2$O molecule.

Similar reasoning leads to an understanding of the bonding in NH$_3$. In the ground state, nitrogen has three electrons in the 2p state. When these three electrons are in the p_x, p_y, and p_z atomic orbitals, they bond to the electrons of hydrogen atoms. Again, because of the repulsion of the hydrogen atoms, the angles between the bonds are somewhat larger than 90°.

The bonding of carbon atoms is somewhat more complicated. Carbon forms a wide variety of different types of molecular bonds, leading to a great diversity in the kinds of organic molecules. The ground-state configuration of carbon is $1s^2 2s^2 2p^2$. From our previous discussion, we might expect carbon to be divalent—that is, bonding only through its two 2p electrons—with the two bonds forming at approximately 90°. However, one of the most important features of the chemistry of carbon is that tetravalent carbon compounds, such as CH_4, are overwhelmingly favored.

The observed valence of 4 for carbon comes about in an interesting way. One of the first excited states of carbon occurs when a 2s electron is excited to a 2p state, giving a configuration of $1s^2 2s^1 2p^3$. In this excited state, we can have four unpaired electrons, one each in the 2s, $2p_x$, $2p_y$, and $2p_z$ atomic orbitals. We might expect there to be three similar bonds corresponding to the three p orbitals and one different bond corresponding to the s orbital. However, when carbon forms tetravalent bonds, these four atomic orbitals become mixed and form four new *equivalent* molecular orbitals called **hybrid orbitals.** This mixing of atomic orbitals, called **hybridization,** is among the most important features involved in the physics of complex molecular bonds. Figure 38-10 shows the tetrahedral structure of the methane molecule (CH_4), and Figure 38-11 shows the structure of the ethane molecule (CH_3–CH_3), which is similar to two joined methane molecules in which one of the C–H bonds is replaced with a C–C bond.

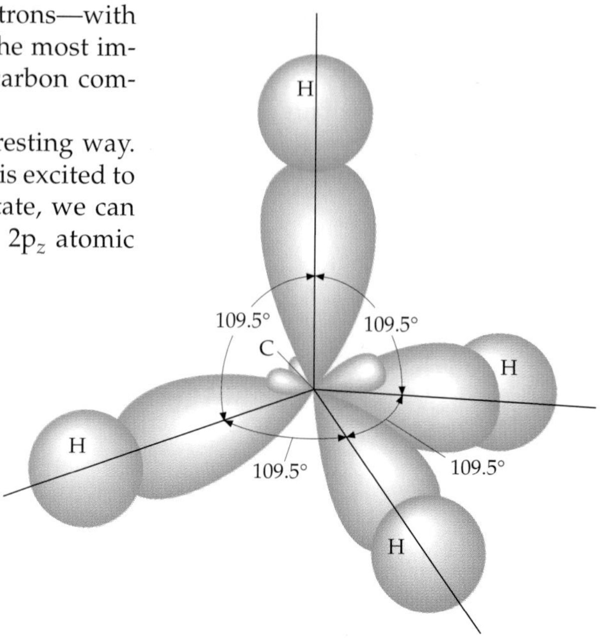

Figure 38-10 Electron charge distribution in the CH_4 (methane) molecule.

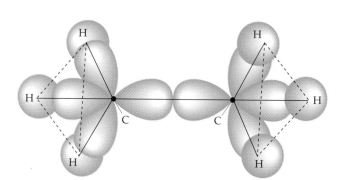

Figure 38-11 Electron charge distribution in the CH_3–CH_3 (ethane) molecule.

Carbon orbitals can also hybridize, with the s, p_x, and p_y orbitals combining to form three hybrid orbitals in the xy plane with 120° bonds and the p_z orbital remaining unmixed. An example of this configuration is graphite, in which the bonds in the xy plane provide the strongly layered structure characteristic of the material.

38-3 Energy Levels and Spectra of Diatomic Molecules

As is the case with an atom, a molecule often emits electromagnetic radiation when it makes a transition from an excited energy state to a state of lower energy. Conversely, a molecule can absorb radiation and make a transition from a lower energy state to a higher energy state. The study of molecular emission and absorption spectra thus provides us with information about the energy states of molecules. For simplicity, we will consider only diatomic molecules here.

The energy of a molecule can be conveniently separated into three parts: electronic, due to the excitation of the electrons of the molecule; vibrational, due to the oscillations of the atoms of the molecule; and rotational, due to the rotation of the molecule about its center of mass. The magnitudes of these energies are sufficiently different that they can be treated separately. The energies due to the electronic excitations of a molecule are of the order of magnitude of 1 eV, the same as for the excitation of an atom. The energies of vibration and rotation are much smaller than this.

Rotational Energy Levels

Figure 38-12 shows a simple schematic model of a diatomic molecule consisting of a mass m_1 and a mass m_2 separated by a distance r and rotating about its center of mass. Classically, the kinetic energy of rotation (see Section 9-5) is

$$E = \tfrac{1}{2}I\omega^2 \qquad\qquad 38\text{-}9$$

where I is the moment of inertia and ω is the angular frequency of rotation. If we write this in terms of the angular momentum $L = I\omega$, we have

$$E = \frac{(I\omega)^2}{2I} = \frac{L^2}{2I} \qquad\qquad 38\text{-}10$$

The solution of the Schrödinger equation for rotation leads to quantization of the angular momentum with values given by

$$L^2 = \ell(\ell + 1)\hbar^2, \qquad \ell = 0, 1, 2, \ldots \qquad\qquad 38\text{-}11$$

where ℓ is the **rotational quantum number.** This is the same quantum condition on angular momentum that holds for the orbital angular momentum of an electron in an atom. Note, however, that L in Equation 38-10 refers to the angular momentum of the entire molecule rotating about its center of mass. The energy levels of a rotating molecule are therefore given by

$$E = \frac{\ell(\ell + 1)\hbar^2}{2I} = \ell(\ell + 1)E_{0r} \qquad \ell = 0, 1, 2, \ldots \qquad\qquad 38\text{-}12$$

Rotational energy levels

where E_{0r} is the characteristic rotational energy of a particular molecule, which is inversely proportional to its moment of inertia:

$$E_{0r} = \frac{\hbar^2}{2I} \qquad\qquad 38\text{-}13$$

Characteristic rotational energy

A measurement of the rotational energy of a molecule from its rotational spectrum can be used to determine the moment of inertia of the molecule, which can then be used to find the separation of the atoms in the molecule. The moment of inertia about an axis through the center of mass of a diatomic molecule (see Figure 38-12) is

$$I = m_1 r_1^2 + m_2 r_2^2$$

Using $m_1 r_1 = m_2 r_2$, which relates the distances r_1 and r_2 from the atoms to the center of mass, and $r_0 = r_1 + r_2$ for the separation of the atoms, we can write the moment of inertia as (see Problem 26)

$$I = \mu r_0^2 \qquad\qquad 38\text{-}14$$

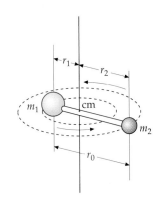

Figure 38-12 Diatomic molecule rotating about an axis through its center of mass.

where μ, called the **reduced mass**, is

$$\mu = \frac{m_1 m_2}{m_1 + m_2}$$ 38-15

Definition—Reduced mass

If the masses are equal ($m_1 = m_2 = m$), as in H_2 and O_2, the reduced mass $\mu = \frac{1}{2}m$ and

$$I = \tfrac{1}{2}mr_0^2$$ 38-16

A unit of mass convenient for discussing atomic and molecular masses is the **unified mass unit,** u, which is defined as one-twelfth the mass of the neutral carbon-12 (^{12}C) atom. The mass of one ^{12}C atom is thus 12 u. The mass of an atom in unified mass units is therefore numerically equal to the molar mass of the atom in grams. The unified mass unit is related to the gram and kilogram by

$$1\,u = \frac{1\,g}{N_A} = \frac{10^{-3}\,kg}{6.0221 \times 10^{23}} = 1.6606 \times 10^{-27}\,kg$$ 38-17

where N_A is Avogadro's number.

Example 38-2

Find the reduced mass of the HCl molecule.

Picture the Problem We find the masses of the hydrogen and chlorine atoms in the periodic table* in Appendix C and use the definition in Equation 38-15.

1. The reduced mass μ is related to the individual masses m_H and m_{Cl}

$$\mu = \frac{m_H m_{Cl}}{m_H + m_{Cl}}$$

2. Find the masses in the periodic table:

$$m_H = 1.01\,u, \qquad m_{Cl} = 35.5\,u$$

3. Substitute to calculate the reduced mass:

$$\mu = \frac{m_H m_{Cl}}{m_H + m_{Cl}} = \frac{(1.01\,u)(35.5\,u)}{1.01\,u + 35.5\,u} = 0.982\,u$$

Remark Note that the reduced mass is less than the mass of either atom in the molecule and that it is approximately equal to the mass of the hydrogen atom. When one atom of a diatomic molecule is much more massive than the other, the center of mass of the molecule is approximately at the center of the more massive atom, and the reduced mass is approximately equal to the mass of the lighter atom.

*The masses in these tables are weighted according to the natural isotopic distribution. Thus, the mass of carbon is given as 12.011 rather than 12.000 because natural carbon consists of about 98.9% ^{12}C and 1.1% ^{13}C. Similarly, natural chlorine consists of about 76% ^{35}Cl and 24% ^{37}Cl.

Example 38-3

Estimate the characteristic rotational energy of an O_2 molecule, assuming that the separation of the atoms is 0.1 nm.

1. The characteristic rotational energy is inversely proportional to the moment of inertia:

$$E_{0r} = \frac{\hbar^2}{2I}$$

2. Calculate the moment of inertia:

$$I = \mu r_0^2 = \tfrac{1}{2}mr_0^2$$

3. Substitute this expression for I into the expression for E_{0r}:

$$E_{0r} = \frac{\hbar^2}{mr_0^2}$$

4. Use $m = 16$ u for the mass of oxygen and the given values of the constants to calculate E_{0r}:

$$E_{0r} = \frac{\hbar^2}{mr_0^2} = \frac{(1.05 \times 10^{-34}\,\text{J}\cdot\text{s})^2}{(16\,\text{u})(10^{-10}\,\text{m})^2} \times \frac{1\,\text{u}}{1.66 \times 10^{-27}\,\text{kg}}$$

$$= 4.15 \times 10^{-23}\,\text{J} = 2.59 \times 10^{-4}\,\text{eV}$$

We can see from Example 38-3 that the rotational energy levels are several orders of magnitude smaller than energy levels due to electron excitation, which have energies of the order of 1 eV or higher. Transitions within a given set of rotational energy levels yield photons in the microwave region of the electromagnetic spectrum. The rotational energies are also small compared with the typical thermal energy kT at normal temperatures. For $T = 300$ K, for example, kT is about 2.6×10^{-2} eV, which is about 100 times the characteristic rotational energy as calculated in Example 38-3, and about 1% of the typical electronic energy. Thus, at ordinary temperatures, a molecule can be easily excited to the lower rotational energy levels by collisions with other molecules. But such collisions cannot excite the molecule to its electronic energy levels above the ground state.

Vibrational Energy Levels

The quantization of energy in a simple harmonic oscillator was one of the first problems solved by Schrödinger in his paper proposing his wave equation. Solving the Schrödinger equation for a simple harmonic oscillator gives

$$E_\nu = (\nu + \tfrac{1}{2})hf, \qquad \nu = 0, 1, 2, \ldots \qquad \text{38-18}$$

Vibrational energy levels

where f is the frequency of the oscillator and ν is the **vibrational quantum number.**[*] An interesting feature of this result is that the energy levels are equally spaced with intervals equal to hf. The frequency of vibration of a diatomic molecule can be related to the force exerted by one atom on the other. Consider two objects of mass m_1 and m_2 connected by a spring of force constant K. The frequency of oscillation of this system can be shown to be (see Problem 30)

$$f = \frac{1}{2\pi}\sqrt{\frac{K}{\mu}} \qquad \text{38-19}$$

where μ is the reduced mass given by Equation 38-15. The effective force constant of a diatomic molecule can thus be determined from a measurement of the frequency of oscillation of the molecule.

A selection rule on transitions between vibrational states (of the same electronic state) requires that ν can change only by ± 1, so the energy of a photon emitted by such a transition is hf and the frequency is f, the same as the frequency of vibration. There is a similar selection rule that ℓ must change by ± 1 for transitions between rotational states.

A typical measured frequency of a transition between vibrational states is 5×10^{13} Hz, which gives for the order of magnitude of vibrational energies

$$E \sim hf = (4.14 \times 10^{-15}\,\text{eV}\cdot\text{s})(5 \times 10^{13}\,\text{s}^{-1}) = 0.2\,\text{eV}$$

[*]We use ν (the Greek letter nu) here rather than n so as not to confuse the vibrational quantum number with the principal quantum number n for electronic energy levels.

This typical vibrational energy is about 1000 times greater than the typical rotational energy E_{0r} of the O_2 molecule we found in Example 38-3 and about 8 times greater than the typical thermal energy $kT = 0.026$ eV at $T = 300$ K. Thus the vibrational levels cannot be excited by molecular collisions at ordinary temperatures.

Example 38-4

The frequency of vibration of the CO molecule is 6.42×10^{13} Hz. What is the effective force constant for this molecule?

Picture the Problem We use Equation 38-19 to relate K to the frequency and reduced mass, and calculate μ from its definition.

1. The effective force constant is related to the frequency and reduced mass by Equation 38-19:

$$f = \frac{1}{2\pi}\sqrt{\frac{K}{\mu}}$$

$$K = (2\pi f)^2 \mu$$

2. Calculate the reduced mass using 12 u for the mass of the carbon atom and 16 u for the mass of the oxygen atom:

$$\mu = \frac{m_1 m_2}{m_1 + m_2} = \frac{(12\text{ u})(16\text{ u})}{12\text{ u} + 16\text{ u}} = 6.86\text{ u}$$

3. Substitute this value of μ into the equation for K in step 1 and convert to SI units:

$$K = (2\pi f)^2 \mu = 4\pi^2 (6.42 \times 10^{13}\text{ Hz})^2 (6.86\text{ u})\frac{1.66 \times 10^{-27}\text{ kg}}{1\text{ u}}$$

$$= 1.85 \times 10^3\text{ N/m}$$

Emission Spectra

Figure 38-13 shows schematically some electronic, vibrational, and rotational energy levels of a diatomic molecule. The vibrational levels are labeled with the quantum number ν and the rotational levels are labeled with ℓ. The lower vibrational levels are evenly spaced, with $\Delta E = hf$. For higher vibrational levels, the approximation that the vibration is simple harmonic is not valid and the levels are not quite evenly spaced. Note that the potential-energy curves representing the force between the two atoms in the molecule do not have exactly the same shape for the electronic ground and excited states. This implies that the fundamental frequency of vibration f is different for different electronic states. For transitions between vibrational states of different electronic states, the selection rule $\Delta \nu = \pm 1$ does not hold. Such transitions result in the emission of photons of wavelengths in or near the visible spectrum, so the emission spectrum of a molecule for electronic transitions is also sometimes called the optical spectrum.

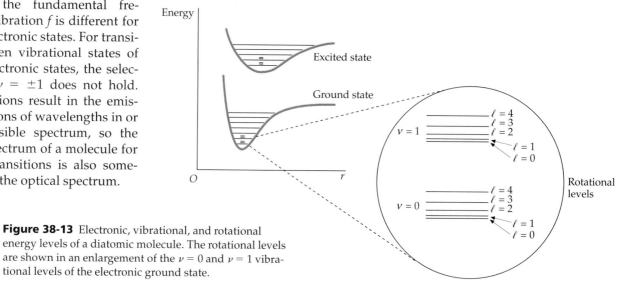

Figure 38-13 Electronic, vibrational, and rotational energy levels of a diatomic molecule. The rotational levels are shown in an enlargement of the $\nu = 0$ and $\nu = 1$ vibrational levels of the electronic ground state.

The spacing of the rotational levels increases with increasing values of ℓ. Since the energies of rotation are so much smaller than those of vibrational or electronic excitation of a molecule, molecular rotation shows up in optical spectra as a fine splitting of the spectral lines. When the fine structure is not resolved, the spectrum appears as bands as shown in Figure 38-14a. Close inspection of these bands reveals that they have a fine structure due to the rotational energy levels, as shown in the enlargement in Figure 38-14b.

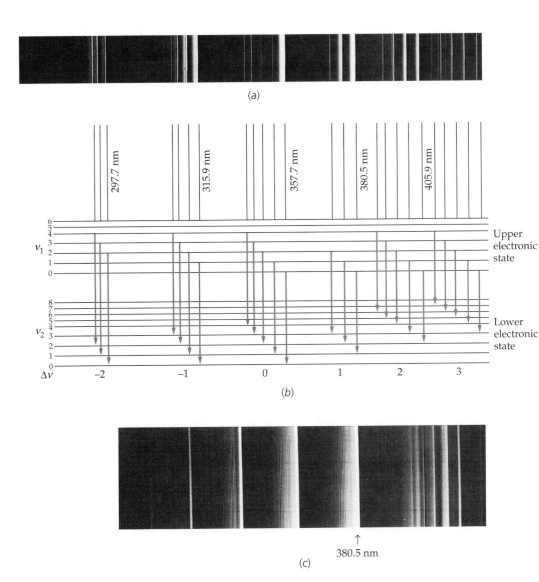

Figure 38-14 (a) Part of the emission spectrum of N_2. The spectral lines are due to transitions between the vibrational levels of two electronic states, as indicated in the energy level diagram (b). (c) An enlargement of part of (a) shows that the apparent lines in (a) are in fact bands with structure caused by rotational levels.

Absorption Spectra

Much molecular spectroscopy is done using infrared absorption techniques in which only the vibrational and rotational energy levels of the ground-state electronic level are excited. For ordinary temperatures, the vibrational energies are sufficiently large in comparison with the thermal energy kT that most of the molecules are in the lowest vibrational state $\nu = 0$, for which the energy is $E_0 = \frac{1}{2}hf$. The transition from $\nu = 0$ to $\nu = 1$ is the predominant transition in absorption. The rotational energies, however, are sufficiently less than kT that the molecules are distributed among several rotational energy states. If the molecule is originally in a vibrational state characterized by

$v = 0$ and a rotational state characterized by the quantum number ℓ, its initial energy is

$$E_\ell = \tfrac{1}{2}hf + \ell(\ell + 1)E_{0r} \qquad\qquad \text{38-20}$$

where E_{0r} is given by Equation 38-13. From this state, two transitions are permitted by the selection rules. For a transition to the next higher vibrational state $v = 1$ and a rotational state characterized by $\ell + 1$, the final energy is

$$E_{\ell+1} = \tfrac{3}{2}hf + (\ell + 1)(\ell + 2)E_{0r} \qquad\qquad \text{38-21}$$

For a transition to the next higher vibrational state and to a rotational state characterized by $\ell - 1$, the final energy is

$$E_{\ell-1} = \tfrac{3}{2}hf + (\ell - 1)\ell E_{0r} \qquad\qquad \text{38-22}$$

The energy differences are

$$\Delta E_{\ell\to\ell+1} = E_{\ell+1} - E_\ell = hf + 2(\ell + 1)E_{0r} \qquad\qquad \text{38-23}$$

where $\ell = 0, 1, 2, \ldots$, and

$$\Delta E_{\ell\to\ell-1} = E_{\ell-1} - E_\ell = hf - 2\ell E_{0r} \qquad\qquad \text{38-24}$$

where $\ell = 1, 2, 3, \ldots$. (In Equation 38-24, ℓ begins at $\ell = 1$ because from $\ell = 0$ only the transition $\ell \to \ell + 1$ is possible.) Figure 38-15 illustrates these transitions. The frequencies of these transitions are given by

$$f_{\ell\to\ell+1} = \frac{\Delta E_{\ell\to\ell+1}}{h}$$

$$= f + \frac{2(\ell + 1)E_{0r}}{h}, \qquad \ell = 0, 1, 2, \ldots \qquad \text{38-25}$$

and

$$f_{\ell\to\ell-1} = \frac{\Delta E_{\ell\to\ell-1}}{h}$$

$$= f - \frac{2\ell E_{0r}}{h}, \qquad \ell = 1, 2, 3, \ldots \qquad \text{38-26}$$

The frequencies for the transitions $\ell \to \ell + 1$ are thus $f + 2(E_{0r}/h)$, $f + 4(E_{0r}/h)$, $f + 6(E_{0r}/h)$, and so forth; those corresponding to the transition $\ell \to \ell - 1$ are $f - 2(E_{0r}/h)$, $f - 4(E_{0r}/h)$, $f - 6(E_{0r}/h)$, and so forth. We thus expect the absorption spectrum to contain frequencies equally spaced by $2E_{0r}/h$ except for a gap of $4E_{0r}/h$ at the vibrational frequency f as shown in Figure 38-16. A measurement of the position of the gap gives f, and a measurement of the spacing of the absorption peaks gives E_{0r}, which is inversely proportional to the moment of inertia of the molecule.

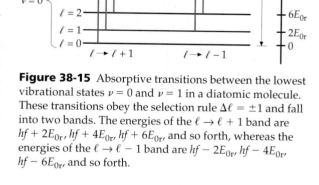

Figure 38-15 Absorptive transitions between the lowest vibrational states $v = 0$ and $v = 1$ in a diatomic molecule. These transitions obey the selection rule $\Delta\ell = \pm 1$ and fall into two bands. The energies of the $\ell \to \ell + 1$ band are $hf + 2E_{0r}$, $hf + 4E_{0r}$, $hf + 6E_{0r}$, and so forth, whereas the energies of the $\ell \to \ell - 1$ band are $hf - 2E_{0r}$, $hf - 4E_{0r}$, $hf - 6E_{0r}$, and so forth.

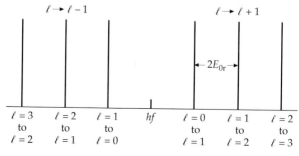

Figure 38-16 Expected absorption spectrum of a diatomic molecule. The right branch corresponds to transitions $\ell \to \ell + 1$ and the left branch to $\ell \to \ell - 1$. The lines are equally spaced by $2E_{0r}$. The energy midway between the branches is hf where f is the frequency of vibration of the molecule.

Figure 38-17 shows the absorption spectrum of HCl. The double-peak structure results from the fact that chlorine occurs naturally in two isotopes, ^{35}Cl and ^{37}Cl, which have different moments of inertia. If all the rotational levels were equally populated initially, we would expect the intensities of each absorption line to be equal. However, the population of a rotational level ℓ is proportional to the degeneracy of the level, that is, to the number of states with the same value of ℓ, which is $2\ell + 1$, and to the Boltzmann factor $e^{-E/kT}$, where E is the energy of the state. For low values of ℓ, the population increases slightly because of the degeneracy factor, whereas for higher values of ℓ, the population decreases because of the Boltzmann factor. The intensities of the absorption lines therefore increase with ℓ for low values of ℓ and then decrease with ℓ for high values of ℓ, as can be seen from the figure.

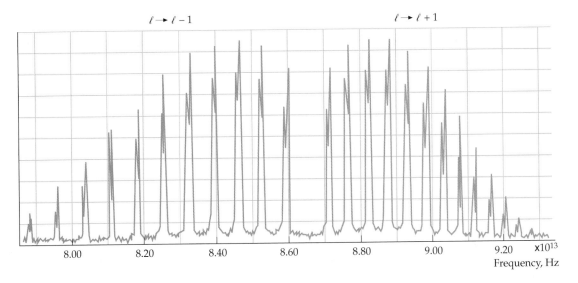

Figure 38-17 Absorption spectrum of the diatomic molecule HCl. The double-peak structure results from the two isotopes of chlorine, ^{35}Cl (abundance 75.5%) and ^{37}Cl (abundance 24.5%). The intensities of the peaks vary because the population of the initial state depends on ℓ.

38-4   The Structure of Solids

The three phases of matter we observe—gas, liquid, and solid—result from the relative strengths of the attractive forces between molecules and the thermal energy of the molecules. Molecules in the gas phase have a high thermal kinetic energy and have little influence on one another except during their frequent but brief collisions. At sufficiently low temperatures (depending on the type of molecule), van der Waals forces will cause practically every substance to condense into a liquid and then a solid. In liquids the molecules are close enough—and their kinetic energy is low enough—that they can develop a temporary **short-range order**. As thermal kinetic energy is further reduced, the molecules form solids, characterized by a lasting order.

If a liquid is cooled slowly so that the kinetic energy of its molecules is reduced slowly, the molecules (or atoms or ions) may arrange themselves in a regular crystalline array, producing the maximum number of bonds and leading to a minimum potential energy. However, if the liquid is cooled rapidly so that its internal energy is removed before the molecules have a chance to arrange themselves, the solid formed is often not crystalline but instead resembles a snapshot of the liquid. Such a solid is called an **amorphous solid.** It displays short-range order but not the **long-range order** (over many molecular diameters) that is characteristic of a crystal. Glass is a typical amorphous solid. A characteristic result of the long-range ordering of a crystal is that it has a well-defined melting point, whereas an amorphous solid merely softens as its temperature is increased. Many materials may solidify into either an amorphous or a crystalline state depending on how they are prepared; others exist only in one form or the other.

Most common solids are polycrystalline; that is, they consist of many single crystals that meet at grain boundaries. The size of a single crystal is typically a fraction of a millimeter. However, large single crystals do occur naturally and can be produced artificially. The most important property of a single crystal is the symmetry and regularity of its structure. It can be thought of as having a single unit structure that is repeated throughout the crystal. This smallest unit of a crystal is called the **unit cell;** its structure depends on the type of bonding—ionic, covalent, metallic, hydrogen, van der Waals—between the atoms, ions, or molecules. If more than one kind of atom is present, the structure will also depend on the atoms' relative sizes.

Figure 38-18 shows the structure of the ionic crystal sodium chloride (NaCl). The Na^+ and Cl^- ions are spherically symmetric, and the Cl^- ion is approximately twice as large as the Na^+ ion. The minimum potential energy for this crystal occurs when an ion of either kind has 6 nearest neighbors of the other kind. This structure is called *face-centered-cubic* (fcc). Note that the Na^+ and Cl^- ions in solid NaCl are *not* paired into NaCl molecules.

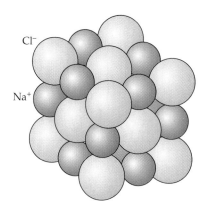

Figure 38-18 Face-centered-cubic structure of the NaCl crystal.

The net attractive part of the potential energy of an ion in a crystal can be written

$$U_{att} = -\alpha\frac{ke^2}{r}$$

38-27

where r is the separation distance between neighboring ions (0.281 nm for the Na^+ and Cl^- ions in crystalline NaCl), and α, called the **Madelung constant,** depends on the geometry of the crystal. If only the 6 nearest neighbors of each ion were important, α would be 6. However, in addition to the 6 neighbors of the opposite charge at a distance r, there are 12 ions of the same charge at a distance $\sqrt{2}r$, 8 ions of opposite charge at a distance $\sqrt{3}r$, and so on. The Madelung constant is thus an infinite sum:

$$\alpha = 6 - \frac{12}{\sqrt{2}} + \frac{8}{\sqrt{3}} - \cdots$$

38-28

The result for face-centered-cubic structures is $\alpha = 1.7476$.*

When Na^+ and Cl^- ions are very close together, they repel each other because of the overlap of their electrons and the exclusion-principle repulsion

---

* A large number of terms are needed to calculate the Madelung constant accurately because the sum converges very slowly.

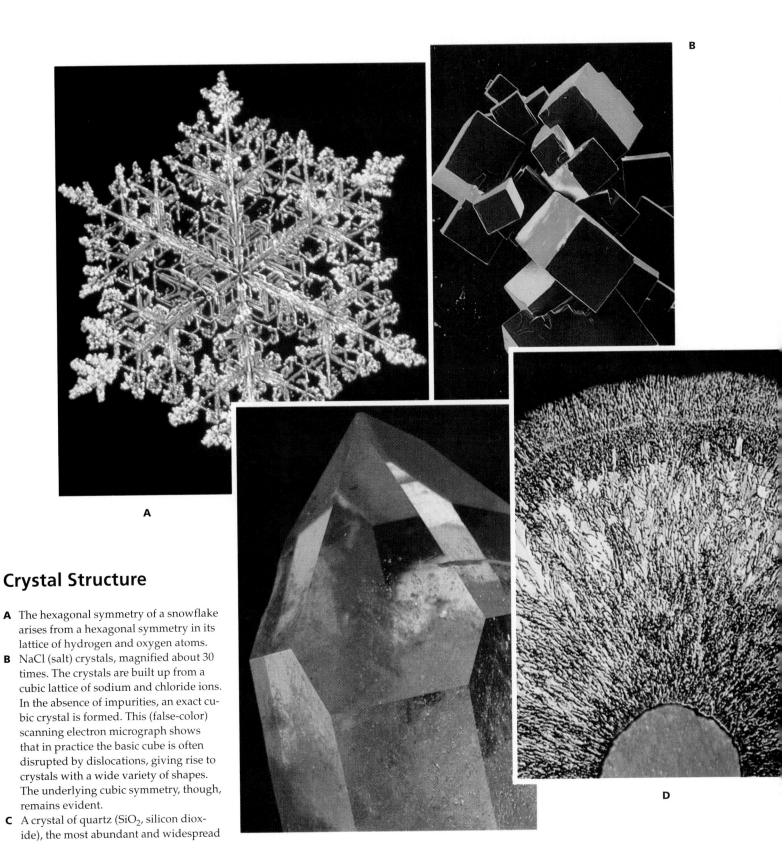

Crystal Structure

A The hexagonal symmetry of a snowflake arises from a hexagonal symmetry in its lattice of hydrogen and oxygen atoms.

B NaCl (salt) crystals, magnified about 30 times. The crystals are built up from a cubic lattice of sodium and chloride ions. In the absence of impurities, an exact cubic crystal is formed. This (false-color) scanning electron micrograph shows that in practice the basic cube is often disrupted by dislocations, giving rise to crystals with a wide variety of shapes. The underlying cubic symmetry, though, remains evident.

C A crystal of quartz (SiO_2, silicon dioxide), the most abundant and widespread mineral on earth. If molten quartz is allowed to solidify without crystallizing, it will form glass.

D A soldering iron tip, ground down to reveal the copper core within its iron sheath. Visible in the iron is its underlying microcrystalline structure.

discussed in Section 38-1. A simple empirical expression for the potential energy associated with this repulsion that works fairly well is

$$U_{rep} = \frac{A}{r^n}$$

where A and n are constants. The total potential energy of an ion is then

$$U = -\alpha \frac{ke^2}{r} + \frac{A}{r^n} \qquad\qquad \text{38-29}$$

The equilibrium separation $r = r_0$ is that at which the force $F = -dU/dr$ is zero. Differentiating and setting $dU/dr = 0$ at $r = r_0$, we obtain

$$A = \frac{\alpha k e^2 r_0^{n-1}}{n} \qquad\qquad \text{38-30}$$

The total potential energy can thus be written

$$U = -\alpha \frac{ke^2}{r_0}\left[\frac{r_0}{r} - \frac{1}{n}\left(\frac{r_0}{r}\right)^n\right] \qquad\qquad \text{38-31}$$

At $r = r_0$, we have

$$U(r_0) = -\alpha \frac{ke^2}{r_0}\left(1 - \frac{1}{n}\right) \qquad\qquad \text{38-32}$$

If we know the equilibrium separation r_0, the value of n can be found approximately from the *dissociation energy* of the crystal, which is the energy needed to break up the crystal into atoms.

Example 38-5

Calculate the equilibrium spacing r_0 for NaCl from the measured density of NaCl, which is $\rho = 2.16 \text{ g/cm}^3$.

Picture the Problem We consider each ion to occupy a cubic volume of side r_0. The mass of 1 mol of NaCl is 58.4 g, which is the sum of the atomic masses of sodium and chlorine. The ions occupy a volume of $2N_A r_0^3$, where $N_A = 6.02 \times 10^{23}$ is Avogadro's number.

1. Relate r_0 to the density ρ:

$$\rho = \frac{m}{V} = \frac{m}{2N_A r_0^3}$$

2. Solve for r_0^3 and substitute the known values:

$$r_0^3 = \frac{m}{2N_A\rho} = \frac{58.4 \text{ g}}{2(6.02 \times 10^{23})(2.16 \text{ g/cm}^3)}$$
$$= 2.25 \times 10^{-23} \text{ cm}^3$$

3. Solve for r_0:

$$r_0 = 2.82 \times 10^{-8} \text{ cm} = 0.282 \text{ nm}$$

The measured dissociation energy of NaCl is 770 kJ/mol. Using $1 \text{ eV} = 1.602 \times 10^{-19}$ J, and the fact that 1 mol of NaCl contains N_A pairs of ions, we can express the dissociation energy in electron volts per ion pair. The conversion between electron volts per ion pair and kilojoules per mole is

$$1\frac{\text{eV}}{\text{ion pair}} \times \frac{6.022 \times 10^{23} \text{ ion pairs}}{\text{mol}} \times \frac{1.602 \times 10^{-19} \text{ J}}{1 \text{ eV}}$$

The result is

$$1\frac{eV}{\text{ion pair}} = 96.47 \frac{kJ}{\text{mol}}$$ 38-33

Thus 770 kJ/mol = 7.98 eV per ion pair. Substituting −7.98 eV for $U(r_0)$, 0.282 nm for r_0, and 1.75 for α in Equation 38-32, we can solve for n. The result is $n = 9.35 \approx 9$.

Most ionic crystals, such as LiF, KF, KCl, KI, and AgCl, have a face-centered-cubic structure. Some elemental solids that have this structure are silver, aluminum, gold, calcium, copper, nickel, and lead.

Figure 38-19 shows the structure of CsCl, which is called the *body-centered-cubic* (bcc) structure. In this structure, each ion has 8 nearest neighbor ions of the opposite charge. The Madelung constant for these crystals is 1.7627. Elemental solids with this structure include barium, cesium, iron, potassium, lithium, molybdenum, and sodium.

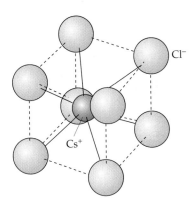 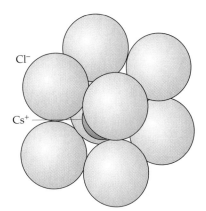

Figure 38-19 Body-centered-cubic structure of the CsCl crystal.

Figure 38-20 shows another important crystal structure: the *hexagonal close-packed* (hcp) structure. It is obtained by stacking identical spheres, such as bowling balls. In the first layer, each ball touches 6 others; thus, the name *hexagonal*. In the next layer, each ball fits into a triangular depression of the first layer. In the third layer, each ball fits into a triangular depression of the second layer, so it lies directly over a ball in the first layer. Elemental solids with hcp structure include beryllium, cadmium, cerium, magnesium, osmium, and zinc.

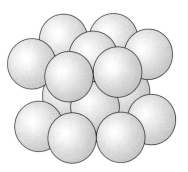

Figure 38-20 Hexagonal close-packed crystal structure.

In some solids with covalent bonding, the crystal structure is determined by the directional nature of the bonds. Figure 38-21 illustrates the diamond structure of carbon, in which each atom is bonded to four others as a result of

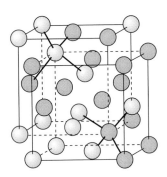

Figure 38-21 Diamond crystal structure. This structure can be considered to be a combination of two interpenetrating face-centered-cubic structures.

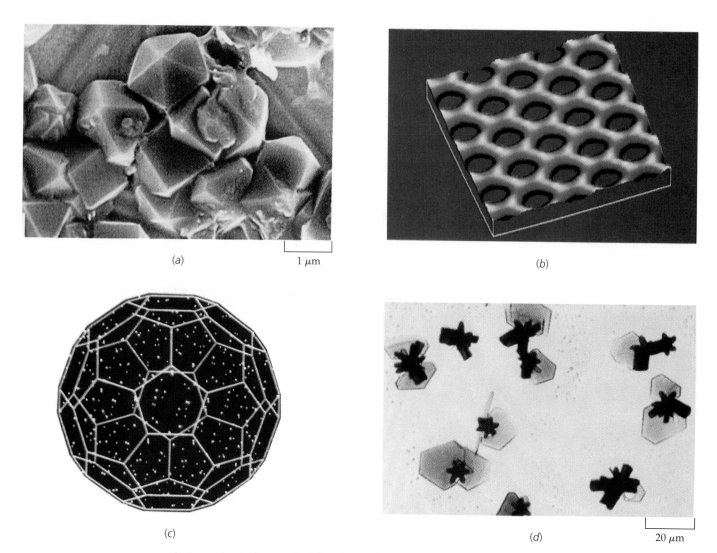

(a)   1 μm

(b)

(c)   (d)   20 μm

Carbon exists in three well-defined crystalline forms: diamond, graphite, and fullerenes (short for "buckminsterfullerenes"), the third of which was predicted and discovered only a few years ago. The forms differ in how the carbon atoms are packed together in a lattice. A fourth form of carbon, in which no well-defined crystalline form exists, is common charcoal. (a) Synthetic diamonds, magnified about 75,000 times. In diamond, each carbon atom is centered in a tetrahedron of four other carbon atoms. The strength of these bonds accounts for the hardness of a diamond. (b) An atomic-force micrograph of graphite. In graphite, carbon atoms are arranged in sheets, each sheet made up of atoms in hexagonal rings. The sheets slide easily across one another, a property that allows graphite to function as a lubricant. (c) A single sheet of carbon rings can be closed on itself if certain rings are allowed to be pentagonal, instead of hexagonal. A computer-generated image of the smallest such structure, C_{60}, is shown here. Each of the 60 vertices corresponds to a carbon atom; 20 of the faces are hexagons and 12 are pentagons. The same geometric pattern is encountered in a soccer ball. (d) Fullerene crystals, in which C_{60} molecules are close-packed. The smaller crystals tend to form thin brownish platelets; larger crystals are usually rod-like in shape. Fullerenes exist in which more than 60 carbon atoms appear. In the crystals shown here, about one-sixth of the molecules are C_{70}.

hybridization, discussed in Section 38-2. This is also the structure of germanium and silicon.

38-5 Semiconductors

In Section 27-4, it was shown that certain materials, such as silicon and germanium, behave as *intrinsic semiconductors* because of the small energy gap between their filled valence band and empty conduction band. The semiconducting property of such materials makes them useful as a basis for electronic circuit components whose resistivity can be controlled by application of an external voltage or current. Most such *solid-state devices*, however, such as the semiconductor diode and the transistor, make use of **impurity semiconductors,** which are created through the controlled addition of certain impurities to intrinsic semiconductors. This process is called **doping.** Figure 38-22*a* is a schematic illustration of silicon doped with a small amount of arsenic such that arsenic atoms replace a few of the silicon atoms in the crystal lattice. Arsenic has five valence electrons rather than the four of silicon. Four of these electrons take part in bonds with the four neighboring silicon atoms, and the fifth electron is very loosely bound to the atom. This extra electron occupies an energy level that is just slightly below the conduction band in the solid, and it is easily excited into the conduction band, where it can contribute to electrical conduction.

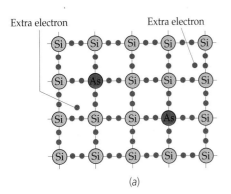

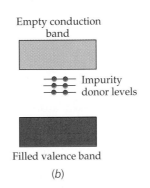

(a)   (b)

Figure 38-22 (*a*) A two-dimensional schematic illustration of silicon doped with arsenic. Because arsenic has five valence electrons, there is an extra, weakly bound electron that is easily excited to the conduction band, where it can contribute to electrical conduction. (*b*) Band structure of an *n*-type semiconductor such as silicon doped with arsenic. The impurity atoms provide filled energy levels that are just below the conduction band. These levels donate electrons to the conduction band.

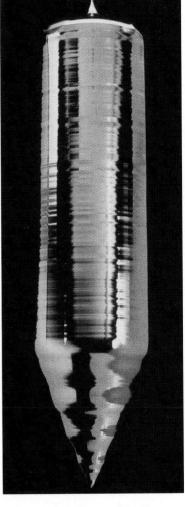

Synthetic crystal silicon is produced beginning with a raw material containing silicon (for instance, common beach sand), separating out the silicon, and melting it. From a seed crystal, the molten silicon grows into a cylindrical crystal, such as the one shown here. The crystals (typically about 1.3 m long) are formed under highly controlled conditions to ensure that they are flawless, and sliced into thousands of thin wafers, onto which the layers of an integrated circuit are etched.

The effect on the band structure of a silicon crystal achieved by doping it with arsenic is shown in Figure 38-22*b*. The levels shown just below the conduction band are due to the extra electrons of the arsenic atoms. These levels are called **donor levels** because they donate electrons to the conduction band without leaving holes in the valence band. Such a semiconductor is called an ***n*-type semiconductor** because the major charge carriers are negative electrons. The conductivity of a doped semiconductor can be controlled by controlling the amount of impurity added. The addition of just one part per million can increase the conductivity by several orders of magnitude.

Another type of impurity semiconductor can be made by replacing a silicon atom with a gallium atom, which has 3 valence electrons (Figure 38-23*a*).

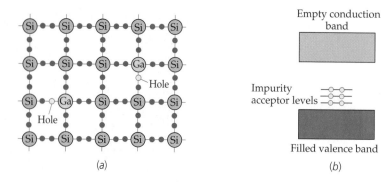

Empty conduction
band

Impurity
acceptor levels

Filled valence band

(a)   (b)

Figure 38-23 (a) A two-dimensional schematic illustration of silicon doped with gallium. Because gallium has only three valence electrons, there is a hole in one of its bonds. As electrons move into the hole, the hole moves about, contributing to the conduction of electrical current. (b) Band structure of a p-type semiconductor such as silicon doped with gallium. The impurity atoms provide empty energy levels just above the filled valence band that accept electrons from the valence band.

The gallium atom accepts electrons from the valence band to complete its four covalent bonds, thus creating a hole in the valence band. The effect on the band structure of silicon achieved by doping it with gallium is shown in Figure 38-23b. The empty levels shown just above the valence band are due to the holes from the ionized gallium atoms. These levels are called **acceptor levels** because they accept electrons from the filled valence band when these electrons are thermally excited to a higher energy state. This creates holes in the valence band that are free to propagate in the direction of an electric field. Such a semiconductor is called a **p-type semiconductor** because the charge carriers are positive holes. The fact that conduction is due to the motion of holes can be verified by the Hall effect.*

---

*The Hall effect is discussed in Chapter 28.

Example 38-6 *try it yourself*

The number of free electrons in pure silicon is about 10^{10} electrons/cm³ at ordinary temperatures. If one silicon atom out of every million atoms is replaced by an arsenic atom, how many free electrons per cubic centimeter are there? (The density of silicon is 2.33 g/cm³ and its molar mass is 28.1 g/mol.)

Picture the Problem The number of silicon atoms per cubic centimeter, n_{Si} can be found from $n_s = N_A \rho / M$. Then, since each arsenic atom contributes one free electron, the number of electrons contributed by the arsenic atoms is $10^{-6} n_{Si}$.

Cover the column to the right and try these on your own before looking at the answers.

Steps   **Answers**

1. Calculate the number of silicon atoms per cubic centimeter.

$$n_{Si} = \frac{\rho N_A}{M} = \frac{(2.33 \text{ g/cm}^3)(6.02 \times 10^{23} \text{ atoms/mol})}{28.1 \text{ g/mol}}$$

$$= 4.99 \times 10^{22} \text{ atoms/cm}^3$$

2. Multiply by 10^{-6} to obtain the number of arsenic atoms per cubic centimeter, which equals the added number of free electrons per cubic centimeter.

$$n_e = 10^{-6} n_s = 4.99 \times 10^{16} \text{ electrons/cm}^3$$

Remark Because silicon has so few free electrons per atom, the number of conduction electrons is increased by a factor of about 5 million by doping silicon with just one arsenic atom per million silicon atoms.

Exercise How many free electrons are there per silicon atom in pure silicon? (*Answer* 2×10^{-13})

38-6 Semiconductor Junctions and Devices

Semiconductor devices such as diodes and transistors make use of *n*-type and *p*-type semiconductors joined together as shown in Figure 38-24. In practice, the two types of semiconductors are often incorporated into a single silicon crystal doped with donor impurities on one side and acceptor impurities on the other. The region in which the semiconductor changes from *p*-type to *n*-type is called a **junction.**

When an *n*-type and a *p*-type semiconductor are placed in contact, the initially unequal concentrations of electrons and holes result in the diffusion of electrons across the junction from the *n* side to the *p* side and holes from the *p* side to the *n* side until equilibrium is established. The result of this diffusion is a net transport of positive charge from the *p* side to the *n* side. Unlike the case when two different metals are in contact, the electrons cannot travel very far from the junction region because the semiconductor is not a particularly good conductor. The diffusion of electrons and holes therefore creates a double layer of charge at the junction similar to that on a parallel-plate capacitor. There is thus a potential difference *V* across the junction, which tends to inhibit further diffusion. In equilibrium, the *n* side with its net positive charge will be at a higher potential than the *p* side with its net negative charge. In the junction region, between the charge layers, there will be very few charge carriers of either type, so the junction region has a high resistance. Figure 38-25 shows the energy level diagram for a *pn* junction. The junction region is also called the **depletion region** because it has been depleted of charge carriers.

Diodes

In Figure 38-26 an external potential difference has been applied across a *pn* junction by connecting a battery and resistor to the semiconductor. When the positive terminal of the battery is connected to the *p* side of the junction, as shown in Figure 38-26*a*, the junction is said to be **forward biased.** Forward biasing lowers the potential across the junction. The diffusion of electrons and holes is thereby increased as they attempt to reestablish equilibrium, resulting in a current in the circuit.

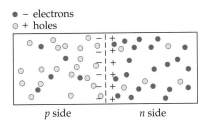

● – electrons
○ + holes

Figure 38-24 A *pn* junction. Because of the difference in their concentrations, holes diffuse from the *p* side to the *n* side and electrons diffuse from the *n* side to the *p* side. As a result, there is a double layer of charge at the junction, with the *p* side being negative and the *n* side being positive.

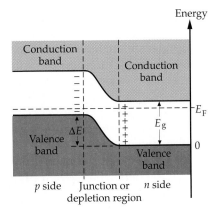

Figure 38-25 Electron energy levels for a *pn* junction.

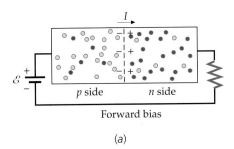

Forward bias

(*a*)

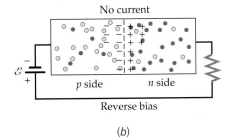

No current

Reverse bias

(*b*)

Figure 38-26 A *pn*-junction diode. (*a*) Forward-biased *pn* junction. The applied potential difference enhances the diffusion of holes from the *p* side to the *n* side and of electrons from the *n* side to the *p* side, resulting in a current *I*. (*b*) Reverse-biased *pn* junction. The applied potential difference inhibits the further diffusion of holes and electrons, so there is no current.

If the positive terminal of the battery is connected to the n side of the junction, as shown in Figure 38-26b, the junction is said to be **reverse biased.** Reverse biasing tends to increase the potential difference across the junction, thereby further inhibiting diffusion. Figure 38-27 shows a plot of current versus voltage for a typical semiconductor junction. Essentially, the junction conducts only in one direction. A single-junction semiconductor device is called a **diode.*** Diodes have many uses. One is to convert alternating current into direct current, a process called rectification.

Note that the current in Figure 38-27 suddenly increases in magnitude at extreme values of reverse bias. In such large electric fields, electrons are stripped from their atomic bonds and accelerated across the junction. These electrons, in turn, cause others to break loose. This effect is called **avalanche breakdown.** Although such a breakdown can be disastrous in a circuit where it is not intended, the fact that it occurs at a sharply defined voltage makes it of use in a special voltage reference standard known as a **Zener diode.**

An interesting effect that we can discuss only qualitatively occurs if both the n side and p side of a pn-junction diode are so heavily doped that the donors on the n side provide so many electrons that the lower part of the conduction band is practically filled and the acceptors on the p side accept so many electrons that the upper part of the valence band is nearly empty. Figure 38-28a shows the energy-level diagram for this situation. Because the depletion region is now so narrow, electrons can easily penetrate the potential barrier across the junction, and tunnel to the other side. The flow of electrons through the barrier is called a **tunneling current,** and such a heavily doped diode is called a **tunnel diode.**

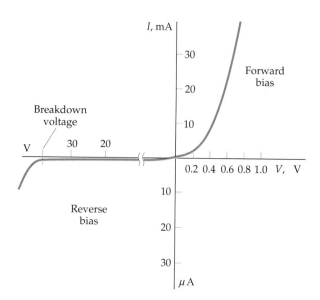

Figure 38-27 Current versus applied voltage across a pn junction. Note the different scales on both axes for the forward and reverse bias conditions.

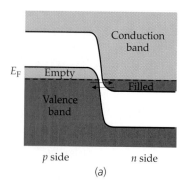 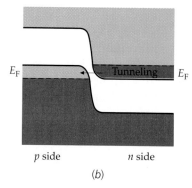 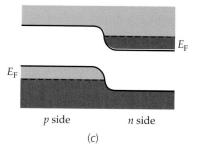

Figure 38-28 Electron energy levels for a heavily doped pn-junction tunnel diode. (a) With no bias voltage, some electrons tunnel in each direction. (b) With a small bias voltage, the tunneling current is enhanced in one direction, making a sizable contribution to the net current. (c) With further increases in the bias voltage, the tunneling current decreases dramatically.

At equilibrium with no bias, there is an equal tunneling current in each direction. When a small bias voltage is applied across the junction, the energy-level diagram is as shown in Figure 38-28b and the tunneling of electrons from the n side to the p side is increased whereas that in the opposite direction is decreased. This tunneling current in addition to the usual current due to diffusion results in a considerable net current. When the bias voltage is increased slightly, the energy-level diagram is as shown in Figure 38-28c and the tunneling current is decreased. Although the diffusion current is in-

---

*The name diode originates from a vacuum tube device consisting of just two electrodes that also conducts electric current in one direction only.

creased, the net current is decreased. At large bias voltages, the tunneling current is completely negligible, and the total current increases with increasing bias voltage due to diffusion as in an ordinary *pn*-junction diode. Figure 38-29 shows the current-versus-voltage curve for a tunnel diode. Such diodes are used in electric circuits because of their very fast response time. When operated near the peak in the current-versus-voltage curve, a small change in bias voltage results in a large change in the current.

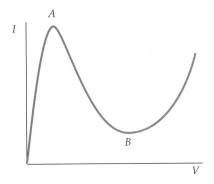

Figure 38-29 Current versus applied voltage for a tunnel diode. Up to point *A*, an increase in the bias voltage enhances tunneling. Between points *A* and *B*, an increase in the bias voltage inhibits tunneling. After point *B*, the tunneling is negligible, and the diode behaves like an ordinary *pn*-junction diode.

Another use for the *pn*-junction semiconductor is the **solar cell,** which is illustrated schematically in Figure 38-30. When a photon of energy greater than the gap energy (1.1 eV in silicon) strikes the *p*-type region, it can excite an electron from the valence band into the conduction band, leaving a hole in the valence band. This region is already rich in holes. Some of the electrons created by the photons will recombine with holes, but some will migrate to the junction. From there they are accelerated into the *n*-type region by the electric field between the double layer of charge. This creates an excess negative charge in the *n*-type region and excess positive charge in the *p*-type region. The result is a potential difference between the two regions, which in practice is about 0.6 V. If a load resistance is connected across the two regions, a charge flows through the resistance. Some of the incident light energy is thus converted into electrical energy. The current in the resistor is proportional to the number of incident photons, which is in turn proportional to the intensity of the incident light.

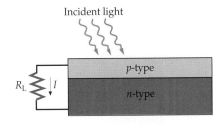

Figure 38-30 A *pn*-junction semiconductor as a solar cell. When light strikes the *p*-type region, electron–hole pairs are created, resulting in a current through the load resistance R_L.

There are many other applications of semiconductors with *pn* junctions. Particle detectors called **surface-barrier detectors** consist of a *pn*-junction semiconductor with a large reverse bias so that there is ordinarily no current. When a high-energy particle, such as an electron, passes through the semiconductor, it creates many electron–hole pairs as it loses energy. The resulting current pulse signals the passage of the particle. **Light-emitting diodes (LEDs)** are *pn*-junction semiconductors with a large forward bias that produces a large excess concentration of electrons on the *p* side and holes on the *n* side of the junction. Under these conditions, the diode emits light as the electrons and holes recombine. This is essentially the reverse of the process that occurs in a solar cell, in which electron–hole pairs are created by the absorption of light. LEDs are commonly used in displays for digital watches and calculators.

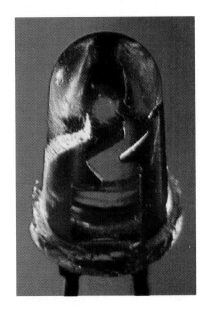

A light-emitting diode (LED).

Transistors

The transistor, a semiconducting device that is used to produce a desired output signal in response to an input signal, was invented in 1948 by William Shockley, John Bardeen, and Walter H. Brattain and has revolutionized the electronics industry and our everyday world. A simple *bipolar junction transistor** consists of three distinct semiconductor regions called the **emitter,** the **base,** and the **collector.** The base is a very thin region of one type of semiconductor sandwiched between two regions of the opposite type. The emitter semiconductor is much more heavily doped than either the base or the col-

---

* Besides the bipolar junction transistor, there are other categories of transistors, notably, the field-effect transistor.

lector. In an *npn* transistor, the emitter and collector are *n*-type semiconductors and the base is a *p*-type semiconductor; in a *pnp* transistor, the base is an *n*-type semiconductor and the emitter and collector are *p*-type semiconductors.

Figures 38-31 and 38-32 show, respectively, a *pnp* transistor and an *npn* transistor with the symbols used to represent each transistor in circuit diagrams. We see that either transistor consists of two *pn* junctions. We will discuss the operation of a *pnp* transistor. The operation of an *npn* transistor is similar.

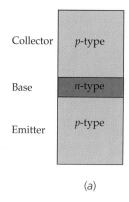

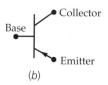

Figure 38-31 A *pnp* transistor. (*a*) The heavily doped emitter emits holes that pass through the thin base to the collector. (*b*) Symbol for a *pnp* transistor in a circuit. The arrow points in the direction of the conventional current, which is the same as that of the emitted holes.

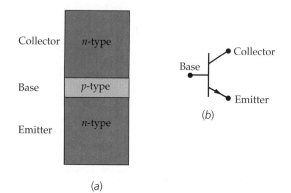

Figure 38-32 An *npn* transistor. (*a*) The heavily doped emitter emits electrons that pass through the thin base to the collector. (*b*) Symbol for an *npn* transistor. The arrow points in the direction of the conventional current, which is opposite the direction of the emitted electrons.

In normal operation, the emitter–base junction is forward biased, and the base–collector junction is reverse biased, as shown in Figure 38-33. The heavily doped *p*-type emitter emits holes that flow across the emitter–base junction into the base. Because the base is very thin, most of these holes flow across the base into the collector. This flow constitutes a current I_c from the emitter to the collector. However, some of the holes recombine in the base producing a positive charge that inhibits the further flow of current. To prevent this, some of the holes that do not reach the collector are drawn off the base as a base current I_b in a circuit connected to the base. In Figure 38-33, therefore, I_c is almost but not quite equal to I_e, and I_b is much smaller than either I_c or I_e. It is customary to express I_c as

$$I_c = \beta I_b \qquad\qquad 38\text{-}34$$

where β is called the **current gain** of the transistor. Transistors can be designed to have values of β as low as 10 or as high as several hundred.

Figure 38-34 shows a simple *pnp* transistor used as an amplifier. A small time-varying input voltage v_S is connected in series with a bias voltage V_{eb}. The base current is then the sum of a steady current I_b produced by the bias voltage V_{eb} and a varying current i_b due to the signal voltage v_s. Because v_s may at any instant be either positive or negative, the bias voltage V_{eb} must be large enough to ensure that there is always a forward bias on the emitter–base junction. The collector current will consist of two parts: a direct current $I_c = \beta I_b$ and an alternating current $i_c = \beta i_b$. We thus have a current amplifier in which the time-varying output current i_c is β times the input current i_b. In such an amplifier, the steady currents I_c and I_b, although essential to the operation of the transistor, are usually not of interest. The input signal voltage v_S is related to the base current by Ohm's law:

$$i_b = \frac{v_s}{R_b + r_b} \qquad\qquad 38\text{-}35$$

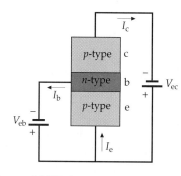

Figure 38-33 A *pnp* transistor biased for normal operation. Holes from the emitter can easily diffuse across the base, which is only tens of nanometers thick. Most of the holes flow to the collector, producing the current I_c.

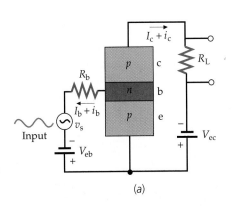 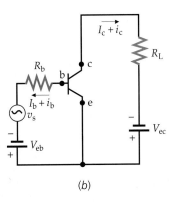

Figure 38-34 (*a*) A *pnp* transistor used as an amplifier. A small change i_b in the base current results in a large change i_c in the collector current. Thus, a small signal in the base circuit results in a large signal across the load resistor R_L in the collector circuit. (*b*) The same circuit as (*a*) with the conventional symbol for the transistor.

where r_b is the internal resistance of the transistor between the base and emitter. Similarly, the collector current i_c produces a voltage v_L across the output or load resistance R_L given by

$$v_L = i_c R_L \qquad\qquad\qquad 38\text{-}36$$

Using Equations 38-34 and 38-35, we have

$$i_c = \beta i_b = \beta \frac{v_s}{R_b + r_b}$$

The output voltage is thus related to the input voltage by

$$v_L = \beta \frac{R_L}{R_b + r_b} v_s \qquad\qquad\qquad 38\text{-}37$$

The ratio of the output voltage to the input voltage is the **voltage gain** of the amplifier:

$$\text{Voltage gain} = \frac{v_L}{v_s} = \beta \frac{R_L}{R_b + r_b} \qquad\qquad\qquad 38\text{-}38$$

A typical amplifier, such as that in a tape player, has several transistors similar to the one in Figure 38-34 connected in series so that the output of one transistor serves as the input for the next. Thus, the very small voltage produced by the passage of the magnetized tape past the pickup heads controls the large amounts of power required to drive the loudspeakers. The power delivered to the speakers is supplied by the dc sources connected to each transistor.

The technology of semiconductors extends well beyond individual transistors and diodes. Many of the electronic devices we now take for granted, such as laptop computers and the processors that govern the operation of vehicles and appliances, rely on large-scale integration of many transistors and other circuit components on a single "chip." Large-scale integration combined with advanced concepts in semiconductor theory has created remarkable new instruments for scientific research; see the Exploring sections that follow.

*e*xploring

Integrated Circuits

Integrated circuits (ICs; often called chips) combine "active" electronic devices (transistors and diodes) with "passive" ones (capacitors and resistors) on a single semiconductor crystal. Chips containing the equivalent of hundreds of thousands of transistors may be as small as a few millimeters square and can be connected to dozens of leads (Figures 1 and 2). Although resistors, capacitors, and conductors (Figure 3) can be incorporated into the chip, no means have been found to directly fabricate inductors (the remaining passive circuit component) on ICs; they are simulated with other circuitry or appended to a chip as discrete components.

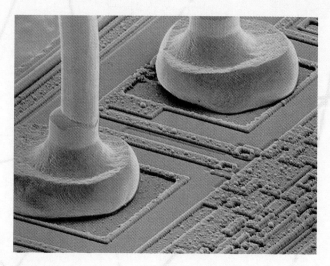

Figure 2 Scanning electron micrograph showing two conductor leads precision bonded to the edge of a chip (magnification: ×163).

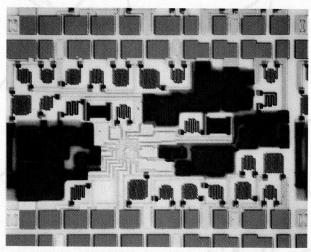

Figure 3 Capacitors (orange blocks), resistors (brown blocks and meandering black lines), and conductors (gold lines) on a ceramic base, formed here by metal films only a few tenths of a micrometer thick.

Figure 1 A chip used to format digitized voice and data signals so they can share a single transmission line. This chip is connected to 44 conductor leads and has an actual size of 6.4 mm squared.

A particular solid-state technology that especially lends itself to very large scale integration is metal-oxide-semiconductor (MOS) technology. MOS transistors are manufactured by heating an original silicon wafer to about 1000°C, causing a layer of silicon dioxide (SiO_2) to form on its surface (Figures 4 and 5). This is coated with a photoresist and exposed to light through a mask. Unexposed (masked) windows of photoresist are etched away with a developer, exposing the silicon dioxide, which is etched away with acid. The exposed (unmasked) areas are resistant to the developer and are not affected. The wafer is again heated and this time doped, via a diffusion process, with a *p*-type impurity, forming *pn* junctions

Integrated circuits making use of the magnetic properties of materials can serve as nonvolatile digital memory. Magnetic bubble memory chips are the integrated-circuit analog to magnetic recording tape and disks (Figure 6). In a thin-film garnet memory crystal, magnetic "bubbles" are created when the garnet is placed between two permanent magnets. They represent regions whose magnetic polarity points in a direction opposite to that of the surrounding crystal. An additional external magnetic field manipulates the position of the bubbles. (Garnet is easy to magnetize, up or down, along a particular axis, but hard to magnetize perpendicular to that axis. This property is necessary for the formation and movement of bubbles.) Storage sites for bubbles are established using a layer of ferromagnetic material deposited on the surface of the crystal; the presence or absence of a bubble at a site can be used to represent a bit of data.

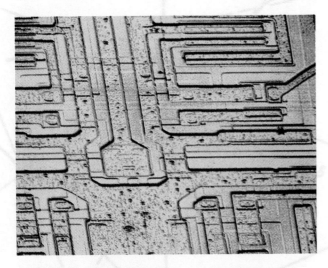

Figure 4 Scanning electron micrograph of metal-oxide-semiconductor (MOS) transistors in patterned layers (magnification ×106).

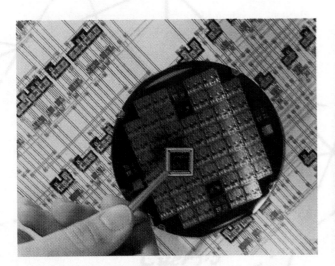

Figure 5 The chip in the tweezers holds 150,000 transistors. Beneath it is a 4-inch wide silicon wafer, awaiting dicing, on which a group of chips have been fabricated simultaneously. In the background is a detail of the "stare plot," the layout of the chip's circuits.

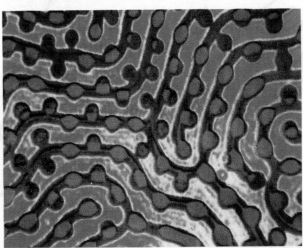

Figure 6 Magnetized domains ("bubbles," actually cylinders seen in cross section), blue in this video micrograph, flow along channels in a thin-film garnet memory crystal. The bubbles are created when the garnet is placed between two permanent magnets. Storage sites for bubbles are established using a layer of ferromagnetic material deposited on the surface of the crystal; the presence or absence of a bubble at a site can be used to represent a bit of data.

in the n-type silicon. The chip is covered with a contact metal (typically aluminum), which bonds to the SiO_2 that has re-formed in windows while the chip was heated and doped. The contact metal itself is patterned in a final photo-etching process. Entire microchips are fabricated by an elaboration, using many masks, of this process.

*e*xploring

Charge-Coupled Devices

Charge-coupled devices (CCDs) are light-sensitive semiconductors at the forefront of imaging technology. They are efficient and fast, and their output is easily stored electronically for processing by computer. Typically 40 to 80% of the photons incident on a CCD surface are converted into a stored electrical signal, allowing for short exposure times and a very low detection threshold. This compares with the 2 or 3% of incoming photons that react with a film's light-sensitive atoms to produce exposed film grains. Also, unlike a photographic film, the response of a CCD is di-

Figure 2 Platinum silicide CCD chip that responds to infrared wavelengths. It contains pixels in a 320 by 244 array.

rectly proportional to the amount of incoming light, making possible a much more precise measurement of data. Arrays of CCDs used as receptors can greatly expand the capabilities of optical telescopes.

A CCD is a three-layer semiconductor: The top layer is a series of metallic electrodes (see Figure 1), the bottom layer is a silicon crystal, and the middle layer is an insulator separating the two. Light striking silicon in the semiconductor frees electrons, which accumulate in potential wells at the surface of the silicon (Figure 2). Each well in the two-dimensional array on the silicon surface stores an amount of charge that is proportional to the number of photons that strike the surface in the region of the well. The charge is dumped electronically into a computer that records the location and amount of charge in each well. A conventional TV monitor can then be used to reconstruct the original image.

Figure 3 shows a CCD image of two galaxies presented with minimal computer processing. In Figures 4 through 6, the same data that yielded the image in Figure 3 have been processed to reveal different degrees and kinds of detail.

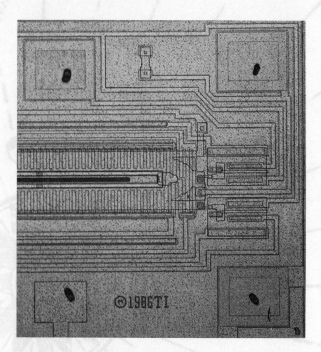

Figure 1 Close-up of part of a CCD. The horizontal bar emerging from the left is the photosensitive area (called a "pixel"). The vertical segments above and below it (called a "transfer register") contain the succession of electrodes that transfer accumulated charge packets along a line of potential wells, from left to right, eventually depositing them in an amplifier located in the central right portion of the chip.

Figure 3 Unprocessed CCD image of spiral galaxy Messier 51 and companion galaxy.

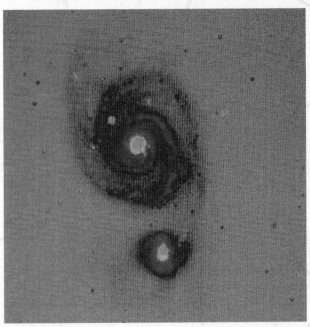

Figure 5 An image that again, like Figure 4, has been generated from the data in Figure 3 and enhanced and colorized by computer.

Figure 4 An image generated from the data contained in Figure 3 in which false colors have been assigned, corresponding to different intensity ranges.

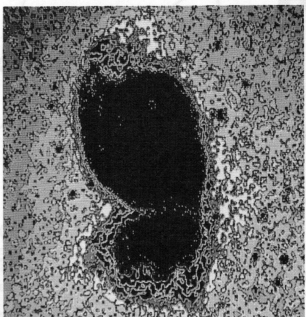

Figure 6 This time the image has been processed for maximum contrast and contoured to show detail in the outer rims of the galaxies.

Summary

1. Atoms are usually found in nature bonded to form molecules or in the lattices of crystalline solids.

2. Ionic bonds and covalent bonds are the principal mechanisms responsible for forming molecules. van der Waals bonds and metallic bonds are important in the formation of liquids and solids. Hydrogen bonds enable large biological molecules to maintain their shape.

3. Like atoms, molecules emit electromagnetic radiation when making a transition from a higher to a lower energy state. The internal energy of a molecule can be separated into three parts: electronic, vibrational, and rotational energy.

4. The molecules in liquids are characterized by a temporary short-range order. The molecules or ions in solids have a more lasting order. Amorphous solids maintain a short-range order similar to the short-range order of a liquid. Crystalline solids display a long-range order determined by their minimum potential energy state.

Topic	Remarks and Relevant Equations
1. Bonding Mechanisms	
Ionic	Ionic bonds result when an electron is transferred from one atom to another, resulting in a positive ion and a negative ion that bond together.
Covalent	The covalent bond is a quantum-mechanical effect that arises from the sharing of one or more electrons by atoms.
van der Waals	The van der Waals bonds are weak bonds that result from the interaction of the instantaneous electric dipole moments of molecules.
Hydrogen	The hydrogen bond results from the sharing of a hydrogen atom by other atoms.
Metallic	In the metallic bond, the positive lattice ions of the metal are held together by a cloud of negative charge comprised of free electrons.
Mixed	A diatomic molecule formed from two identical atoms, such as O_2, must bond by covalent bonding. The bonding of two nonidentical atoms is often a mixture of covalent and ionic bonding. The percentage of ionic bonding can be found from the ratio of the measured electric dipole moment to the ionic electric dipole moment defined by $$p_{\text{ionic}} = er_0 \qquad \textbf{38-5}$$ where r_0 is the equilibrium separation of the ions.
2. Polyatomic Molecules (optional)	The shapes of such polyatomic molecules as H_2O and NH_3 can be understood from the spatial distribution of the atomic-orbital or molecular-orbital wave functions. The tetravalent nature of the carbon atom is a result of the hybridization of the 2s and 2p atomic orbitals.
3. Diatomic Molecules	
Moment of inertia	$$I = \mu r_0^2 \qquad \textbf{38-14}$$ where r_0 is the equilibrium separation, and μ is the reduced mass:
Reduced mass	$$\mu = \frac{m_1 m_2}{m_1 + m_2} \qquad \textbf{38-15}$$

Rotational energy levels	$E = \dfrac{\ell(\ell + 1)\hbar^2}{2I} = \ell(\ell + 1)E_{0r} \qquad \ell = 0, 1, 2, \ldots$	38-12
	where $E_{0r} = \hbar^2/2I$.	38-13
Vibrational energy levels	$E_\nu = (\nu + \frac{1}{2})hf \qquad \nu = 0, 1, 2, 3, \ldots$	38-18
Effective force constant K	$f = \dfrac{1}{2\pi}\sqrt{\dfrac{K}{\mu}}$	38-19

4. Molecular Spectra

The optical spectra of molecules have a band structure due to transitions between rotational levels. Information about the structure and bonding of a molecule can be found from its rotational and vibrational absorption spectrum involving transitions from one vibrational–rotational level to another. These transitions obey the selection rules

$$\Delta \nu = \pm 1, \qquad \Delta \ell = \pm 1$$

5. Crystal Structure of Solids

Solids are often found in crystalline form in which a small structure called the unit cell is repeated over and over. A crystal may have a face-centered-cubic, body-centered-cubic, hexagonal-close-packed, or other structure depending on the type of bonding between the atoms, ions, or molecules in the crystal and on the relative sizes of the atoms if there are more than one kind as in NaCl.

Potential energy	$U = -\alpha\dfrac{ke^2}{r} + \dfrac{A}{r^n}$	38-29

where r is the separation distance between neighboring ions, α is the Madelung constant, which depends on the geometry of the crystal and is of the order of 1.8, and n is about 9.

6. Impurity Semiconductors

The conductivity of a semiconductor can be greatly increased by doping. In an n-type semiconductor, the doping adds electrons just below the conduction band. In a p-type semiconductor, holes are added just above the valence band.

7. Semiconductor Devices (optional)

Junction

Semiconductor devices such as diodes and transistors make use of n-type and p-type semiconductors joined together. The two types of semiconductors are often a single silicon crystal doped with donor impurities on one side and acceptor impurities on the other. The region in which the semiconductor changes from a p-type to an n-type is called a junction. Junctions are used in diodes, solar cells, surface barrier detectors, LEDs, and transistors

Diode

A diode is a single-junction device that carries current in one direction only.

Zener diode

A Zener diode is a diode with a very high reverse bias. It breaks down suddenly at a distinct voltage and is therefore used as a voltage reference standard.

Tunnel diode

A tunnel diode is a diode that is heavily doped so that electrons tunnel through the depletion barrier. At normal operation, a small change in bias voltage results in a large change in current.

Transistors

A transistor consists of a very thin semiconductor of one type sandwiched between two semiconductors of the opposite type. Transistors are used in amplifiers because a small variation in the base current results in a large variation in the collector current.

Problem-Solving Guide

Summary of Worked Examples

Type of Calculation	Procedure and Relevant Examples
1. Molecular Bonding	
Calculate the energy due to repulsion of two ions in a ionic molecule.	The dissociation energy equals the total potential energy, which is the sum of the electrostatic energy of attraction plus the energy of repulsion plus the energy required to form the ions. **Example 38-1**
2. Diatomic Molecules	
Find the reduced mass.	Use $\mu = m_1 m_2/(m_1 + m_2)$. **Example 38-2**
Find the characteristic rotational energy.	Use $E_{0r} = \hbar^2/2I$ with $I = \frac{1}{2}m_0^2$. **Example 38-3**
Find the effective force constant given the vibration frequency.	Use $f = \dfrac{1}{2\pi}\sqrt{\dfrac{K}{\mu}}$. **Example 38-4**
3. Solids	
Estimate the equilibrium spacing of ions in an ionic crystal.	Assume each ion occupies a cubic volume of side r_0. Then calculate r_0 from the density and molecular mass. **Example 38-5**
Find the number of free electrons per cubic centimeter in a doped semiconductor.	Calculate the number of doped atoms per cubic centimeter. **Example 38-6**

Problems

In a few problems, you are given more data than you actually need; in a few other problems, you are required to supply data from your general knowledge, outside sources, or informed estimates.

  Conceptual Problems

  Problems from Optional and Exploring sections

•   Single-concept, single-step, relatively easy
••   Intermediate-level, may require synthesis of concepts
•••   Challenging, for advanced students

Molecular Bonding

1 • Would you expect the NaCl molecule to be polar or nonpolar?

2 • Would you expect the N_2 molecule to be polar or nonpolar?

3 • Does neon occur naturally as Ne or Ne_2? Why?

4 • What type of bonding mechanism would you expect for (a) HF, (b) KBr, (c) N_2?

5 • What kind of bonding mechanism would you expect for (a) the N_2 molecule, (b) the KF molecule, (c) Ag atoms in a solid?

6 • Calculate the separation of Na^+ and Cl^- ions for which the potential energy is -1.52 eV.

7 • The dissociation energy of Cl_2 is 2.48 eV. Consider the formation of an NaCl molecule according to the reaction

$$Na + \tfrac{1}{2}Cl_2 \rightarrow NaCl$$

Does this reaction absorb energy or release energy? How much energy per molecule is absorbed or released?

8 • The dissociation energy is sometimes expressed in kilocalories per mole (kcal/mol). (a) Find the relation between the units eV/molecule and kcal/mol. (b) Find the dissociation energy of molecular NaCl in kcal/mol.

9 • The equilibrium separation of the HF molecule is 0.0917 nm and its measured electric dipole moment is 6.40×10^{-30} C·m. What percentage of the bonding is ionic?

10 • Do Problem 9 for CsCl, for which the equilibrium separation is 0.291 nm and the measured electric dipole moment is 3.48×10^{-29} C·m.

11 •• The dissociation energy of LiCl is 4.86 eV and the equilibrium separation is 0.202 nm. The electron affinity of chlorine is 3.62 eV, and the ionization energy of lithium is 5.39 eV. Determine the core-repulsion energy of LiCl.

12 •• The equilibrium separation of the K^+ and Cl^- ions in KCl is about 0.267 nm. (a) Calculate the potential energy of attraction of the ions assuming them to be point charges at this separation. (b) The ionization energy of potassium is 4.34 eV and the electron affinity of Cl is 3.62 eV. Find the dissociation energy neglecting any energy of repulsion. (See Figure 38-1.) The measured dissociation energy is 4.49 eV. What is the energy due to repulsion of the ions at the equilibrium separation?

13 •• Indicate the mean value of r for two vibration levels in the potential-energy curve for a diatomic molecule and show that because of the asymmetry in the curve, r_{av} increases with increasing vibration energy, and therefore solids expand when heated.

14 •• Calculate the potential energy of attraction between the Na^+ and Cl^- ions at the equilibrium separation $r_0 = 0.236$ nm and compare this result with the dissociation energy given in Figure 38-1. What is the energy due to repulsion of the ions at the equilibrium separation?

15 •• The equilibrium separation of the K^+ and F^- ions in KF is about 0.217 nm. (a) Calculate the potential energy of attraction of the ions, assuming them to be point charges at this separation. (b) The ionization energy of potassium is 4.34 eV and the electron affinity of F is 3.40 eV. Find the dissociation energy neglecting any energy of repulsion. (c) The measured dissociation energy is 5.07 eV. Calculate the energy due to repulsion of the ions at the equilibrium separation.

16 ••• Assume that the core repulsion can be represented by a potential energy of the form $U_{rep} = C/r^n$ so the total potential energy is

$$U = -\frac{ke^2}{r} + \frac{C}{r^n} + \Delta E$$

Use the fact that $dU/dr = 0$ at $r = r_0$, and the results for U_{rep} at $r = r_0$ from Problem 11, to calculate C and n.

17 ••• (a) Find U_{rep} at $r = r_0$ for NaCl. (b) Assume $U_{rep} = C/r^n$ and find C and n for NaCl. (See Problem 16.)

Polyatomic Molecules (optional)

18 • Find other elements with the same subshell electron configuration in the two outermost orbitals as carbon. Would you expect the same type of hybridization for these elements as for carbon?

Energy Levels and Spectra of Diatomic Molecules

19 • How does the effective force constant calculated for the CO molecule in Example 38-4 compare with the force constant of an ordinary spring?

20 • Explain why the moment of inertia of a diatomic molecule increases slightly with increasing angular momentum.

21 • The characteristic rotational energy E_{0r} for the rotation of the N_2 molecule is 2.48×10^{-4} eV. From this find the separation distance of the N atoms in N_2.

22 • The separation of the O atoms in O_2 is actually slightly greater than the 0.1 nm used in Example 38-3, and the characteristic energy of rotation E_{0r} is 1.78×10^{-4} eV rather than the result obtained in that example. Use this value to calculate the separation distance of the O atoms in O_2.

23 •• Show that the reduced mass is smaller than either mass in a diatomic molecule, and calculate it for (a) H_2, (b) N_2, (c) CO, and (b) HCl. Express your answers in unified mass units.

24 • The equilibrium separation between the nuclei of the LiH molecule is 0.16 nm. Determine the energy separation between the $\ell = 3$ and $\ell = 2$ rotational levels of this diatomic molecule.

25 •• Repeat Problem 24 for LiD, where D is the symbol for deuterium. Note that replacing the proton by the deuteron does not change the equilibrium separation between the nuclei of the molecule.

26 •• Derive Equations 38-14 and 38-15 for the moment of inertia in terms of the reduced mass of a diatomic molecule.

27 •• Use the separation of the K^+ and Cl^- ions given in Problem 12 and the reduced mass of KCl to calculate the characteristic rotational energy E_{0r}.

28 •• The central frequency for the absorption band of HCl shown in Figure 38-17 is at $f = 8.66 \times 10^{13}$ Hz, and the absorption peaks are separated by about $\Delta f = 6 \times 10^{11}$ Hz. Use this information, to find (a) the lowest (zero-point) vibrational energy for HCl, (b) the moment of inertia of HCl, and (c) the equilibrium separation of the atoms.

29 •• Calculate the effective force constant for HCl from its reduced mass and the fundamental vibrational frequency obtained from Figure 38-17.

30 •• Two objects of mass m_1 and m_2 are attached to a spring of force constant K and equilibrium length r_0. (a) Show that when m_1 is moved a distance Δr_1 from the center of mass, the force exerted by the spring is

$$F = -K\left(\frac{m_1 + m_2}{m_2}\right)\Delta r_1$$

(b) Show that the angular frequency of oscillation is $f = (1/2\pi)\sqrt{K/\mu}$, where μ is the reduced mass.

31 ••• Calculate the reduced mass for the $H^{35}Cl$ and $H^{37}Cl$ molecules and the fractional difference $\Delta\mu/\mu$. Show that the mixture of isotopes in HCl leads to a fractional difference in the frequency of a transition from one rotational state to another given by $\Delta f/f = -\Delta\mu/\mu$. Compute $\Delta f/f$ and compare your result with Figure 38-17.

32 ••• In calculating the rotational energy levels of a diatomic molecule, we did not consider rotation of the molecule about the line joining the atoms. (a) Estimate the moment of inertia of the H_2 molecule about this line. (b) Use your results of (a) to estimate the typical rotational energy E_{0r} for rotation about the line joining the atoms. (c) Compare your answer in (b) with the typical thermal energy kT at $T = 300$ K.

The Structure of Solids

33 • Suppose that hard spheres of radius R are located at the corners of a unit cell with a simple cubic structure. (a) If the hard spheres touch so as to take up the minimum volume possible, what is the size of the unit cell? (b) What fraction of the volume of the cubic structure is occupied by the hard spheres?

34 • Calculate the distance r_0 between the K^+ and the Cl^- ions in KCl, assuming that each ion occupies a cubic volume of side r_0. The molar mass of KCl is 74.55 g/mol and its density is 1.984 g/cm³.

35 • The distance between the Li^+ and Cl^- ions in LiCl is 0.257 nm. Use this and the molecular mass of LiCl (42.4 g/mol) to compute the density of LiCl.

36 • Find the value of n in Equation 38-32 that gives the measured dissociation energy of 741 kJ/mol for LiCl, which has the same structure as NaCl and for which $r_0 = 0.257$ nm.

37 •• Suppose identical bowling balls of radius R are packed into a hexagonal close-packed structure. What fraction of the available volume of the unit cell is filled by the bowling balls?

Semiconductors

38 • Which of the following elements are most likely to act as acceptor impurities in germanium?

(a) Bromine   (b) Gallium
(c) Silicon   (d) Phosphorus
(e) Magnesium

39 • Which of the following elements are most likely to serve as donor impurities in germanium?

(a) Bromine   (b) Gallium
(c) Silicon   (d) Phosphorus
(e) Magnesium

40 • What type of semiconductor is obtained if silicon is doped with (a) aluminum and (b) phosphorus? (See Table 37-1 for the electron configurations of these elements.)

41 • What type of semiconductor is obtained if silicon is doped with (a) indium and (b) antimony? (See Table 37-1 for the electron configurations of these elements.)

42 • The donor energy levels in an n-type semiconductor are 0.01 eV below the conduction band. Find the temperature for which $kT = 0.01$ eV.

43 •• The relative binding of the extra electron in the arsenic atom that replaces an atom in silicon or germanium can be understood from a calculation of the first Bohr orbit of this electron in these materials. Four of arsenic's outer electrons form covalent bonds, so the fifth electron sees a singly charged center of attraction. This model is a modified hydrogen atom. In the Bohr model of the hydrogen atom, the electron moves in free space at a radius a_0 given by

$$a_0 = \frac{\epsilon_0 h^2}{\pi m_e e^2}$$

When an electron moves in a crystal, we can approximate the effect of the other atoms by replacing ϵ_0 with $\kappa\epsilon_0$ and m_e with an effective mass for the electron. For silicon, κ is 12 and the effective mass is about $0.2m_e$. For germanium, κ is 16 and the effective mass is about $0.1m_e$. Estimate the Bohr radii for the outer electron as it orbits the impurity arsenic atom in silicon and germanium.

44 •• The ground-state energy of the hydrogen atom is given by

$$E_1 = -\frac{mk^2 e^4}{2\hbar^2} = -\frac{e^4 m_e}{8\epsilon_0^2 h^2}$$

Modify this equation in the spirit of Problem 43 by replacing ϵ_0 by $\kappa\epsilon_0$ and m_e by an effective mass for the electron to estimate the binding energy of the extra electron of an impurity arsenic atom in (a) silicon and (b) germanium.

45 •• A doped n-type silicon sample with 10^{16} electrons per cubic centimeter in the conduction band has a resistivity of 5×10^{-3} Ω·m at 300 K. Find the mean free path of the electrons. Use the effective mass of $0.2m_e$ for the mass of the electrons. (See Problem 43.) Compare this mean free path with that of conduction electrons in copper at 300 K.

46 •• The measured Hall coefficient of a doped silicon sample is 0.04 V·m/A·T at room temperature. If all the doping impurities have contributed to the total charge carriers of the sample, find (a) the type of impurity (donor or acceptor) used to dope the sample and (b) the concentration of these impurities.

Semiconductor Junctions and Devices (optional)

47 • When a pnp junction transistor is used as an amplifier, a small signal in the _____ current results in a large signal in the _____ current.

(a) collector; base   (b) base; collector
(c) emitter; base   (d) emitter; collector
(e) collector; base

48 • When light strikes the p-type semiconductor in a pn junction solar cell,

(a) only free electrons are created.
(b) only positive holes are created.
(c) both electrons and holes are created.
(d) positive protons are created.
(e) none of these is correct.

49 •• Simple theory for the current versus the bias voltage across a *pn* junction yields the equation

$$I = I_0(e^{eV_b/kT} - 1)$$

Sketch I versus V_b for both positive and negative values of V_b using this equation.

50 •• For a temperature of 300 K, use the equation in Problem 49 to find the bias voltage V_b for which the exponential term has the value (a) 10 and (b) 0.1.

51 •• In Figure 38-34 for the *pnp*-transistor amplifier, suppose $R_b = 2$ kΩ and $R_L = 10$ kΩ. Suppose further that a 10-μA ac base current generates a 0.5-mA ac collector current. What is the voltage gain of the amplifier?

52 •• Germanium can be used to measure the energy of incident particles. Consider a 660-keV gamma ray emitted from ^{137}Cs. (a) Given that the band gap in germanium is 0.72 eV, how many electron–hole pairs can be generated as this gamma ray travels through germanium? (b) The number of pairs N in part (a) will have statistical fluctuations given by $\pm\sqrt{N}$. What then is the energy resolution of this detector in this photon energy region?

53 •• Make a sketch showing the valence and conduction band edges and Fermi energy of a *pn*-junction diode when biased (a) in the forward direction and (b) in the reverse direction.

54 •• A "good" silicon diode has the current–voltage characteristic given in Problem 49. Let $kT = 0.025$ eV (room temperature) and the saturation current $I_0 = 1$ nA. (a) Show that for small reverse-bias voltages, the resistance is 25 MΩ. *Hint:* Do a Taylor expansion of the exponential function, or use your calculator and enter small values for V_b. (b) Find the dc resistance for a reverse bias of 0.5 V. (c) Find the dc resistance for a 0.5-V forward bias. What is the current in this case? (d) Calculate the ac resistance dV/dI for a 0.5-V forward bias.

55 •• A slab of silicon of thickness $t = 1.0$ mm and width $w = 1.0$ cm is placed in a magnetic field $B = 0.4$ T. The slab is in the xy plane, and the magnetic field points in the positive z direction. When a current of 0.2 A flows through the sample in the positive x direction, a voltage difference of 5 mV develops across the width of the sample with the electric field in the sample pointing in the positive y direction. Determine the semiconductor type (n or p) and the concentration of charge carriers.

General Problems

56 • Why would you expect the separation distance between the two protons to be smaller in the H_2^+ ion than in the H_2 molecule?

57 • What kind of bonding mechanism would you expect for (a) the HCl molecule (b) the O_2 molecule, and (c) Cu atoms in a solid?

58 • Why does an atom usually absorb radiation only from the ground state, whereas a diatomic molecule can absorb radiation from many different rotational states?

59 • The equilibrium separation of CsF is 0.2345 nm. If its bonding is 70% ionic, what is its electric dipole moment.

60 • Show that when one atom in a diatomic molecule is much more massive than the other the reduced mass is approximately equal to the mass of the lighter atom.

61 •• The equilibrium separation between the nuclei of the CO molecule is 0.113 nm. Determine the energy difference between the $\ell = 2$ and $\ell = 1$ rotational energy levels of this molecule.

62 •• When a thin slab of semiconducting material is illuminated with monochromatic light most of the light is transmitted through the slab if the wavelength is greater than 1.85 μm. For wavelengths less than 1.85 μm, most of the incident light is absorbed. Determine the energy gap of this semiconductor.

63 •• Show that when an intrinsic semiconductor carries a current in a transverse magnetic field no Hall voltage is developed across the sample.

64 •• The semiconducting compound CdSe is widely used for light emitting diodes (LEDs). The energy gap in CdSe is 1.8 eV. What is the frequency of the light emitted by a CdSe LED?

65 •• The resistivity of a sample of pure silicon diminishes drastically when it is irradiated with infrared light of wavelength less than 1.13 μm. What is the energy gap of silicon?

66 •• The effective force constant for the HF molecule is 970 N/m. Find the frequency of vibration for this molecule.

67 •• The frequency of vibration of the NO molecule is 5.63×10^{13} Hz. Find the effective force constant for NO.

68 •• The force constant of the hydrogen bond in the H_2 molecule is 580 N/m. Obtain the energies of the four lowest vibrational levels of H_2, HD, and D_2 molecules, and the wavelengths of photons resulting from transitions between adjacent vibrational levels of these molecules.

69 •• The potential energy between two atoms in a molecule can often be described rather well by the Lenard-Jones potential, which can be written

$$U = U_0\left[\left(\frac{a}{r}\right)^{12} - 2\left(\frac{a}{r}\right)^6\right]$$

where U_0 and a are constants. Find the interatomic separation r_0 in terms of a for which the potential energy is a minimum. Find the corresponding value of U_{min}. Use Figure 38-4 to obtain numerical values of r_0 and U_0 for the H_2 molecule and express your answers in nanometers and electron volts.

70 •• In this problem you are to find how the van der Waals force between a polar and a nonpolar molecule depends on the distance between the molecules. Let the dipole moment of the polar molecule be in the x direction and the nonpolar molecule be a distance x away. (a) How does the electric field due to an electric dipole depend on distance x? (b) Use the fact that the potential energy of an electric dipole of moment $\vec{p}$ in an electric field $\vec{E}$ is $U = -\vec{p}\cdot\vec{E}$ and that the induced dipole moment of the nonpolar molecule is proportional to E to find how the potential energy of interaction of the two molecules depends on separation distance. (c) Using $F_x = -dU/dx$, find the x dependence of the force between the two molecules.

71 •• Find the dependence of the force on separation distance between two polar molecules. (See Problem 70.)

72 •• Use the infrared absorption spectrum of HCl in Figure 38-17 to obtain (a) the characteristic rotational energy E_{0r} (in eV) and (b) the vibrational frequency f and the vibrational energy hf (in eV).

73 •• For a molecule such as CO, which has a permanent electric dipole moment, radiative transitions obeying the selection rule $\Delta \ell = \pm 1$ between two rotational energy levels of the same vibrational level are allowed. (That is, the selection rule $\Delta \nu = \pm 1$ does not hold.) (a) Find the moment of inertia of CO and calculate the characteristic rotational energy E_{0r} (in eV). (b) Make an energy level diagram for the rotational levels for $\ell = 0$ to $\ell = 5$ for some vibrational level. Label the energies in electron volts starting with $E = 0$ for $\ell = 0$. (c) Indicate on your diagram transitions that obey $\Delta \ell = -1$ and calculate the energy of the photon emitted. (d) Find the wavelength of the photons emitted for each transition in (c). In what region of the electromagnetic spectrum are these photons?

74 ••• Use the results of Problem 16 to calculate the vibrational frequency of the LiCl molecule. To do this, expand the potential about $r = r_0$, where r_0 is the equilibrium separation, in a Taylor series. Retain only the term proportional to $(r - r_0)^2$. Recall that the potential energy of a simple harmonic oscillator is given by $V = \frac{1}{2}m\omega^2 x^2$. What is the wavelength resulting from transitions between adjacent harmonic oscillator levels of this molecule?

75 ••• Repeat Problem 74 for the NaCl molecule.

76 ••• A one-dimensional model of an ionic crystal consists of a line of alternating positive and negative ions with distance r_0 between each ion. (a) Show that the potential energy of attraction of one ion in the line is

$$V = -\frac{2ke^2}{r_0}\left(1 - \frac{1}{2} + \frac{1}{3} - \frac{1}{4} + \frac{1}{5} - \cdots\right)$$

(b) Using the result that

$$\ln(1 + x) = x - \frac{x^2}{2} + \frac{x^3}{3} - \frac{x^4}{4} + \cdots$$

show that the Madelung constant for this one-dimensional model is $\alpha = 2 \ln 2 = 1.386$.

CHAPTER 39

Relativity

Albert Einstein in 1916.

The theory of relativity consists of two rather different theories, the special theory and the general theory. The special theory, developed by Einstein and others in 1905, concerns the comparison of measurements made in different inertial reference frames moving with constant velocity relative to one another. Its consequences, which can be derived with a minimum of mathematics, are applicable in a wide variety of situations encountered in physics and engineering. On the other hand, the general theory, also developed by Einstein and others around 1916, is concerned with accelerated reference frames and gravity. A thorough understanding of the general theory requires sophisticated mathematics, and the applications of this theory are chiefly in the area of gravitation. It is of great importance in cosmology, but it is rarely encountered in other areas of physics or in engineering. In this chapter we concentrate on the special theory (often referred to as *special relativity*). General relativity will be discussed briefly in an Exploring section near the end of the chapter.

39-1   Newtonian Relativity

Newton's first law does not distinguish between a particle at rest and one moving with constant velocity. If there is no net external force acting, the particle will remain in its initial state—either at rest or moving with its initial velocity. A particle at rest relative to you is moving with constant velocity relative to an observer who is moving with constant velocity relative to you. How might we distinguish whether you and the particle are at rest and the second observer is moving with constant velocity, or the second observer is at rest and you and the particle are moving?

Let us consider some simple experiments. Suppose we have a railway boxcar moving along a straight, flat track with a constant velocity V. We note that a ball at rest in the boxcar remains at rest. If we drop the ball, it falls straight down relative to the boxcar with an acceleration g due to gravity. Of course, when viewed from the track the ball moves along a parabolic path because it has an initial velocity V to the right. No mechanics experiment that we can do—measuring the period of a pendulum, observing the collisions between two objects, or whatever—will tell us whether the boxcar is moving and the track is at rest or the track is moving and the boxcar is at rest. If we have a coordinate system attached to the track and another attached to the boxcar, Newton's laws hold in either system.

A set of coordinate systems at rest relative to each other is called a *reference frame*.* A reference frame in which Newton's laws hold is called an *inertial reference frame*. All reference frames moving at constant velocity relative to an inertial reference frame are also inertial reference frames. If we have two inertial reference frames moving with constant velocity relative to each other, there are no mechanics experiments that can tell us which is at rest and which is moving or if they are both moving. This result is known as the principle of **Newtonian relativity:**

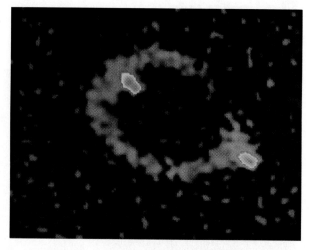

This ringlike structure of the radio source MG1131 + 0456 is thought to be due to "gravitational lensing," first proposed by Einstein in 1936, in which a source is imaged into a ring by a large, massive object in the foreground.

> Absolute motion cannot be detected.

Principle of Newtonian relativity

This principle was well known by Galileo, Newton, and others in the seventeenth century. By the late nineteenth century, however, this view had changed. It was then generally thought that Newtonian relativity was not valid and that absolute motion could be detected in principle by a measurement of the speed of light.

Ether and the Speed of Light

We saw in Chapter 15 that the velocity of a wave depends on the properties of the medium in which the wave travels and not on the velocity of the source of the waves. For example, the velocity of sound relative to still air depends on the temperature of the air. Light and other electromagnetic waves (radio, X rays, etc.) travel through a vacuum with a speed $c \approx 3 \times 10^8$ m/s that is predicted by Maxwell's equations for electricity and magnetism. But what is this speed relative to? What is the equivalent of still air for a vacuum? A proposed medium for the propagation of light was called the *ether*; it was

---

* Inertial reference frames were discussed in Section 4-1.

thought to pervade all space. The velocity of light relative to the ether was assumed to be c as predicted by Maxwell's equations. The velocity of any object relative to the ether was considered its absolute velocity.

Albert Michelson, first in 1881 and then again with Edward Morley in 1887, set out to measure the velocity of the earth relative to the ether by an ingenious experiment in which the velocity of light relative to the earth was compared for two light beams, one in the direction of the earth's motion relative to the sun and the other perpendicular to the direction of the earth's motion. Despite painstakingly careful measurements, they could detect no difference. The experiment has since been repeated under various conditions by a number of people, and no difference has ever been found. The absolute motion of the earth relative to the ether cannot be detected.

39-2 Einstein's Postulates

In 1905, at the age of 26, Albert Einstein published a paper on the electrodynamics of moving bodies.* In this paper, he postulated that absolute motion cannot be detected by any experiment. That is, there is no ether. The earth can be considered to be at rest and the velocity of light will be the same in any direction.[†] His theory of special relativity can be derived from two postulates. Simply stated, these postulates are:

> Postulate 1. Absolute uniform motion cannot be detected.
>
> Postulate 2. The speed of light is independent of the motion of the source.

Einstein's postulates

Postulate 1 is merely an extension of the Newtonian principle of relativity to include all types of physical measurements (not just those that are mechanical). Postulate 2 describes a common property of all waves. For example, the speed of sound waves does not depend on the motion of the sound source. The sound waves from a car horn travel through the air with the same velocity independent of whether the car is moving or not. The speed of the waves depends only on the properties of the air, such as its temperature.

Although each postulate seems quite reasonable, many of the implications of the two together are quite surprising and contradict what is often called common sense. For example, one important implication of these postulates is that every observer measures the same value for the speed of light independent of the relative motion of the source and the observer. Consider a light source S and two observers, R_1 at rest relative to S and R_2 moving toward S with speed v, as shown in Figure 39-1a. The speed of light measured by R_1 is $c = 3 \times 10^8$ m/s. What is the speed measured by R_2? The answer is *not* $c + v$. By postulate 1, Figure 39-1a is equivalent to Figure 39-1b, in which R_2 is at rest and the source S and R_1 are moving with speed v. That is, since absolute motion cannot be detected, it is not possible to say

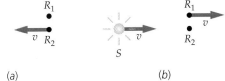

(a)   (b)

Figure 39-1 (a) A stationary light source S and a stationary observer R_1, with a second observer R_2 moving toward the source with speed v. (b) In the reference frame in which the observer R_2 is at rest, the light source S and observer R_1 move to the right with speed v. If absolute motion cannot be detected, the two views are equivalent. Since the speed of light does not depend on the motion of the source, observer R_2 measures the same value for that speed as observer R_1.

---

* *Annalen der Physik*, vol. 17, 1905, p. 841. For a translation from the original German, see W. Perrett and G.B. Jeffery (trans.), *The Principle of Relativity: A Collection of Original Memoirs on the Special and General Theory of Relativity* by H. A. Lorentz, A. Einstein, H. Minkowski, and W. Weyl, Dover, New York, 1923.

† Einstein did not set out to explain the results of the Michelson–Morley experiment. His theory arose from his considerations of the theory of electricity and magnetism and the unusual property of electromagnetic waves that they propagate in a vacuum. In his first paper, which contains the complete theory of special relativity, he made only a passing reference to the Michelson–Morley experiment, and in later years he could not recall whether he was aware of the details of this experiment before he published his theory.

which is really moving and which is at rest. By postulate 2, the speed of light from a moving source is independent of the motion of the source. Thus, looking at Figure 39-1b, we see that R_2 measures the speed of light to be c, just as R_1 does. This result is often considered as an alternative to Einstein's second postulate:

> **Postulate 2 (Alternate).** Every observer measures the same value c for the speed of light.

This result contradicts our intuitive ideas about relative velocities. If a car moves at 50 km/h away from an observer and another car moves at 80 km/h in the same direction, the velocity of the second car relative to the first car is 30 km/h. This result is easily measured and conforms to our intuition. However, according to Einstein's postulates, if a light beam is moving in the direction of the cars, observers in both cars will measure the same speed for the light beam. Our intuitive ideas about the combination of velocities are approximations that hold only when the speeds are very small compared with the speed of light. Even in an airplane moving with the speed of sound, it is not possible to measure the speed of light accurately enough to distinguish the difference between the results c and $c + v$, where v is the speed of the plane. To perceive such a distinction, we would have to either move with a very great velocity (much greater than that of sound) or make extremely accurate measurements.

39-3   The Lorentz Transformation

Einstein's postulates have important consequences for measuring time intervals and space intervals as well as relative velocities. Throughout this chapter we will be comparing measurements of the positions and times of events (such as lightning flashes) made by observers who are moving relative to each other. We will use a rectangular coordinate system xyz with origin O, called the S reference frame, and another system $x'y'z'$ with origin O', called the S' frame, that is moving with a constant velocity $\vec{V}$ relative to the S frame. Relative to the S' frame, the S frame is moving with a constant velocity $-\vec{V}$. For simplicity, we will consider the S' frame to be moving along the x axis in the positive x direction relative to S. In each frame, we will assume that there are as many observers as are needed who are equipped with measuring devices, such as clocks and metersticks, that are identical when compared at rest (see Figure 39-2).

We will use Einstein's postulates to find the general relation between the coordinates x, y, and z and the time t of an event as seen in reference frame S and the coordinates x', y', and z' and the time t' of the same event as seen in reference frame S', which is moving with uniform velocity relative to S. We assume that the origins are coincident at time $t = t' = 0$. The classical relation, called the **Galilean transformation,** is

$$x = x' + Vt', \qquad y = y', \quad z = z', \quad t = t' \qquad\qquad \text{39-1}a$$

Galilean transformation

The inverse transformation is

$$x' = x - Vt, \qquad y' = y, \quad z' = z, \quad t' = t \qquad\qquad \text{39-1}b$$

Galilean transformation

These equations are consistent with experimental observations as long as V is much less than c. They lead to the familiar classical addition law for veloc-

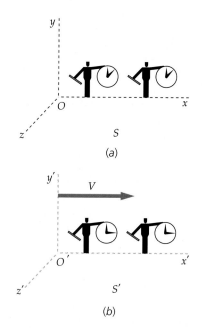

Figure 39-2 Coordinate reference frames S and S' moving with relative speed V. In each frame, there are observers with metersticks and clocks that are identical when compared at rest.

ities. If a particle has velocity $u_x = dx/dt$ in frame S, its velocity in frame S' is

$$u'_x = \frac{dx'}{dt'} = \frac{dx'}{dt} = \frac{dx}{dt} - V = u_x - V \qquad \text{39-2}$$

If we differentiate this equation again, we find that the acceleration of the particle is the same in both frames:

$$a_x = \frac{du_x}{dt} = \frac{du'_x}{dt'} = a'_x$$

It should be clear that the Galilean transformation is not consistent with Einstein's postulates of special relativity. If light moves along the x axis with speed $u'_x = c$ in S', these equations imply that the speed in S' is $u_x = c + V$ rather than $u_x = c$, which is consistent with Einstein's postulates and with experiment. The classical transformation equations must therefore be modified to make them consistent with Einstein's postulates. We will give a brief outline of one method of obtaining the relativistic transformation.

We assume that the relativistic transformation equation for x is the same as the classical equation (Equation 39-1a) except for a constant multiplier on the right side. That is, we assume the equation is of the form

$$x = \gamma(x' + Vt') \qquad \text{39-3}$$

where γ is a constant that can depend on V and c but not on the coordinates. The inverse transformation must look the same except for the sign of the velocity:

$$x' = \gamma(x - Vt) \qquad \text{39-4}$$

Let us consider a light pulse that starts at the origin of S at $t = 0$. Since we have assumed that the origins are coincident at $t = t' = 0$, the pulse also starts at the origin of S' at $t' = 0$. Einstein's postulates require that the equation for the x component of the wave front of the light pulse is $x = ct$ in frame S and $x' = ct'$ in frame S'. Substituting ct for x and ct' for x' in Equations 39-3 and 39-4, we obtain

$$ct = \gamma(ct' + Vt') = \gamma(c + V)t' \qquad \text{39-5}$$

and

$$ct' = \gamma(ct - Vt) = \gamma(c - V)t \qquad \text{39-6}$$

We can eliminate either t' or t from these two equations and determine γ. We get

$$\gamma^2 = \left(1 - \frac{V^2}{c^2}\right)^{-1}$$

$$\gamma = \frac{1}{\sqrt{1 - V^2/c^2}} \qquad \text{39-7}$$

Note that γ is always greater than 1, and that when V is much less than c, $\gamma \approx 1$. The relativistic transformation for x and x' is therefore given by Equations 39-3 and 39-4 with γ given by Equation 39-7. We can obtain equations for t and t' by combining Equation 39-3 with the inverse transformation given by Equation 39-4. Substituting $x = \gamma(x' + Vt')$ for x in Equation 39-4, we obtain

$$x' = \gamma[\gamma(x' + Vt') - Vt] \qquad \text{39-8}$$

which can be solved for t in terms of x' and t'. The complete relativistic transformation is

$$x = \gamma(x' + Vt'), \qquad y = y', \quad z = z' \tag{39-9}$$

$$t = \gamma\left(t' + \frac{Vx'}{c^2}\right) \tag{39-10}$$

Lorentz transformation

The inverse transformation is

$$x' = \gamma(x - Vt), \qquad y' = y, \quad z' = z \tag{39-11}$$

$$t' = \gamma\left(t - \frac{Vx}{c^2}\right) \tag{39-12}$$

The transformation described by Equations 39-9 through 39-12 is called the **Lorentz transformation.** It relates the space and time coordinates $x, y, z,$ and t of an event in frame S to the coordinates $x', y', z',$ and t' of the same event as seen in frame S', which is moving along the x axis with speed V relative to frame S.

We will now look at some applications of the Lorentz transformation.

Time Dilation

Consider two events that occur at a single point x_0' at times t_1' and t_2' in frame S'. We can find the times t_1 and t_2 for these events in S from Equation 39-10. We have

$$t_1 = \gamma\left(t_1' + \frac{Vx_0'}{c^2}\right)$$

and

$$t_2 = \gamma\left(t_2' + \frac{Vx_0'}{c^2}\right)$$

so

$$t_2 - t_1 = \gamma(t_2' - t_1')$$

The time between events that happen at the *same place* in a reference frame is called **proper time** t_{p}. In this case, the time interval $t_2' - t_1'$ measured in frame S' is proper time. The time interval Δt measured in any other reference frame is always longer than the proper time. This expansion is called **time dilation:**

$$\Delta t = \gamma \,\Delta t_{\mathrm{p}} \tag{39-13}$$

Time dilation

Example 39-1

Two events occur at the same point x_0' at times t_1' and t_2' in frame S', which is traveling at speed V relative to frame S. **What is the spatial separation of these events in frame S?**

Picture the Problem The spatial separation in S is $x_2 - x_1$, where x_2 and x_1 are the coordinates of the events in S, which are found using Equation 39-9.

1. The position x_1 in S at time t_1' is given by Equation 39-9: $\qquad x_1 = \gamma(x_0' + Vt_1')$

2. Similarly, at time t_2', the position is x_2, given by:

$$x_2 = \gamma(x_0' + Vt_2')$$

3. Subtract to find the spatial separation:

$$x_2 - x_1 = \gamma V(t_2' - t_1') = V(t_2 - t_1)$$

Remarks The spatial separation of these events in S is the distance a single point, such as x_0' in S', moves in S during the time interval between the events.

We can understand time dilation directly from Einstein's postulates without using the Lorentz transformation. Figure 39-3a shows an observer A' a distance D from a mirror. The observer and the mirror are in a spaceship that is at rest in frame S'. The observer explodes a flash gun and measures the time interval $\Delta t'$ between the original flash and his seeing the return flash from the mirror. Since light travels with speed c, this time is

$$\Delta t' = \frac{2D}{c}$$

We now consider these same two events, the original flash of light and the receiving of the return flash, as observed in reference frame S, in which observer A' and the mirror are moving to the right with speed V as shown in Figure 39-3b. The events happen at two different places x_1 and x_2 in frame S.

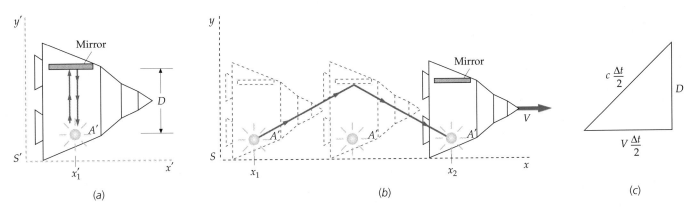

(a)　　　　　　　　　　　　　　(b)　　　　　　　　　　　　　　(c)

Figure 39-3 (a) Observer A' and the mirror are in a spaceship at rest in frame S'. The time it takes for the light pulse to reach the mirror and return is measured by A' to be $2D/c$. (b) In frame S, the spaceship is moving to the right with speed V. If the speed of light is the same in both frames, the time it takes for the light to reach the mirror and return is longer than $2D/c$ in S because the distance traveled is greater than $2D$. (c) A right triangle for computing the time Δt in frame S.

During the time interval Δt (as measured in S) between the original flash and the return flash, observer A' and his spaceship have moved a horizontal distance $V \Delta t$. In Figure 39-3b, we can see that the path traveled by the light is longer in S than in S'. However, by Einstein's postulates, light travels with the same speed c in frame S as it does in frame S'. Since it travels farther in S at the same speed, it takes longer in S to reach the mirror and return. The time interval in S is thus longer than it is in S'. From the triangle in Figure 39-3c, we have

$$\left(\frac{c \, \Delta t}{2}\right)^2 = D^2 + \left(\frac{V \, \Delta t}{2}\right)^2$$

or

$$\Delta t = \frac{2D}{\sqrt{c^2 - V^2}} = \frac{2D}{c} \frac{1}{\sqrt{1 - V^2/c^2}}$$

Using $\Delta t' = 2D/c$, we obtain

$$\Delta t = \frac{\Delta t'}{\sqrt{1 - V^2/c^2}} = \gamma \, \Delta t'$$

Example 39-2   *try it yourself*

Astronauts in a spaceship traveling at $V = 0.6c$ relative to the earth sign off from space control, saying that they are going to nap for 1 h and then call back. How long does their nap last as measured on earth?

Picture the Problem Since the astronauts go to sleep and wake up at the same place in their reference frame, the time interval for their nap of 1 h as measured by them is proper time. In the earth's reference frame, they move a considerable distance between these two events. The time interval measured in the earth's frame (using two clocks located at those events) is longer by the factor γ.

Cover the column to the right and try these on your own before looking at the answers.

Steps	*Answers*
1. Relate the time interval measured on earth Δt to the proper time Δt_p.	$\Delta t = \gamma \, \Delta t_p$
2. Calculate γ for $V = 0.6c$.	$\gamma = 1.25$
3. Substitute to calculate the time of the nap in the earth's frame.	$\Delta t = \gamma \, \Delta t_p = 1.25 \text{ h}$

Exercise If the spaceship is moving at $V = 0.8c$, how long would a 1-h nap last as measured on earth? (*Answer*   1.67 h)

Length Contraction

A phenomenon closely related to time dilation is **length contraction.** The length of an object measured in the reference frame in which the object is at rest is called its **proper length** L_p. In a reference frame in which the object is moving, the measured length is shorter than its proper length. Consider a rod at rest in frame S' with one end at x_2' and the other end at x_1'. The length of the rod in this frame is its proper length $L_p = x_2' - x_1'$. Some care must be taken to find the length of the rod in frame S. In this frame, the rod is moving to the right with speed V, the speed of frame S'. The length of the rod in frame S is *defined* as $L = x_2 - x_1$, where x_2 is the position of one end at some time t_2, and x_1 is the position of the other end *at the same time* $t_1 = t_2$ as measured in frame S. Equation 39-11 is convenient to use to calculate $x_2 - x_1$ at some time t because it relates x and x' to t, whereas Equation 39-9 is not convenient because it relates x and x' to t':

$$x_2' = \gamma \, (x_2 - Vt_2)$$

and

$$x_1' = \gamma(x_1 - Vt_1)$$

Since $t_2 = t_1$, we obtain

$$x_2' - x_1' = \gamma(x_2 - x_1)$$

$$x_2 - x_1 = \frac{1}{\gamma}(x_2' - x_1') = \sqrt{1 - V^2/c^2}\,(x_2' - x_1')$$

or

$$L = \frac{1}{\gamma}L_p = \sqrt{1 - V^2/c^2}\,L_p \qquad\qquad \text{39-14}$$

Length contraction

Thus, the length of a rod is smaller when it is measured in a frame in which it is moving. Before Einstein's paper was published, Lorentz and FitzGerald tried to explain the null result of the Michelson–Morley experiment by assuming that distances in the direction of motion contracted by the amount given in Equation 39-14. This contraction is now known as the **Lorentz–FitzGerald contraction.**

Example 39-3

A stick that has a proper length of 1 m moves in a direction along its length with speed V relative to you. The length of the stick as measured by you is 0.914 m. What is the speed V?

Picture the Problem Since both L and L_p are given, we can find V directly from Equation 39-14.

1. Equation 39-14 relates the lengths L and L_p and the speed V:

$$L = \sqrt{1 - V^2/c^2}\,L_p$$

2. Solve for V:

$$1 - V^2/c^2 = (L/L_p)^2 = (0.914\ \text{m}/1\ \text{m})^2 = 0.835$$

$$V = c\sqrt{1 - 0.835} = 0.406c$$

An interesting example of time dilation or length contraction is afforded by the appearance of muons as secondary radiation from cosmic rays. Muons decay according to the statistical law of radioactivity:

$$N(t) = N_0\,e^{-t/\tau} \qquad\qquad \text{39-15}$$

where N_0 is the original number of muons at time $t = 0$, $N(t)$ is the number remaining at time t, and τ is the mean lifetime, which is about 2 μs for muons at rest. Since muons are created (from the decay of pions) high in the atmosphere, usually several thousand meters above sea level, few muons should reach sea level. A typical muon moving with speed $0.9978c$ would travel only about 600 m in 2 μs. However, the lifetime of the muon measured in the earth's reference frame is increased by the factor $1/\sqrt{1 - V^2/c^2}$, which is 15 for this particular speed. The mean lifetime measured in the earth's reference frame is therefore 30 μs, and a muon with speed $0.9978c$ travels about 9000 m in this time. From the muon's point of view, it lives only 2 μs, but the atmosphere is rushing past it with a speed of $0.9978c$. The distance of 9000 m in the earth's frame is thus contracted to only 600 m in the muon's frame, as indicated in Figure 39-4.

It is easy to distinguish experimentally between the classical and relativistic predictions of the observation of muons at sea level. Suppose that we observe 10^8 muons at an altitude of 9000 m in some time interval with a muon detector. How many would we expect to observe at sea level in the same time interval? According to the nonrelativistic prediction, the time it takes for these muons to travel 9000 m is $(9000 \text{ m})/0.998c \approx 30 \ \mu s$, which is 15 lifetimes. Substituting $N_0 = 10^8$ and $t = 15\tau$ into Equation 39-15, we obtain

$$N = 10^8 e^{-15} = 30.6$$

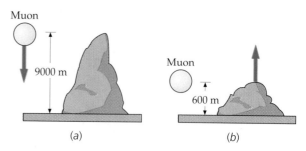

Figure 39-4 Although muons are created high above the earth and their mean lifetime is only about 2 μs when at rest, many appear at the earth's surface. (a) In the earth's reference frame, a typical muon moving at 0.998c has a mean lifetime of 30 μs and travels 9000 m in this time. (b) In the reference frame of the muon, the distance traveled by the earth is only 600 m in the muon's lifetime of 2 μs.

We would thus expect all but about 31 of the original 100 million muons to decay before reaching sea level.

According to the relativistic prediction, the earth must travel only the contracted distance of 600 m in the rest frame of the muon. This takes only 2 $\mu s = 1\tau$. Therefore, the number of muons expected at sea level is

$$N = 10^8 e^{-1} = 3.68 \times 10^7$$

Thus, relativity predicts that we would observe 36.8 million muons in the same time interval. Experiments of this type have confirmed the relativistic predictions.

The Relativistic Doppler Effect

For light or other electromagnetic waves in a vacuum a distinction between motion of source and receiver cannot be made. Therefore, the expressions we derived in Chapter 15 for the Doppler effect cannot be correct for light. The reason is that in that derivation, we assumed the time intervals in the reference frames of the source and receiver to be the same.

Consider a source moving toward a receiver with velocity V, relative to the receiver. If the source emits N electromagnetic waves in a time Δt_R (measured in the frame of the receiver), the first wave will travel a distance $c \, \Delta t_R$ and the source will travel a distance $V \, \Delta t_R$ measured in the frame of the receiver. The wavelength will be

$$\lambda' = \frac{c \, \Delta t_R - V \, \Delta t_R}{N}$$

The frequency f' observed by the receiver will therefore be

$$f' = \frac{c}{\lambda'} = \frac{c}{c - V} \frac{N}{\Delta t_R} = \frac{1}{1 - V/c} \frac{N}{\Delta t_R}$$

If the frequency of the source is f_0, it will emit $N = f_0 \, \Delta t_S$ waves in the time Δt_S measured by the source. Then

$$f' = \frac{1}{1 - V/c} \frac{N}{\Delta t_R} = \frac{1}{1 - V/c} \frac{f_0 \Delta t_S}{\Delta t_R} = \frac{f_0}{1 - V/c} \frac{\Delta t_S}{\Delta t_R}$$

Here Δt_S is the proper time interval (the first wave and the Nth wave are emitted at the same place in the source's reference frame). Times Δt_S and Δt_R are related by Equation 39-13 for time dilation:

$$\Delta t_R = \gamma \Delta t_S = \frac{\Delta t_S}{\sqrt{1 - V^2/c^2}}$$

Thus, when the source and receiver are moving toward one another we obtain

$$f' = \frac{f_0}{1 - V/c}\frac{1}{\gamma} = \frac{\sqrt{1 - V^2/c^2}}{1 - V/c}f_0 = \sqrt{\frac{1 + V/c}{1 - V/c}}f_0 \qquad \text{approaching} \quad \text{39-16}a$$

This differs from our classical equation only in the time-dilation factor.

When the source and receiver are moving away from one another, the same analysis shows that the observed frequency is given by

$$f' = \frac{\sqrt{1 - V^2/c^2}}{1 + V/c}f_0 = \sqrt{\frac{1 - V/c}{1 + V/c}}f_0 \qquad \text{receding} \qquad \text{39-16}b$$

It is left as a problem (Problem 36) for you to show that the same results are obtained if the calculations are done in the reference frame of the source.

An application of the relativistic Doppler effect is the **redshift** observed in the light from distant galaxies. Because the galaxies are moving away from us, the light they emit is shifted toward the longer, red wavelengths. The speed of the galaxies relative to us can be determined by measuring this shift.

Example 39-4   *try it yourself*

The longest wavelength of light emitted by hydrogen in the Balmer series is $\lambda_0 = 656$ nm. In light from a distant galaxy, this wavelength is measured to be $\lambda' = 1458$ nm. Find the speed at which the distant galaxy is receding from the earth.

Cover the column to the right and try these on your own before looking at the answers.

Steps	*Answers*
1. Use Equation 39-16b to relate the speed V to the received frequency f' and the emitted frequency f_0.	$f' = \sqrt{\dfrac{1 - V/c}{1 + V/c}}f_0$
2. Substitute $f' = c/\lambda'$ and $f_0 = c/\lambda_0$ and solve for V/c.	$\dfrac{1 - V/c}{1 + V/c} = \left(\dfrac{f'}{f_0}\right)^2 = \left(\dfrac{\lambda_0}{\lambda'}\right)^2 = 0.202$ $V/c = 0.664$

39-4   Clock Synchronization and Simultaneity

We saw in Section 39-3 that proper time is the time interval between two events that occur at the same point in some reference frame. It can therefore be measured on a single clock. However, in another reference frame moving relative to the first, the same two events occur at different places, so two clocks are needed to record the times. The time of each event is measured on a different clock, and the interval is found by subtraction. This procedure requires that the clocks be **synchronized.** We will show in this section that

> Two clocks that are synchronized in one reference frame are not synchronized in any other frame moving relative to the first frame.

A corollary to this result is:

> Two events that are simultaneous in one reference frame are not si-
> multaneous in another frame moving relative to the first.*

Comprehension of these facts usually resolves all relativity paradoxes. Unfortunately, the intuitive (and incorrect) belief that simultaneity is an absolute relation is difficult to overcome.

Suppose we have two clocks at rest at points A and B a distance L apart in
frame S. How can we synchronize these two clocks? If an observer at A looks
at the clock at B and sets her clock to read the same time, the clocks will not
be synchronized because of the time L/c it takes light to travel from one clock
to another. To synchronize the clocks, the observer at A must set her clock
ahead by the time L/c. Then she will see that the clock at B reads a time that is
L/c behind the time on her clock, but she will calculate that the clocks are synchronized when she allows for the time L/c for the light to reach her. Any
other observers except those equidistant from the clocks will see the clocks
reading different times, but they will also calculate that the clocks are synchronized when they correct for the time it takes the light to reach them. An
equivalent method for synchronizing two clocks would be for an observer C
at a point midway between the clocks to send a light signal and for observers
at A and B to set their clocks to some prearranged time when they receive the
signal.

We now examine the question of **simultaneity.** Suppose A and B agree to
explode flashguns at t_0 (having previously synchronized their clocks). Observer C will see the light from the two flashes at the same time, and because
he is equidistant from A and B, he will conclude that the flashes were simultaneous. Other observers in frame S will see the light from A or B first, depending on their location, but after correcting for the time the light takes to
reach them, they also will conclude that the flashes were simultaneous. We
can thus define simultaneity as follows:

> Two events in a reference frame are simultaneous if light signals
> from the events reach an observer halfway between the events at
> the same time.

Definition—Simultaneity

To show that two events that are simultaneous
in frame S are not simultaneous in another frame
S' moving relative to S, we will use an example
introduced by Einstein. A train is moving with
speed V past a station platform. We will consider
the train to be at rest in S' and the platform to be
at rest in S. We have observers A', B', and C' at the
front, back, and middle of the train. We now suppose that the train and platform are struck by
lightning at the front and back of the train and
that the lightning bolts are simultaneous in the
frame of the platform S (Figure 39-5). That is, an
observer C on the platform halfway between the
positions A and B, where the lightning strikes,
sees the two flashes at the same time. It is convenient to suppose that the lightning scorches the

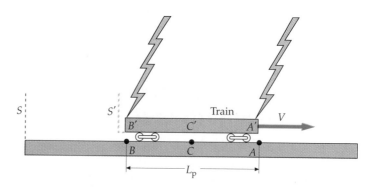

Figure 39-5 Simultaneous lightning bolts strike the ends
of a train traveling with speed V in frame S attached to the
platform. The light from these simultaneous events
reaches observer C midway between the events at the
same time. The distance between the bolts is $L_{p,\text{platform}}$.

---

* This is true unless the events and clocks are in the same plane and are perpendicular to the relative motion.

train and platform so that the events can be easily located. Since C' is in the middle of the train, halfway between the places on the train that are scorched, the events are simultaneous in S' only if C' sees the flashes at the same time. However, the flash from the front of the train is seen by C' before the flash from the back of the train. We can understand this by considering the motion of C' as seen in frame S (Figure 39-6). By the time the light from the front flash reaches C', C' has moved some distance toward the front flash and some distance away from the back flash. Thus, the light from the back flash has not yet reached C' as indicated in the figure. Observer C' must therefore conclude that the events are not simultaneous and that the front of the train was struck before the back. Furthermore, all observers in S' on the train will agree with C' when they have corrected for the time it takes the light to reach them.

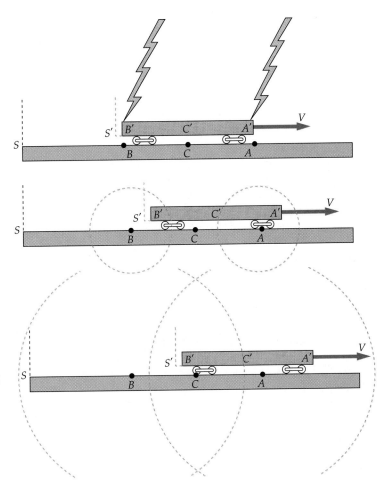

Figure 39-6 The light from the lightning bolt at the front of the train reaches observer C' at the middle of the train before that from the bolt at the back of the train. Since C' is midway between the events (which occur at the front and rear of the train), these events are not simultaneous for him.

Figure 39-7 shows the events of the lightning bolts as seen in the reference frame of the train (S'). In this frame, the platform is moving so the distance between the burns on the platform is contracted. The platform is shorter than it is in S, and, since the train is at rest, the train is longer than its contracted length in S. When the lightning bolt strikes the front of the train at A', the front of the train is at point A, and the back of the train has not yet reached point B. Later, when the lightning bolt strikes the back of the train at B', the back has reached point B on the platform.

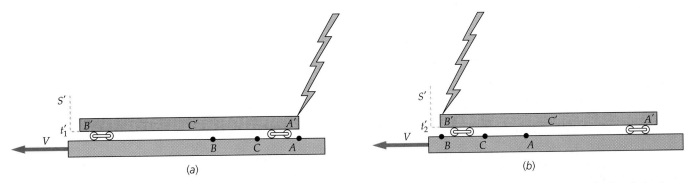

(a)   (b)

Figure 39-7 The lightning bolts of Figure 39-5 as seen in frame S' of the train. In this frame, the distance between A and B on the platform is less than $L_{p,platform}$, and the proper length of the train $L_{p,train}$ is longer than $L_{p,platform}$. The first lightning bolt strikes the front of the train when A' and A are coincident. The second bolt strikes the rear of the train when B' and B are coincident.

The time discrepancy of two clocks that are synchronized in frame S as seen in frame S' can be found from the Lorentz transformation equations. Suppose we have clocks at points x_1 and x_2 that are synchronized in S. What are the times t_1 and t_2 on these clocks as observed from frame S' at a time t_0'? From Equation 39-12, we have

$$t_0' = \gamma\left(t_1 - \frac{Vx_1}{c^2}\right)$$

and

$$t_0' = \gamma\left(t_2 - \frac{Vx_2}{c^2}\right)$$

Then

$$t_2 - t_1 = \frac{V}{c^2}(x_2 - x_1)$$

Note that the chasing clock (at x_2) leads the other (at x_1) by an amount that is proportional to their proper separation $L_p = x_2 - x_1$.

If two clocks are synchronized in the frame in which they are at rest, they will be out of synchronization in another frame. In the frame in which they are moving, the chasing clock leads (shows a later time) by an amount

$$\Delta t_S = L_p \frac{V}{c^2}$$

39-17

where L_p is the proper distance between the clocks.

A numerical example should help clarify time dilation, clock synchronization, and the internal consistency of these results.

Example 39-5

An observer in a spaceship has a flash gun and a mirror as shown in Figure 39-3. The distance from the gun to the mirror is 15 light-minutes (written 15$c\cdot$min) and the spaceship in frame S' travels with speed $V = 0.8c$ relative to a very long space platform in frame S that has two synchronized clocks, one at the position x_1 of the spaceship when the observer explodes the flash gun and the other at the position x_2 of the spaceship when the light returns to the gun from the mirror. Find the time intervals between the events (exploding the flash gun and receiving the return flash from the mirror) (a) in the frame of the ship and (b) in the frame of the platform. (c) Find the distance traveled by the ship, and (d) the amount by which the clocks on the platform are out of synchronization as viewed by the ship.

(a)1. In the spaceship, the light travels from the gun to the mirror and back, a total distance $D = 30\ c\cdot$min. The time required is D/c: $\Delta t' = \dfrac{D}{c} = \dfrac{30\ c\cdot\text{min}}{c} = 30$ min

2. Since these events happen at the same place in the spaceship, the time interval is proper time: $\Delta t_p = 30$ min

(b)1. In frame S, the time between the events is longer by the factor γ:

$$\Delta t = \gamma \, \Delta t' = \gamma(30 \text{ min})$$

2. Calculate γ:

$$\gamma = \frac{1}{\sqrt{1 - V^2/c^2}} = \frac{1}{\sqrt{1 - (0.8)^2}} = \frac{1}{\sqrt{0.36}} = \frac{5}{3}$$

3. Use this value of γ to calculate the time between the events as observed in frame S:

$$\Delta t = \gamma \, \Delta t_\text{p} = \frac{5}{3}(30 \text{ min}) = 50 \text{ min}$$

(c)1. The distance traveled by the ship in S is $V \, \Delta t$:

$$D = V \, \Delta t = (0.8c)(50 \text{ min}) = 40 \, c \cdot \text{min}$$

2. This distance is the proper distance between the clocks on the platform:

$$L_\text{p} = D = 40 \, c \cdot \text{min}$$

(d) The amount that the clocks on the platform are out of synchronization is related to the proper distance between the clocks L_p:

$$\Delta t_\text{s} = L_\text{p} \frac{V}{c^2} = \frac{(40 \, c \cdot \text{min})(0.8c)}{c^2} = 32 \text{ min}$$

Remarks Observers on the platform would say that the spaceship's clock is running slow because it records a time of only 30 min between the events, whereas the time measured on the platform is 50 min.

Figure 39-8 shows the situation viewed from the spaceship in S'. The platform is traveling past the ship with speed $0.8c$. There is a clock at point x_1, which coincides with the ship when the flash gun is exploded, and another at point x_2, which coincides with the ship when the return flash is received from the mirror. We assume that the clock at x_1 reads 12:00 noon at the time of the light flash. The clocks at x_1 and x_2 are synchronized in S but not in S'. In S', the clock at x_2, which is chasing the one at x_1, leads by 32 min; it would thus read 12:32 to an observer in S'. When the spaceship coincides with x_2, the clock there reads 12:50. The time between the events is therefore 50 min in S. Note that according to observers in S', this clock ticks off 50 min $-$ 32 min $= 18$ min for a trip that takes 30 min in S'. Thus, observers in S' see this clock run slow by the factor $30/18 = 5/3$.

Every observer in one frame sees the clocks in the other frame run slow. According to observers in S, who measure 50 min for the time interval, the time interval in S' (30 min) is too small, so they see the single clock in S' run too slow by the factor $5/3$. According to the observers in S', the observers in S measure a time that is too *long* despite the fact that their clocks run too slow because the clocks in S are out of synchronization. The clocks tick off only 18 min, but the second one leads the first by 32 min, so the time interval is 50 min.

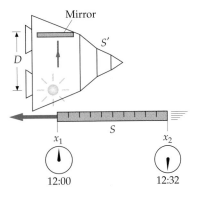

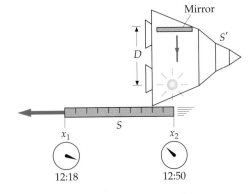

Figure 39-8 Clocks on a platform as observed from the spaceship's frame of reference S'. During the time $\Delta t' = 30$ min it takes for the platform to pass the spaceship, the clocks on the platform run slow and tick off $(30 \text{ min})/\gamma = 18$ min. But the clocks are unsynchronized, with the chasing clock leading by $L_\text{p}V/c^2$, which for this case is 32 min. The time it takes for the spaceship to pass as measured on the platform is therefore 32 min + 18 min = 50 min.

$\mathcal{e}$xploring

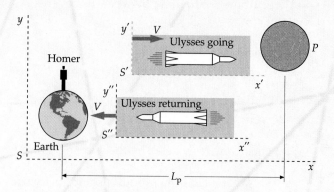

The Twin Paradox

Homer and Ulysses are identical twins. Ulysses travels at high speed to a planet beyond the solar system and returns while Homer remains at home. When they are together again, which twin is older, or are they the same age? The correct answer is that Homer, the twin who stays at home, is older. This problem, with variations, has been the subject of spirited debate for decades, though there are very few who disagree with the answer. The problem is a paradox because of the seemingly symmetric roles played by the twins with the asymmetric result in their aging. The paradox is resolved when the asymmetry of the twins' roles is noted. The relativistic result conflicts with common sense based on our strong but incorrect belief in absolute simultaneity. We will consider a particular case with some numerical magnitudes that, though impractical, make the calculations easy.

Let planet P and Homer on earth be at rest in reference frame S a distance L_p apart, as illustrated in Figure 1. We neglect the motion of the earth. Reference frames S' and S'' are moving with speed V toward and away from the planet, respectively. Ulysses quickly accelerates to speed V, then coasts in S' until he reaches the planet, where he stops and is momentarily at rest in S. To return he quickly accelerates to speed V toward earth and then coasts in S'' until he reaches earth, where he stops. We can assume that the acceleration times are negligible compared with the coasting times. We use the following values for illustration: $L_p = 8$ light-years and $V = 0.8c$. Then $\sqrt{1 - V^2/c^2} = 3/5$ and $\gamma = 5/3$.

It is easy to analyze the problem from Homer's point of view on earth. According to Homer's clock, Ulysses coasts in S' for a time $L_p/V = 10$ y and in S'' for an equal time. Thus, Homer is 20 y older when Ulysses returns. The time interval in S' between Ulysses' leaving earth and his arriving

Figure 1 The twin paradox. The earth and a distant planet are fixed in frame S. Ulysses coasts in frame S' to the planet and then coasts back in frame S''. His twin Homer stays on earth. When Ulysses returns, he is younger than his twin. The roles played by the twins are not symmetric. Homer remains in one inertial reference frame, but Ulysses must accelerate if he is to return home.

at the planet is shorter because it is proper time. The time it takes to reach the planet by Ulysses' clock is

$$\Delta t' = \frac{\Delta t}{\gamma} = \frac{10\ \text{y}}{5/3} = 6\ \text{y}$$

Since the same time is required for the return trip, Ulysses will have recorded 12 y for the round trip and will be 8 y younger than Homer upon his return.

From Ulysses' point of view, the distance from the earth to the planet is contracted and is only

$$L' = \frac{L_p}{\gamma} = \frac{8\ \text{light-years}}{5/3} = 4.8\ \text{light-years}$$

At $V = 0.8c$, it takes only 6 y each way.

The real difficulty in this problem is for Ulysses to understand why his twin aged 20 y during his absence. If we consider Ulysses as being at rest and Homer as moving away, Homer's clock should run slow and measure only $3/5(6) = 3.6$ y. Then why shouldn't Homer age only 7.2 y during the round trip? This, of course, is the paradox. The difficulty with the analysis from the point of view of Ulysses is that he does not remain in an inertial frame. What happens while Ulysses is stopping and starting? To investigate this problem in detail, we would need to treat accelerated reference frames, a subject dealt with in the study of general relativity and beyond the scope of this book. However, we can get some insight into the problem by having the twins send regular signals to each other so that they can record the other's age continuously. If they arrange to send a signal once a year, each can determine the age of the other merely by counting the signals received. The arrival frequency of the signals will not be 1 per year because of the Doppler shift. The frequency observed will be given by Equations 39-16a and b. Using $V/c = 0.8$ and $V^2/c^2 = 0.64$, we have for the case in which the twins are receding from each other

$$f' = \frac{\sqrt{1 - V^2/c^2}}{1 + V/c} f_0 = \frac{\sqrt{1 - 0.64}}{1 + 0.8} f_0 = \frac{1}{3} f_0$$

When they are approaching, Equation 39-16a gives $f' = 3f_0$.

Consider the situation first from the point of view of Ulysses. During the 6 y it takes him to reach the planet (remember that the distance is contracted in his frame), he receives signals at the rate of $\frac{1}{3}$ signal per year, and so he receives 2 signals. As soon as he turns around and starts back to earth, he begins to receive 3 signals per year. In the 6 y it takes him to return he receives 18 signals, giving a total of 20 for the trip. He accordingly expects his twin to have aged 20 years.

We now consider the situation from Homer's point of view. He receives signals at the rate of $\frac{1}{3}$ signal per year not only for the 10 y it takes Ulysses to reach the planet but also for the time it takes for the last signal sent by Ulysses before he turns around to get back to earth. (He cannot know that Ulysses has turned around until the signals begin reaching him with increased frequency.) Since the planet is 8 light-years away, there is an additional 8 y of receiving signals at the rate of $\frac{1}{3}$ signal per year. During the first 18 y, Homer receives 6 signals. In the final 2 y before Ulysses arrives, Homer receives 6 signals, or 3 per year. (The first signal sent after Ulysses turns around takes 8 y to reach earth, whereas Ulysses, traveling at 0.8c, takes 10 y to return and therefore arrives just 2 y after Homer begins to receive signals at the faster rate.) Thus, Homer expects Ulysses to have aged 12 y. In this analysis, the asymmetry of the twins' roles is apparent. When they are together again, both twins agree that the one who has been accelerated will be younger than the one who stayed home.

The predictions of the special theory of relativity concerning the twin paradox have been tested using small particles that can be accelerated to such large speeds that γ is appreciably greater than 1. Unstable particles can be accelerated and trapped in circular orbits in a magnetic field, for example, and their lifetimes can then be compared with those of identical particles at rest. In all such experiments, the accelerated particles live longer on the average than those at rest, as predicted. These predictions have also been confirmed by the results of an experiment in which high-precision atomic clocks were flown around the world in commercial airplanes, but the analysis of this experiment is complicated due to the necessity of including gravitational effects treated in the general theory of relativity.

39-5   The Velocity Transformation

We can find how velocities transform from one reference frame to another by differentiating the Lorentz transformation equations. Suppose a particle has velocity $u'_x = dx'/dt'$ in frame S', which is moving to the right with speed V relative to frame S. Its velocity in frame S is

$$u_x = \frac{dx}{dt}$$

From the Lorentz transformation equations (Equations 39-9 and 39-10), we have

$$dx = \gamma(dx' + V\,dt')$$

and

$$dt = \gamma\left(dt' + \frac{V\,dx'}{c^2}\right)$$

The velocity in S is thus

$$u_x = \frac{dx}{dt} = \frac{\gamma(dx' + V\,dt')}{\gamma\left(dt' + \dfrac{V\,dx'}{c^2}\right)} = \frac{\dfrac{dx'}{dt'} + V}{1 + \dfrac{V}{c^2}\dfrac{dx'}{dt'}} = \frac{u'_x + V}{1 + \dfrac{Vu'_x}{c^2}}$$

If a particle has components of velocity along the y or z axes, we can use the same relation between dt and dt', with $dy = dy'$ and $dz = dz'$, to obtain

$$u_y = \frac{dy}{dt} = \frac{dy'}{\gamma\left(dt' + \dfrac{V\,dx'}{c^2}\right)} = \frac{\dfrac{dy'}{dt'}}{\gamma\left(1 + \dfrac{V}{c^2}\dfrac{dx'}{dt'}\right)} = \frac{u'_y}{\gamma\left(1 + \dfrac{Vu'_x}{c^2}\right)}$$

and

$$u_z = \frac{u'_z}{\gamma\left(1 + \dfrac{Vu'_x}{c^2}\right)}$$

The complete relativistic velocity transformation is

$$u_x = \frac{u'_x + V}{1 + Vu'_x/c^2} \tag{39-18a}$$

$$u_y = \frac{u'_y}{\gamma(1 + Vu'_x/c^2)} \tag{39-18b}$$

$$u_z = \frac{u'_z}{\gamma(1 + Vu'_x/c^2)} \tag{39-18c}$$

Relativistic velocity transformation

The inverse velocity transformation equations are

$$u'_x = \frac{u_x - V}{1 - Vu_x/c^2} \tag{39-19a}$$

$$u'_y = \frac{u_y}{\gamma(1 - Vu_x/c^2)} \tag{39-19b}$$

$$u'_z = \frac{u_z}{\gamma(1 - Vu_x/c^2)} \tag{39-19c}$$

Relativistic velocity transformation

These equations differ from the classical and intuitive result $u_x = u'_x + V$, $u_y = u'_y$, and $u_z = u'_z$ because the denominators in the equations are not equal to 1. When V and u'_x are small compared with the speed of light c, $\gamma \approx 1$ and $Vu'_x/c^2 \ll 1$. Then the relativistic and classical expressions are the same.

Example 39-6

A supersonic plane moves away from you along the x axis with speed 1000 m/s (about 3 times the speed of sound) relative to you. Another plane moves along the x axis away from you and away from the second plane at speed 500 m/s relative to the first plane. How fast is the second plane moving relative to you?

Picture the Problem These speeds are so small compared with c that we expect the classical equations for combining velocities to be accurate. We show this by calculating the correction term in the denominator of Equation 39-18a. Let frame S be your rest frame and frame S' be moving with velocity $V = 1000$ m/s. The first plane is then at rest in frame S' and the second has velocity $u'_x = 500$ m/s in S'.

1. The classical formula for combining velocities, gives for the velocity of the second plane relative to you:

$$u_x = u'_x + V = 500 \text{ m/s} + 1000 \text{ m/s} = 1500 \text{ m/s}$$

2. Calculate the correction term in the denominator of Equation 39-18a:

$$\frac{Vu'_x}{c^2} = \frac{(1000)(500)}{(3 \times 10^8)^2} \approx 5.6 \times 10^{-12}$$

Remark This correction term is so small that the classical and relativistic results are essentially the same.

Example 39-7

Work Example 39-6 if the first plane moves with speed $V = 0.8c$ relative to you and the second plane moves with the same speed $0.8c$ relative to the first plane.

Picture the Problem These speeds are not small compared with c so we use the relativistic expression (Equation 39-18a). We again assume that you are at rest in frame S and the first plane is at rest in frame S' that is moving at $V = 0.8c$ relative to you. The velocity of the second plane in S' is $u'_x = 0.8c$.

1. Use Equation 39-18a to calculate the speed of the second plane relative to you:

$$u_x = \frac{u'_x + V}{1 + Vu'_x/c^2} = \frac{0.8c + 0.8c}{1 + (0.8c)(0.8c)/c^2} = \frac{1.6c}{1.64} = 0.98c$$

The result in Example 39-7 is quite different from the classically expected result of $0.8c + 0.8c = 1.6c$. In fact, it can be shown from Equation 39-18 that if the speed of an object is less than c in one frame, it is less than c in all other frames moving relative to that frame with a speed less than c. (See Problem 20.) We will see in Section 39-7 that it takes an infinite amount of energy to accelerate a particle to the speed of light. The speed of light c is thus an upper, unattainable limit for the speed of a particle having mass.*

---

* There are massless particles, such as photons, that always move at the speed of light.

Example 39-8

Light moves along the x axis in frame S' with speed $u'_x = c$. What is its speed in frame S?

1. The speed in S' is given by Equation 39-18a:

$$u_x = \frac{u'_x + V}{1 + Vu'_x/c^2} = \frac{c + V}{1 + Vc/c^2} = \frac{c(1 + V/c)}{1 + V/c} = c$$

Remark The speed in both frames is c as required by Einstein's postulates.

Example 39-9

Two spaceships, each 100 m long when measured at rest, travel toward each other with speeds of 0.85c relative to the earth. (*a*) How long is each ship as measured by someone on earth? (*b*) How fast is each ship traveling as measured by an observer on the other? (*c*) How long is one ship when measured by an observer on the other? (*d*) At time $t = 0$ on earth, the fronts of the ships are together as they just begin to pass each other. At what time on earth are their ends together?

Figure 39-9

Picture the Problem (*a*) The length of each ship as measured on earth is the contracted length $\sqrt{1 - V^2/c^2}\, L_\mathrm{p}$ (Equation 39-14). To solve part (*b*) let the earth be in frame S, and the ship on the left be in frame S' moving with velocity $V = 0.85c$ relative to the earth. Then the ship on the right moves with velocity $u_x = -0.85c$ as shown in Figure 39-9. (*c*) The length of one ship as seen by the other is $\sqrt{1 - V_2^2/c^2}\, L_\mathrm{p}$, where V_2 is the velocity of one ship relative to the other.

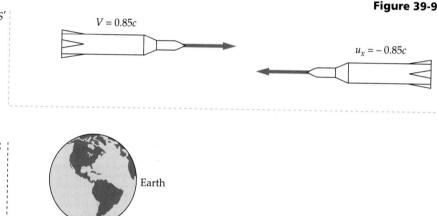

(*a*) The length of each ship in the earth's frame is the proper length divided by γ:

$$L = \sqrt{1 - V^2/c^2}\, L_\mathrm{p} = \sqrt{1 - (0.85c)^2/c^2}\,(100\text{ m})$$
$$= 52.7\text{ m}$$

(*b*) Use the velocity transformation formula (Equation 39-19a) to find the velocity u'_x of the ship on the right as seen in frame S':

$$u'_x = \frac{u_x - V}{1 - Vu_x/c^2} = \frac{-0.85c - 0.85c}{1 - (0.85c)(-0.85c)/c^2}$$
$$= \frac{-1.70c}{1 + 0.7225} = -0.987c$$

(*c*) In the frame of the left ship, the right ship is moving with velocity $V_2 = -0.987c$. Use this to calculate the contracted length of the ship on the right:

$$L = \sqrt{1 - V_2^2/c^2}\, L_\mathrm{p} = \sqrt{1 - (0.987c)^2/c^2}\,(100\text{ m})$$
$$= 16.1\text{ m}$$

(*d*) If the heads of the ships are together at $t = 0$ on earth, their ends will be together after the time it takes either ship to move the length of the ship as seen on earth:

$$t = \frac{L}{V} = \frac{52.7\text{ m}}{0.85c} = \frac{52.7\text{ m}}{(0.85)(3 \times 10^8\text{ m/s})} = 2.07 \times 10^{-7}\text{ s}$$

39-6   Relativistic Momentum

We have seen in previous sections that Einstein's postulates require important modifications in our ideas of simultaneity and in our measurements of time and length. Perhaps more importantly, they also require modifications in our concepts of mass, momentum, and energy. In classical mechanics, the momentum of a particle is defined as the product of its mass and its velocity, $\vec{p} = m\vec{u}$, where $\vec{u}$ is the velocity. In an isolated system of particles, with no net force acting on the system, the total momentum of the system remains constant.

We can see from a simple thought experiment that the quantity $\vec{p} = m\vec{u}$ is not conserved in an isolated system. We consider two observers: observer A in reference frame S and observer B in frame S', which is moving to the right in the x direction with speed V with respect to frame S. Each has a ball of mass m. The two balls are identical when compared at rest. One observer throws his ball up with a speed u_0 relative to him and the other throws his ball down with a speed u_0 relative to him, so that each ball travels a distance L, makes an elastic collision with the other ball, and returns. Figure 39-10 shows how the collision looks in each reference frame. Classically, each ball has vertical momentum of magnitude mu_0. Since the vertical components of the momenta are equal and opposite, the total vertical component of momentum is zero before the collision. The collision merely reverses the momentum of each ball, so the total vertical momentum is zero after the collision.

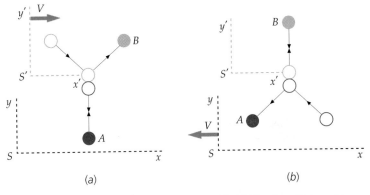

(a)   (b)

Figure 39-10 (a) Elastic collision of two identical balls as seen in frame S. The vertical component of the velocity of ball B is u_0/γ in S if it is u_0 in S'. (b) The same collision as seen in S'. In this frame, ball A has vertical component of velocity equal to u_0/γ.

Relativistically, however, the vertical components of the velocities of the two balls as seen by either observer are not equal and opposite. Thus, when they are reversed by the collision, classical momentum is not conserved. Consider the collision as seen by A in frame S. The velocity of his ball is $u_{Ay} = +u_0$. Since the velocity of B's ball in frame S' is $u'_{Bx} = 0$, $u'_{By} = -u_0$, the y component of the velocity of B's ball in frame S is $u_{By} = -u_0/\gamma$ (Equation 39-18b). Thus, if the classical expression $\vec{p} = m\vec{u}$ is taken as the definition of momentum, the vertical components of momentum of the two balls are not equal and opposite as seen by observer A. Since the balls are reversed by the collision, classical momentum is not conserved. Of course, the same result is observed by B. In the classical limit, when u is much less than c, γ is approximately 1, and the momentum of the system is conserved as seen by either observer.

The reason that the total momentum of a system is important in classical mechanics is that it is conserved when there are no external forces acting on the system, as is the case in collisions. But we have just seen that $\Sigma m\vec{u}$ is conserved only in the approximation that $u \ll c$. We will define the relativistic momentum $\vec{p}$ of a particle to have the following properties:

1. In collisions, $\vec{p}$ is conserved.

2. As u/c approaches zero, $\vec{p}$ approaches $m\vec{u}$.

We will show below that the quantity

$$\vec{p} = \frac{m\vec{u}}{\sqrt{1 - u^2/c^2}} \qquad\qquad 39\text{-}20$$

is conserved in the elastic collision shown in Figure 39-10. Since this quantity also approaches $m\vec{u}$ as u/c approaches zero, we take this equation for the definition of the **relativistic momentum** of a particle.

One interpretation of Equation 39-20 is that the mass of an object increases with speed. The quantity $m/\sqrt{1 - u^2/c^2}$ is called the **relativistic mass** of a particle. The mass of a particle when it is at rest in some reference frame is called its **rest mass** m_0. The mass thus increases from m_0 at rest to $m_r = m_0/\sqrt{1 - u^2/c^2}$ when it is moving at speed u. To avoid confusion, we will label the rest mass m_0 and use $m_0/\sqrt{1 - u^2/c^2}$ for the relativistic mass in this chapter. The rest mass of a particle is the same in all reference frames. Using this notation, the relativistic momentum of a particle is written

$$\vec{p} = \frac{m_0\vec{u}}{\sqrt{1 - u^2/c^2}}$$

39-21

Relativistic momentum

Illustration of Conservation of Relativistic Momentum

We will compute the y component of the relativistic momentum of each particle in the reference frame S for the collision of Figure 39-10 and show that the y component of the total relativistic momentum is zero. The speed of ball A in S is u_0, so the y component of its relativistic momentum is

$$p_{Ay} = \frac{mu_0}{\sqrt{1 - u_0^2/c^2}}$$

The speed of ball B in S is more complicated. Its x component is V and its y component is $-u_0/\gamma$. Thus,

$$u_B^2 = u_{Bx}^2 + u_{By}^2 = V^2 + (-u_0\sqrt{1 - V^2/c^2})^2 = V^2 + u_0^2 - \frac{u_0^2V^2}{c^2}$$

Using this result to compute $\sqrt{1 - u_B^2/c^2}$, we obtain

$$1 - \frac{u_B^2}{c^2} = 1 - \frac{V^2}{c^2} - \frac{u_0^2}{c^2} + \frac{u_0^2V^2}{c^4} = \left(1 - \frac{V^2}{c^2}\right)\left(1 - \frac{u_0^2}{c^2}\right)$$

and

$$\sqrt{1 - u_B^2/c^2} = \sqrt{1 - V^2/c^2}\,\sqrt{1 - u_0^2/c^2} = (1/\gamma)\sqrt{1 - u_0^2/c^2}$$

The y component of the relativistic momentum of ball B as seen in S is therefore

$$p_{By} = \frac{mu_{By}}{\sqrt{1 - u_B^2/c^2}} = \frac{-mu_0/\gamma}{(1/\gamma)\sqrt{1 - u_0^2/c^2}} = \frac{-mu_0}{\sqrt{1 - u_0^2/c^2}}$$

Since $p_{By} = -p_{Ay}$, the y component of the total momentum of the two balls is zero. If the speed of each ball is reversed by the collision, the total momentum will remain zero and momentum will be conserved.

39-7   Relativistic Energy

In classical mechanics, the work done by an unbalanced force acting on a particle equals the change in the kinetic energy of the particle. In relativistic mechanics, we equate the unbalanced force to the rate of change of the relativistic momentum. The work done by such a force can then be calculated and set equal to the change in kinetic energy.

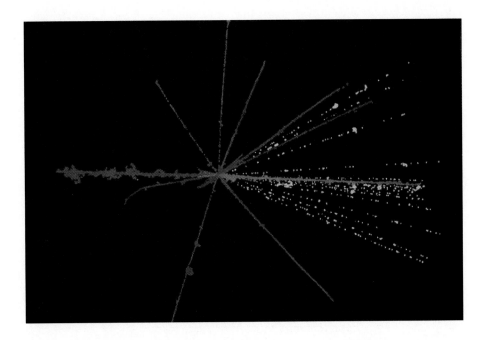

The creation of elementary particles demonstrates the conversion of kinetic energy to rest energy. In this 1950 photograph of a cosmic ray shower, a high-energy sulfur nucleus (red) collides with a nucleus in a photographic emulsion and produces a spray of particles, including a fluorine nucleus (green), other nuclear fragments (blue), and about 16 pions (yellow).

As in classical mechanics, we will define kinetic energy as the work done by an unbalanced force in accelerating a particle from rest to some velocity. Considering one dimension only, we have

$$K = \int_{u=0}^{u} \sum F \, ds = \int_{0}^{u} \frac{dp}{dt} \, ds = \int_{0}^{u} u \, dp$$

$$= \int_{0}^{u} u \, d\left(\frac{m_0 u}{\sqrt{1 - u^2/c^2}}\right) \qquad \text{39-22}$$

where we have used $u = ds/dt$. It is left as a problem (Problem 51) for you to show that

$$d\left(\frac{m_0 u}{\sqrt{1 - u^2/c^2}}\right) = m_0\left(1 - \frac{u^2}{c^2}\right)^{-3/2} du$$

If we substitute this expression into the integrand in Equation 39-22, we obtain

$$K = \int_{0}^{u} u \, d\left(\frac{m_0 u}{\sqrt{1 - u^2/c^2}}\right) = \int_{0}^{u} m_0\left(1 - \frac{u^2}{c^2}\right)^{-3/2} u \, du$$

$$= m_0 c^2\left(\frac{1}{\sqrt{1 - u^2/c^2}} - 1\right)$$

or

$$K = \frac{m_0 c^2}{\sqrt{1 - u^2/c^2}} - m_0 c^2 \qquad \text{39-23}$$

Relativistic kinetic energy

The expression for kinetic energy consists of two terms. The first term depends on the speed of the particle. The second, $m_0 c^2$, is independent of the speed. The quantity $m_0 c^2$ is called the **rest energy** E_0 of the particle. The rest energy is the product of the rest mass and c^2:

$$E_0 = m_0 c^2 \qquad \text{39-24}$$

Rest energy

The total **relativistic energy** E is then defined to be the sum of the kinetic energy and the rest energy:

$$E = K + m_0c^2 = \frac{m_0c^2}{\sqrt{1 - u^2/c^2}}$$ 39-25

Relativistic energy

Thus, the work done by an unbalanced force increases the energy from the rest energy m_0c^2 to the final energy $m_0c^2/\sqrt{1 - u^2/c^2} = m_rc^2$ where $m_r = m_0/\sqrt{1 - u^2/c^2}$ is the relativistic mass. We can obtain a useful expression for the velocity of a particle by multiplying Equation 39-21 for the relativistic momentum by c^2 and comparing the result with Equation 39-25 for the relativistic energy. We have

$$pc^2 = \frac{m_0c^2u}{\sqrt{1 - u^2/c^2}} = Eu$$

or

$$\frac{u}{c} = \frac{pc}{E}$$ 39-26

Energies in atomic and nuclear physics are usually expressed in units of electron volts (eV) or mega-electron-volts (MeV):

$$1\text{ eV} = 1.6 \times 10^{-19}\text{ J}$$

A convenient unit for the masses of atomic particles is eV/c^2 or MeV/c^2, which is just the rest energy of the particle divided by c^2. The rest energies of some elementary particles and light nuclei are given in Table 39-1.

Table 39-1

Rest Energies of Some Elementary Particles and Light Nuclei

Particle	Symbol	Rest energy, MeV
Photon	γ	0
Electron (positron)	e or e^- (e^+)	0.5110
Muon	$\mu^\pm$	105.7
Pion	π^0	135
	$\pi^\pm$	139.6
Proton	p	938.280
Neutron	n	939.573
Deuteron	^{2}H or d	1875.628
Triton	^{3}H or t	2808.944
Helium-3	^{3}He	2808.41 MeV
Alpha particle	^{4}He or α	3727.409

Example 39-10

An electron (rest energy 0.511 MeV) moves with speed $u = 0.8c$. Find (*a*) its total energy, (*b*) its kinetic energy, and (*c*) the magnitude of its momentum.

(*a*) The total energy is given by Equation 39-25:

$$E = \frac{m_0c^2}{\sqrt{1 - u^2/c^2}} = \frac{m_0c^2}{\sqrt{1 - 0.64}} = \frac{m_0c^2}{0.6}$$

$$= \frac{0.511\text{ MeV}}{0.6} = 0.852\text{ MeV}$$

(*b*)1. The kinetic energy is the total energy minus the rest energy:

$$K = E - m_0c^2 = 0.852\text{ MeV} - 0.511\text{ MeV} = 0.341\text{ MeV}$$

2. The magnitude of the momentum is found from Equation 39-21:

$$p = \frac{m_0u}{\sqrt{1 - u^2/c^2}} = \frac{m_0(0.8c)}{0.6}$$

3. We can simplify by multiplying both numerator and denominator by c:

$$p = \frac{(0.8)m_0c^2}{(0.6)c} = \frac{(1.33)(0.511\text{ MeV})}{c} = 0.680\text{ MeV}/c$$

Remark The unit MeV/c is a convenient unit for momentum.

The expression for kinetic energy given by Equation 39-23 doesn't look much like the classical expression $\frac{1}{2}m_0u^2$. However, when u is much less than c, we can approximate $1/\sqrt{1 - u^2/c^2}$ using the binomial expansion

$$(1 + x)^n = 1 + nx + n(n - 1)\frac{x^2}{2} + \cdots \approx 1 + nx \qquad \text{39-27}$$

Then

$$\frac{1}{\sqrt{1 - u^2/c^2}} = \left(1 - \frac{u^2}{c^2}\right)^{-1/2} \approx 1 + \frac{1}{2}\frac{u^2}{c^2}$$

From this result, when u is much less than c, the expression for relativistic kinetic energy becomes

$$K = m_0c^2\left(\frac{1}{\sqrt{1 - u^2/c^2}} - 1\right) \approx m_0c^2\left(1 + \frac{1}{2}\frac{u^2}{c^2} - 1\right) = \frac{1}{2}m_0u^2$$

Thus, at low speeds, the relativistic expression is the same as the classical expression.

We note from Equation 39-25 that as the speed u approaches the speed of light c, the energy of the particle becomes very large because $1/\sqrt{1 - u^2/c^2}$ becomes very large. At $u = c$, the energy becomes infinite. For u greater than c, $\sqrt{1 - u^2/c^2}$ is the square root of a negative number and is therefore imaginary. A simple interpretation of the result that it takes an infinite amount of energy to accelerate a particle to the speed of light is that no particle that is ever at rest in any inertial reference frame can travel as fast or faster than the speed of light c. As we noted in Example 39-7, if the speed of a particle is less than c in one reference frame, it is less than c in all other reference frames moving relative to that frame at speeds less than c.

In practical applications, the momentum or energy of a particle is often known rather than the speed. Equation 39-21 for the relativistic momentum and Equation 39-25 for the relativistic energy can be combined to eliminate the speed u. (See Problem 52.) The result is

$$E^2 = p^2c^2 + (m_0c^2)^2 \qquad \text{39-28}$$

Relation for total energy, momentum, and rest energy

$$E^2 = (pc)^2 + (m_0c^2)^2$$

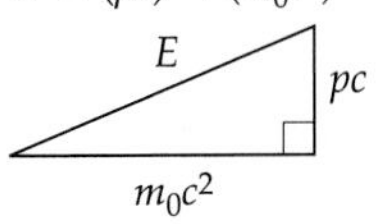

Figure 39-11 Right triangle for remembering Equation 39-28.

This useful equation can be conveniently remembered from the right triangle shown in Figure 39-11. If the energy of a particle is much greater than its rest energy m_0c^2, the second term on the right side of Equation 39-28 can be neglected, giving the useful approximation

$$E \approx pc \qquad \text{for } E \gg m_0c^2 \qquad \text{39-29}$$

Equation 39-29 is an exact relation between energy and momentum for particles with no rest mass, such as photons.

Exercise   A proton (rest mass 938 MeV/c^2) has a total energy of 1400 MeV. Find (a) $1/\sqrt{1 - u^2/c^2}$, (b) the momentum of the proton, and (c) the speed u of the proton. (*Answers*   (a) 1.49, (b) $p = 1.04 \times 10^3$ MeV/c, (c) $u = 0.74c$)

Rest Mass and Energy

Einstein considered Equation 39-24 relating the energy of a particle to its mass to be the most significant result of the theory of relativity. Energy and inertia, which were formerly two distinct concepts, are related through this famous equation. As discussed in Chapter 7, the conversion of rest energy to kinetic energy with a corresponding loss in rest mass is a common occurrence in radioactive decay and nuclear reactions, including nuclear fission and nuclear fusion. We illustrated this in Chapter 7 with the deuteron, whose rest mass is 2.22 MeV/c^2 less than the rest mass of its parts—a proton and neutron. When a neutron and proton combine to form a deuteron, 2.22 MeV of energy is released. The breaking up of a deuteron into a neutron and proton requires 2.22 MeV of energy input. The proton and neutron are thus bound together in a deuteron by a binding energy of 2.22 MeV. Any stable composite particle, such as a deuteron or a helium nucleus (2 neutrons plus 2 protons), that is made up of other particles has a rest mass and rest energy that are less than the sum of the rest masses and rest energies of its parts. The difference in rest energy is the binding energy of the composite particle. The binding energies of atoms and molecules are of the order of a few electron volts, which leads to a negligible difference in mass between the composite particle and its parts. The binding energies of nuclei are of the order of several MeV, which leads to a noticeable difference in mass. Some very heavy nuclei, such as radium, are radioactive and decay into a lighter nucleus plus an alpha particle. In this case, the original nucleus has a rest energy greater than that of the decay particles. The excess energy appears as the kinetic energy of the decay products.

To further illustrate the interrelation of rest mass and energy, we consider a perfectly inelastic collision of two particles. Classically, kinetic energy is lost in such a collision. Relativistically, this loss in kinetic energy shows up as an increase in rest energy of the system, that is, the total energy of the system is conserved. Consider a particle of rest mass $m_{0,1}$ moving with initial speed u_1 that collides with a particle of rest mass $m_{0,2}$ moving with initial speed u_2. The particles collide and stick together, forming a particle of rest mass M_0 that moves with speed u_f, as shown in Figure 39-12. The initial total energy of particle 1 is

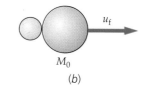

(a)

(b)

Figure 39-12 A perfectly inelastic collision between two particles. One particle of rest mass m_{10} collides with another particle of rest mass m_{20}. After the collision, the particles stick together, forming a composite particle of rest mass M_0 that moves

with speed u_f such that relativistic momentum is conserved. Kinetic energy is lost in this process. If we assume that the total energy is conserved, the loss in kinetic energy must equal c^2 times the increase in the rest mass of the system.

$$E_1 = K_1 + m_{0,1}c^2$$

where K_1 is its initial kinetic energy. Similarly the initial total energy of particle 2 is

$$E_2 = K_2 + m_{0,2}c^2$$

The total initial energy of the system is

$$E_i = E_1 + E_2 = K_1 + m_{0,1}c^2 + K_2 + m_{0,2}c^2 = K_i + (m_{0,1} + m_{0,2})c^2$$

where $K_i = K_1 + K_2$ is the initial kinetic energy of the system. The final total energy of the system is

$$E_f = K_f + M_0c^2$$

If we set the final total energy equal to the initial total energy, we obtain

$$K_f + M_0c^2 = K_i + (m_{0,1} + m_{0,2})c^2$$

The change in kinetic energy is thus

$$K_f - K_i = [M_0 - (m_{0,1} + m_{0,2})]c^2 = (\Delta m_0)c^2 \qquad\qquad 39\text{-}30$$

where $\Delta m_0 = M_0 - (m_{0,1} + m_{0,2})$ is the increase in rest mass of the system.

Example 39-11

A particle of rest mass 2 MeV/c^2 and kinetic energy 3 MeV collides with a stationary particle of rest mass 4 MeV/c^2. After the collision, the two particles stick together. Find (*a*) the initial momentum of the system, (*b*) the final velocity of the two-particle system, and (*c*) the rest mass of the two-particle system.

Picture the Problem (*a*) The initial momentum of the system is that of the incoming particle, which can be found from the total energy of the particle. (*b*) The final velocity of the system can be found from its total energy and momentum using $u/c = pc/E$ (Equation 39-26). The energy is found from conservation of energy, and the momentum from conservation of momentum. (*c*) Since the final energy and momentum are known, the final rest mass can be found from $E^2 = p^2c^2 + (M_0c^2)^2$.

(*a*)1. The initial momentum of the incoming particle is related to its energy and rest mass:	$E^2 = p^2c^2 + (m_0c^2)^2$ $pc = \sqrt{E_1^2 - (m_0c^2)^2}$
2. The total energy of the moving particle is the sum of its kinetic energy and rest energy:	$E_1 = 3\ \text{MeV} + 2\ \text{MeV} = 5\ \text{MeV}$
3. Use this total energy to calculate the momentum:	$pc = \sqrt{E_1^2 - (m_0c^2)^2} = \sqrt{(5\ \text{MeV})^2 - (2\ \text{MeV})^2} = \sqrt{21}\ \text{MeV}$ $p = 4.58\ \text{MeV}/c$
(*b*)1. We can find the final velocity of the two-particle system from its total energy E and its momentum p using Equation 39-26:	$\dfrac{u}{c} = \dfrac{pc}{E}$
2. By the conservation of total energy, the final energy of the system equals the initial total energy of the two particles:	$E_f = E_i = E_1 + E_2 = 5\ \text{MeV} + 4\ \text{MeV} = 9\ \text{MeV}$
3. By the conservation of momentum, the final momentum of the two-particle system equals the initial momentum:	$p = 4.58\ \text{MeV}/c$
4. Calculate the velocity of the two-particle system from its total energy and momentum using $u/c = pc/E$:	$\dfrac{u}{c} = \dfrac{pc}{E} = \dfrac{4.58\ \text{MeV}}{9\ \text{MeV}} = 0.509$ $u = 0.509c$
(*c*) We can find the rest mass of the final two-particle system from Equation 39-28 using $pc = 4.58\ \text{MeV}$ and $E = 9\ \text{MeV}$:	$E^2 = (pc)^2 + (M_0c^2)^2$ $(9\ \text{MeV})^2 = (4.58\ \text{MeV})^2 + (M_0c^2)^2$ $M_0 = 7.75\ \text{MeV}/c^2$

Remarks Note that the rest mass of the system increased from 6 MeV/c^2 to 7.75 MeV/c^2. This increase times c^2 equals the loss in kinetic energy of the system, as you will show in the exercise below.

Exercise (*a*) Find the final kinetic energy of the two-particle system in Example 39-11. (*b*) Find the loss in kinetic energy, K_{loss}, in the collision. (*c*) Show that $K_{\text{loss}} = \Delta Mc^2$, where ΔM is the increase in rest mass of the system found in part (*c*) of that example. (*Answers* (*a*) $K_f = E - M_0c^2 = 9\ \text{MeV} - 7.75\ \text{MeV} = 1.25\ \text{MeV}$, (*b*) $K_{\text{loss}} = K_i - K_f = 3\ \text{MeV} - 1.25\ \text{MeV} = 1.75\ \text{MeV}$, (*c*) $\Delta Mc^2 = M_0c^2 - M_i\,c^2 = 7.75\ \text{MeV} - (2\ \text{MeV} + 4\ \text{MeV}) = 1.75\ \text{MeV} = K_{\text{loss}}$)

Example 39-12

A rocket of mass $m_r = 10^6$ kg is coasting through space when it suddenly becomes necessary to accelerate. The rocket ejects 10^3 kg of burned fuel in a very short time at a speed of $c/2$ relative to the rocket. (*a*) Neglecting any change in the rest mass of the system, calculate the speed of the rocket u_r in the frame in which it was initially at rest. (*b*) Use your results from (*a*) to estimate the change in the rest mass of the system. (*c*) Calculate the speed of the rocket using classical, newtonian mechanics.

Picture the Problem The speed of the rocket is calculated from its relativistic momentum. In the frame in which the rocket is initially at rest the total momentum of rocket plus fuel is zero. After the burn, the magnitude of the momentum of the rocket equals that of the ejected fuel. Let m_r be the mass of the rocket *after* ejecting the burned fuel and m_f be the mass of the fuel. Since $m_r \approx 1000 m_f$, the loss in fuel has a negligible effect on the mass of the rocket. In part (*b*) we can neglect the change in rest mass of the ejected fuel compared to the change in rest mass of the remaining rocket.

(*a*)1. The speed of the rocket is related to its momentum:

$$p_r = \frac{m_r u_r}{\sqrt{1 - u_r^2/c^2}}$$

2. Set the magnitude of the momentum of the rocket equal to the magnitude of the momentum of the ejected fuel:

$$\frac{m_r u_r}{\sqrt{1 - u_r^2/c^2}} = \frac{m_f u_f}{\sqrt{1 - u_f^2/c^2}}$$
$$= \frac{m_f(0.5c)}{\sqrt{1 - (0.5c)^2/c^2}} = \frac{m_f(0.5c)}{\sqrt{1 - 0.25}} = \frac{m_f c}{\sqrt{3}}$$

3. Solve for u_r:

$$\sqrt{3}\, m_r u_r = \sqrt{1 - u_r^2/c^2}\, m_f c = m_f \sqrt{c^2 - u_r^2}$$
$$c^2 - u_r^2 = 3(m_r/m_f)^2 u_r^2 = 3 \times 10^6 u_r^2$$
$$u_r = (1/\sqrt{3}) \times 10^{-3} c = 1.73 \times 10^5\,\text{m/s}$$

(*b*)1. The rest mass is related to the total energy. Write the initial energy E_i in terms of m_r and m_f:

$$E_i = m_r c^2 + m_f c^2$$

2. Write the final energy in terms of the final rest mass of the rocket minus fuel m_r' and the final rest mass of the fuel m_f':

$$E_f = \frac{m_r' c^2}{\sqrt{1 - u_r^2/c^2}} + \frac{m_f' c^2}{\sqrt{1 - (0.5c)^2/c^2}}$$

3. Simplify using $u_r \ll c$ and $m_f' \approx m_f$:

$$E_f \approx m_r' c^2 + \frac{2}{\sqrt{3}} m_f c^2$$

4. Apply conservation of energy and solve for m_r':

$$E_f = E_i$$
$$m_r' c^2 + \frac{2}{\sqrt{3}} m_f c^2 = m_r c^2 + m_f c^2$$
$$m_r' = m_r - \left(\frac{2}{\sqrt{3}} - 1\right) m_f = m_r - 0.155 m_f$$
$$\Delta m_r = -0.155 m_f = -155\,\text{kg}$$

(*c*)1. Set the magnitude of the classical momentum of the rocket equal to the magnitude of the classical momentum of the ejected fuel and solve for u_r:

$$m_r u_r = m_f u_f$$
$$u_r = \frac{m_f}{m_r} u_f = \frac{10^3\,\text{kg}}{10^6\,\text{kg}}(0.5c) = 1.5 \times 10^5\,\text{m/s}$$

Remarks The decrease in rest mass is extremely small as it usually is in macroscopic problems. Note that the classical calculation of the speed gives an error of about 13% in this case.

xploring

General Relativity

The generalization of the theory of relativity to noninertial reference frames by Einstein in 1916 is known as the general theory of relativity. It is much more difficult mathematically than the special theory of relativity, and there are fewer situations in which it can be tested. Nevertheless, its importance calls for a brief qualitative discussion.

The basis of the general theory of relativity is the **principle of equivalence:**

A homogeneous gravitational field is completely equivalent to a uniformly accelerated reference frame.

Principle of equivalence

This principle arises in Newtonian mechanics because of the apparent identity of gravitational mass and inertial mass. In a uniform gravitational field, all objects fall with the same acceleration $\vec{g}$ independent of their mass because the gravitational force is proportional to the (gravitational) mass, whereas the acceleration varies inversely with the (inertial) mass. Consider a compartment in space undergoing a uniform acceleration $\vec{a}$, as shown in Figure 1a. No mechanics experiment can be performed *inside* the compartment that will distinguish whether the compartment is actually accelerating in space or is at rest (or is moving with uniform velocity) in the presence of a uniform gravitational field $\vec{g} = -\vec{a}$, as shown in Figure 1b. If objects are dropped in the compartment, they will fall to the "floor" with an acceleration $\vec{g} = -\vec{a}$. If people stand on a spring scale, it will read their "weight" of magnitude ma.

Einstein assumed that the principle of equivalence applies to all physics and not just to mechanics. In effect, he assumed that there is no experiment of any kind that can distinguish uniformly accelerated motion from the presence of a gravitational field.

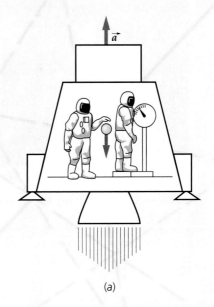

(a)

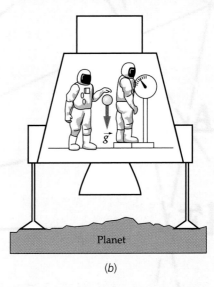

(b)

Figure 1 The results of experiments in a uniformly accelerated reference frame (*a*) cannot be distinguished from those in a uniform gravitational field (*b*) if the acceleration $\vec{a}$ and the gravitational field $\vec{g}$ have the same magnitude.

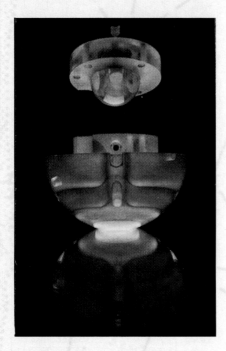

The quartz sphere in the top part of the container is probably the world's most perfectly round object. It is designed to spin as a gyroscope in a satellite orbiting the earth. General relativity predicts that the rotation of the earth will cause the axis of rotation of the gyroscope to precess in a circle at a rate of about 1 revolution in 100,000 years.

One consequence of the principle of equivalence—the deflection of a light beam in a gravitational field—was one of the first to be tested experimentally. Figure 2 shows a beam of light entering a compartment that is accelerating. Suc-

cessive positions of the compartment at equal time intervals are shown in Figure 2a. Because the compartment is accelerating, the distance it moves in each time interval increases with time. The path of the beam of light as observed from inside the compartment is therefore a parabola, as shown in Figure 2b. But according to the principle of equivalence, there is no way to distinguish between an accelerating compartment and one moving with uniform velocity in a uniform gravitational field. We conclude, therefore, that a beam of light will accelerate in a gravitational field, just like objects that have mass. For example, near the surface of the earth, light will fall with an acceleration of 9.81 m/s^2. This is difficult to observe because of the enormous speed of light. For example, in a distance of 3000 km, which takes light about 0.01 s to traverse, a beam of light should fall about 0.5 mm. Einstein pointed out that the deflection of a light beam in a gravitational field might be observed when light from a distant star passes close to the sun, as illustrated in Figure 3. Because of the brightness of the sun, such a star cannot ordinarily be seen. Such a deflection was first observed in 1919 during an eclipse of the sun. This well-publicized observation brought instant worldwide fame to Einstein.

A second prediction from Einstein's theory of general relativity, which we will not discuss in detail, is the excess precession of the perihelion of the orbit of Mercury of about 0.01° per century. This effect had been known and unexplained for some time, so, in a sense, explaining it constituted an immediate success of the theory.

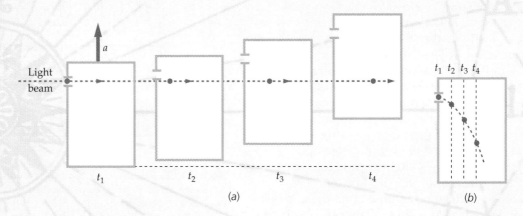

Figure 2 (a) A light beam moving in a straight line through a compartment that is undergoing uniform acceleration. The position of the beam is shown at equally spaced times t_1, t_2, t_3, and t_4. (b) In the reference frame of the compartment, the light travels in a parabolic path as a ball would if it were projected horizontally. The vertical displacements are greatly exaggerated in both (a) and (b) for emphasis.

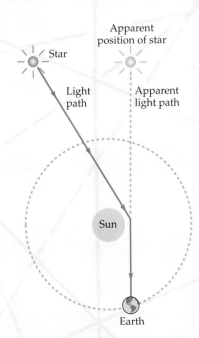

Figure 3 The deflection (greatly exaggerated) of a beam of light due to the gravitational attraction of the sun.

This extremely accurate hydrogen maser clock was launched in a satellite in 1976, and its time was compared to that of an identical clock on earth. In accordance with the prediction of general relativity, the clock on earth, where the gravitational potential was lower, "lost" about 4.3×10^{-10} s each second compared with the clock orbiting the earth at an altitude of about 10,000 km.

A third prediction of general relativity concerns the change in time intervals and frequencies of light in a gravitational field. In Chapter 11, we found that the gravitational potential energy between two masses M and m a distance r apart is

$$U = -\frac{GMm}{r}$$

where G is the universal gravitational constant, and the point of zero potential energy has been chosen to be when the separation of the masses is infinite. The potential energy per unit mass near a mass M is called the *gravitational potential* ϕ:

$$\phi = -\frac{GM}{r} \qquad\qquad 1$$

According to the general theory of relativity, clocks run more slowly in regions of low gravitational potential. (Since the gravitational potential is negative, as can be seen from Equation 1, low gravitational potential occurs near the mass, where the *magnitude* of the potential is large.) If Δt_1 is a time interval between two events measured by a clock where the gravitational potential is ϕ_1 and Δt_2 is the interval between the same events as measured by a clock where the gravitational potential is ϕ_2, general relativity predicts

that the fractional difference between these times will be approximately*

$$\frac{\Delta t_2 - \Delta t_1}{\Delta t} = \frac{1}{c^2}(\phi_2 - \phi_1) \qquad\qquad 2$$

A clock in a region of low gravitational potential will therefore run slower than one in a region of high potential. Since a vibrating atom can be considered to be a clock, the frequency of vibration of an atom in a region of low potential, such as near the sun, will be lower than that of the same atom on earth. This shift toward a lower frequency and therefore a longer wavelength is called the **gravitational redshift.**

As our final example of the predictions of general relativity, we mention **black holes,** which were first predicted by Oppenheimer and Snyder in 1939. According to the general theory of relativity, if the density of an object such as a star is great enough, its gravitational attraction will be so great that once inside a critical radius, nothing

---

*Since this shift is usually very small, it does not matter by which interval we divide on the left side of the equation.

can escape, not even light or other electromagnetic radiation. (The effect of a black hole on objects outside the critical radius is the same as that of any other mass.) A remarkable property of such an object is that nothing that happens inside it can be communicated to the outside. As sometimes occurs in physics, a simple but incorrect calculation gives the correct results for the relation between the mass and the critical radius of a black hole. In Newtonian mechanics, the speed needed for a particle to escape from the surface of a planet or star of mass M and radius R is given by Equation 11-19:

$$v_e = \sqrt{\frac{2GM}{R}}$$

If we set the escape speed equal to the speed of light and solve for the radius, we obtain the critical radius R_S, called the **Schwarzschild radius:**

$$R_S = \frac{2GM}{c^2} \qquad\qquad 3$$

For an object with a mass equal to that of our sun to be a black hole, its radius would have to be about 3 km. Since no radiation is emitted from a black hole and its radius is expected to be small, the detection of a black hole is not easy. The best chance of detection would occur if a black hole were a companion to a normal star in a binary star system. Then both stars would revolve around their center of mass, and the existence of the black hole could be inferred from the motion of the normal star. At present there are several excellent candidates—one in the constellation Cygnus, one in the Small Magellanic Cloud, and perhaps one in our own galaxy—but the evidence is not conclusive.

Summary

Topic	Remarks and Relevant Equations
1. Einstein Postulates	The special theory of relativity is based on two postulates of Albert Einstein. All of the results of special relativity can be derived from these postulates.
	Postulate 1. Absolute, uniform motion cannot be detected.
	Postulate 2. The speed of light is independent of the motion of the source.
	An important implication of these postulates is
	Postulate 2 (Alternate). Every observer measures the same value for the speed of light independent of the relative motion of the source and observer.
2. Lorentz Transformation	$x = \gamma(x' + Vt'), \qquad y = y', \quad z = z'$ **39-9**
	$t = \gamma\left(t' + \dfrac{Vx'}{c^2}\right)$ **39-10**
Inverse transformation	$x' = \gamma(x - Vt), \qquad y' = y, \quad z' = z$ **39-11**
	$t' = \gamma\left(t - \dfrac{Vx}{c^2}\right)$ **39-12**
3. Proper Time and Time Dilation	The time interval measured between two events that occur at the same point in space in some reference frame is called the proper time. In another reference frame in which the events occur at different places, the time interval between the events is longer by the factor γ:
	$\Delta t = \gamma\,\Delta t_{\text{p}}$ **39-13**
4. Length Contraction	The length of an object measured in a frame in which it is at rest is called its proper length L_{p}. When measured in another reference frame, the length of the object is
	$L = \dfrac{L_{\text{p}}}{\gamma}$ **39-14**
5. The Relativistic Doppler Effect	$f' = \dfrac{\sqrt{1 - V^2/c^2}}{1 - V/c}\,f_0 \qquad$ approaching **39-16a**
	$f' = \dfrac{\sqrt{1 - V^2/c^2}}{1 + V/c}\,f_0 \qquad$ receding **39-16b**
6. Clock Synchronization	Two events that are simultaneous in one reference frame are not simultaneous in another frame that is moving relative to the first. If two clocks are synchronized in the frame in which they are at rest, they will be out of synchronization in another frame. In the frame in which they are moving, the chasing clock leads by an amount
	$\Delta t_{\text{S}} = L_{\text{p}}(V/c^2)$ **39-17**
	where L_{p} is the proper distance between the clocks.
7. Velocity Transformation	$u_x = \dfrac{u_x' + V}{1 + Vu_x'/c^2}$ **39-18a**
	$u_y = \dfrac{u_y'}{\gamma(1 + Vu_x'/c^2)}$ **39-18b**
	$u_z = \dfrac{u_z'}{\gamma(1 + Vu_x'/c^2)}$ **39-18c**

Inverse velocity transformation	$u'_x = \dfrac{u_x - V}{1 - Vu_x/c^2}$	**39-19a**
	$u'_y = \dfrac{u_y}{\gamma(1 - Vu_x/c^2)}$	**39-19b**
	$u'_z = \dfrac{u_z}{\gamma(1 - Vu_x/c^2)}$	**39-19c**

8. Relativistic Momentum

$$\vec{p} = \frac{m_0 \vec{u}}{\sqrt{1 - u^2/c^2}}$$ **39-21**

where m_0 is the rest mass of the particle.

9. Relativistic Energy

Kinetic energy

$$K = \frac{m_0 c^2}{\sqrt{1 - u^2/c^2}} - m_0 c^2 = \frac{m_0 c^2}{\sqrt{1 - u^2/c^2}} - E_0$$ **39-23**

Rest energy

$$E_0 = m_0 c^2$$ **39-24**

Total energy

$$E = K + E_0 = \frac{m_0 c^2}{\sqrt{1 - u^2/c^2}}$$ **39-25**

10. Useful Formulas for Speed, Energy, and Momentum

$$\frac{u}{c} = \frac{pc}{E}$$ **39-26**

$$E^2 = p^2 c^2 + (m_0 c^2)^2$$ **39-28**

$$E \approx pc \quad \text{for } E \gg m_0 c^2$$ **39-29**

11. Binding Energy

The total rest mass of bound systems of particles, such as nuclei or atoms, is less than the sum of the rest masses of the particles making up the system. The difference in mass times c^2 equals the binding energy of the system. The binding energy is the energy that must be added to break up the system into its parts. The binding energies of electrons in atoms are of the order of eV or keV, leading to a negligible difference in rest mass. The binding energies of nuclei are of the order of several MeV, and the difference in rest mass is noticeable.

Problem-Solving Guide

Summary of Worked Examples

Type of Calculation	Procedure and Relevant Examples
1. Lorentz Transformation	
Find the spatial separation or time interval between two events in another reference frame.	Use the Lorentz transformation. If the events happen at the same place in one frame, the time interval in that frame is proper time and in another frame $\Delta t = \gamma \Delta t_p$. **Examples 39-1, 39-2, 39-3, 39-5**
2. Doppler Effect	
Find the speed of a moving source from its Doppler shift.	Use $$f' = \frac{\sqrt{1 - V^2/c^2}}{1 \pm V/c} f_0$$ and solve for V/c. **Example 39-4**

3. Clock Synchronization

Find the amount by which two clocks are unsynchronized.	The chasing clock leads the other by $\Delta t_s = \gamma V L_p / c^2$, where L_p is the proper separation of the clocks. **Example 39-5**

4. Velocity Transformation

Find the velocity of an object relative to you given its velocity in another frame.	Use the velocity transformation equations. **Examples 39-6, 39-7, 39-8, 39-9**

5. Momentum and Energy

Find the total energy, kinetic energy, and momentum of a particle given its rest energy and speed.	Use $E = m_0 c^2 / \sqrt{1 - u^2/c^2}$, $K = E - E_0$, $p = m_0 u / \sqrt{1 - u^2/c^2}$. **Examples 39-10, 39-11**
Find the velocity and rest energy of a system after an inelastic collision.	Use $E^2 = p^2 c^2 + (m_0 c^2)^2$ to find the initial momentum of the particles. Then use conservation of momentum and energy. The velocity can be found from $u/c = pc/E$. **Example 39-11**
Find the speed of a rocket given the mass and speed of its exhaust.	Use conservation of momentum. **Example 39-12**

Problems

Conceptual Problems

Problems from Optional and Exploring sections

In a few problems, you are given more data than you actually need; in a few other problems, you are required to supply data from your general knowledge, outside sources, or informed estimates.

• Single-concept, single-step, relatively easy
•• Intermediate-level, may require synthesis of concepts
••• Challenging, for advanced students

Time Dilation and Length Contraction

1 • You are standing on a corner and a friend is driving past in an automobile. Both of you note the times when the car passes two different intersections and determine from your watch readings the time that elapses between the two events. Which of you has determined the proper time interval?

2 • The proper mean lifetime of pions is 2.6×10^{-8} s. If a beam of pions has a speed of $0.85c$, (a) what would their mean lifetime be as measured in the laboratory? (b) How far would they travel, on average, before they decay? (c) What would your answer be to part (b) if you neglect time dilation?

3 • (a) In the reference frame of the pion in Problem 2, how far does the laboratory travel in a typical lifetime of 2.6×10^{-8} s? (b) What is this distance in the laboratory's frame?

4 • The proper mean lifetime of a muon is 2 μs. Muons in a beam are traveling at $0.999c$. (a) What is their mean lifetime as measured in the laboratory? (b) How far do they travel, on average, before they decay?

5 • (a) In the reference frame of the muon in Problem 4, how far does the laboratory travel in a typical lifetime of 2 μs? (b) What is this distance in the laboratory's frame?

6 • Jay has been posted to a remote region of space to monitor traffic. Toward the end of a quiet shift, a spacecraft goes by, and he measures its length using a laser device, which reports a length of 85 m. He flips open his handy reference catalogue and identifies the craft as a CCCNX-22, which has a proper length of 100 m. When he phones in his report, what speed should Jay give for this spacecraft?

7 • A spaceship travels to a star 95 light-years away at a speed of 2.2×10^8 m/s. How long does it take to get there (a) as measured on earth and (b) as measured by a passenger on the spaceship?

8 • The mean lifetime of a pion traveling at high speed is measured to be 7.5×10^{-8} s. Its lifetime when measured at rest is 2.6×10^{-8} s. How fast is the pion traveling?

9 • A meterstick moves with speed $V = 0.8c$ relative to you in the direction parallel to the stick. (a) Find the length of the stick as measured by you. (b) How long does it take for the stick to pass you?

10 • The half-life of charged pions, π^+ and π^-, is 1.8×10^{-8} s; i.e., in the rest frame of the pions if there are N pions at time $t = 0$, there will be only $N/2$ pions at time $t = 1.8 \times 10^{-8}$ s. Pions are produced in an accelerator and emerge with a speed of $0.998c$. How far do these particles travel in the laboratory before half of them have decayed?

11 •• A friend of yours who is the same age as you travels to the star Alpha Centauri, which is 4 light-years away and returns immediately. He claims that the entire trip took just 6 y. How fast did he travel?

12 •• Two spaceships pass each other traveling in opposite directions. A passenger in ship A, which she knows to be 100 m long, notes that ship B is moving with a speed of $0.92c$ relative to A and that the length of B is 36 m. What are the lengths of the two spaceships as measured by a passenger in ship B?

13 •• In the Stanford linear collider, small bundles of electrons and positrons are fired at each other. In the laboratory's frame of reference, each bundle is about 1 cm long and 10 μm in diameter. In the collision region, each particle has an energy of 50 GeV, and the electrons and positrons are moving in opposite directions. (a) How long and how wide is each bundle in its own reference frame? (b) What must be the minimum proper length of the accelerator for a bundle to have both its ends simultaneously in the accelerator in its own reference frame? (The actual length of the accelerator is less than 1000 m.) (c) What is the length of a positron bundle in the reference frame of the electron bundle?

The Lorentz Transformation

14 • Use the binomial expansion

$$(1 + x)^n = 1 + nx + n(n - 1)\frac{x^2}{2} + \cdots \approx 1 + nx$$

to derive the following results for the case when V is much less than c, and use the results when applicable in the following problems:

(a) $\gamma \approx 1 + \dfrac{1}{2}\dfrac{V^2}{c^2}$

(b) $\dfrac{1}{\gamma} \approx 1 - \dfrac{1}{2}\dfrac{V^2}{c^2}$

(c) $\gamma - 1 \approx 1 - \dfrac{1}{\gamma} \approx \dfrac{1}{2}\dfrac{V^2}{c^2}$

15 •• Show that when $V \ll c$ the transformation equations for x, t, and u reduce to the Galilean equations.

16 •• Supersonic jets achieve maximum speeds of about $(3 \times 10^{-6})c$. (a) By what percentage would you see a jet traveling at this speed contracted in length? (b) During a time of 1 y $= 3.15 \times 10^7$ s on your clock, how much time would elapse on the pilot's clock? How many minutes are lost by the pilot's clock in 1 y of your time?

17 •• How great must the relative speed of two observers be for the time-interval measurements to differ by 1%? (See Problem 14.)

18 •• A spaceship of proper length $L' = 400$ m moves past a transmitting station at a speed of $0.76c$. At the instant that the nose of the ship passes the transmitter, clocks at the transmitter and in the nose of the ship are synchronized to $t = t' = 0$. The instant that the tail of the ship passes the transmitter a signal is sent and subsequently detected by the receiver in the nose of the ship. (a) When, according to the clock in the ship, is the signal sent? (b) When, according to the clock at the transmitter, is the signal received by the spaceship? (c) When, according to the clock in the ship, is the signal received? (d) Where, according to an observer at the transmitter, is the nose of the spaceship when the signal is received?

19 •• A beam of unstable particles emerges from the exit slit of an accelerator with a speed of $0.89c$. Particle detectors 3.0 and 6.0 m from the exit slit measure beam intensities of 2×10^8 particles/cm^2·s and 5×10^7 particles/cm^2·s, respectively. (a) Find the proper half-life of the particles. (b) Determine the beam intensity at the exit slit of the accelerator. (c) The accelerator is adjusted so that the particles emerge from the exit slit with a speed of $0.96c$. The beam intensity at the farther detector is again 5×10^7 particles/cm^2·s. Find the beam intensity at the exit slit of the accelerator.

20 •• Show that if u'_x and V in Equation 39-18a are both less than c, then u_x is less than c. (Hint: Let $u'_x = (1 - \epsilon_1)c$ and $V = (1 - \epsilon_2)c$, where ϵ_1 and ϵ_2 are small positive numbers that are less than 1.)

21 ••• Two events in S are separated by a distance $D = x_2 - x_1$ and a time $T = t_2 - t_1$. (a) Use the Lorentz transformation to show that in frame S', which is moving with speed V relative to S, the time separation is $t'_2 - t'_1 = \gamma(T - VD/c^2)$. (b) Show that the events can be simultaneous in frame S' only if D is greater than cT. (c) If one of the events is the *cause* of the other, the separation D must be less than cT, since D/c is the smallest time that a signal can take to travel from x_1 to x_2 in frame S. Show that if D is less than cT, t'_2 is greater than t'_1 in all reference frames. This shows that if the cause precedes the effect in one frame, it must precede it in all reference frames. (d) Suppose that a signal could be sent with speed $c' > c$ so that in frame S the cause precedes the effect by the time $T = D/c'$. Show that there is then a reference frame moving with speed V less than c in which the effect precedes the cause.

Clock Synchronization and Simultaneity

22 • If event A occurs before event B in some frame, might it be possible for there to be a reference frame in which event B occurs before event A?

23 • Two events are simultaneous in a frame in which they also occur at the same point in space. Are they simultaneous in other reference frames?

24 •• Two observers are in relative motion. In what circumstances can they agree on the simultaneity of two different events?

Problems 25 through 29 refer to the following situation: An observer in S' lays out a distance $L' = 100$ light-minutes between points A' and B' and places a flashbulb at the midpoint C'. She arranges for the bulb to flash and for clocks at A' and B' to be started at zero when the light from the flash reaches them (see Figure 39-13). Frame S' is moving to the right with speed $0.6c$ relative *to an observer C in S who is at the midpoint between A′ and B′ when the bulb flashes. At the instant he sees the flash, observer C sets his clock to zero.*

Figure 39-13
Problems 25 through 29

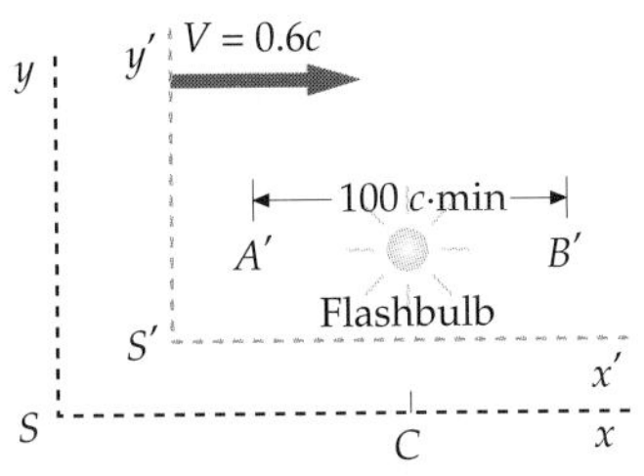

25 •• What is the separation distance between clocks A′ and B′ according to the observer in S?

26 •• As the light pulse from the flashbulb travels toward A′ with speed c, A′ travels toward C with speed 0.6c. Show that the clock in S reads 25 min when the flash reaches A′. (*Hint*: In time t, the light travels a distance ct and A′ travels 0.6ct. The sum of these distances must equal the distance between A′ and the flashbulb as seen in S.)

27 •• Show that the clock in S reads 100 min when the light flash reaches B′, which is traveling away from C with speed 0.6c. (See the hint for Problem 26.)

28 •• The time interval between the reception of the flashes at A′ and B′ in Problems 26 and 27 is 75 min according to the observer in S. How much time does he expect to have elapsed on the clock at A′ during this 75-min interval?

29 •• The time interval calculated in Problem 28 is the amount that the clock at A′ leads that at B′ according to the observer in S. Compare this result with $L_p V/c^2$.

30 •• In frame S, event B occurs 2 μs after event A, which occurs at $\Delta x = 1.5$ km from event A. How fast must an observer be moving along the +x axis so that events A and B occur simultaneously? Is it possible for event B to precede event A for some observer?

31 •• Observers in reference frame S see an explosion located at $x_1 = 480$ m. A second explosion occurs 5 μs later at $x_2 = 1200$ m. In reference frame S′, which is moving along the +x axis at speed V, the explosions occur at the same point in space. What is the separation in time between the two explosions as measured in S′?

The Doppler Effect

32 • How fast must you be moving toward a red light ($\lambda = 650$ nm) for it to appear green ($\lambda = 525$ nm)?

33 • A distant galaxy is moving away from us at a speed of 1.85×10^7 m/s. Calculate the fractional redshift $(\lambda′ - \lambda_0)/\lambda_0$ in the light from this galaxy.

34 • Sodium light of wavelength 589 nm is emitted by a source that is moving toward the earth with speed V. The wavelength measured in the frame of the earth is 620 nm. Find V.

35 • A student on earth hears a tune on her radio that seems to be coming from a record that is being played too fast. She has a 33-rev/min record of that tune and determines that the tune sounds the same as when her record is played at 78 rev/min, that is, the frequencies are all too high by a factor of 78/33. If the tune is being played correctly, but is being broadcast by a spaceship that is approaching the earth at speed V, determine V.

36 •• Derive Equation 39-16a for the frequency received by an observer moving with speed V toward a stationary source of electromagnetic waves.

Exploring . . . The Twin Paradox

37 • Herb and Randy are twin jazz musicians who perform as a trombone–saxophone duo. At the age of twenty, however, Randy got an irresistible offer to join a road trip to perform on a star 15 light-years away. To celebrate his bounteous luck, he bought a new vehicle for the trip—a deluxe space-coupe which could do 0.999c. Each of the twins promises to practice diligently, so they can reunite afterward. Randy's gig goes so fabulously well, however, that he stays for a full 10 years before returning to Herb. After their reunion, (a) how many years of practice will Randy have? (b) how many years of practice will Herb have?

38 •• A clock is placed in a satellite that orbits the earth with a period of 90 min. By what time interval will this clock differ from an identical clock on earth after 1 y? (Assume that special relativity applies and neglect general relativity.)

39 •• A and B are twins. A travels at 0.6c to Alpha Centauri (which is 4 c·y from earth as measured in the reference frame of the earth) and returns immediately. Each twin sends the other a light signal every 0.01 y as measured in her own reference frame. (a) At what rate does B receive signals as A is moving away from her? (b) How many signals does B receive at this rate? (c) How many total signals are received by B before A has returned? (d) At what rate does A receive signals as B is receding from her? (e) How many signals does A receive at this rate? (f) How many total signals are received by A? (g) Which twin is younger at the end of the trip, and by how many years?

The Velocity Transformation

40 • A light beam moves along the y′ axis with speed c in frame S′, which is moving to the right with speed V relative to frame S. (a) Find the x and y components of the velocity of the light beam in frame S. (b) Show that the magnitude of the velocity of the light beam in S is c.

41 • A spaceship is moving east at speed 0.90c relative to the earth. A second spaceship is moving west at speed 0.90c relative to the earth. What is the speed of one spaceship relative to the other?

42 •• Two spaceships are approaching each other. (a) If the speed of each is 0.6c relative to the earth, what is the speed of one relative to the other? (b) If the speed of each relative to the earth is 30,000 m/s (about 100 times the speed of sound), what is the speed of one relative to the other?

43 •• A particle moves with speed $0.8c$ along the x'' axis of frame S'', which moves with speed $0.8c$ along the x' axis relative to frame S'. Frame S' moves with speed $0.8c$ along the x axis relative to frame S. (a) Find the speed of the particle relative to frame S'. (b) Find the speed of the particle relative to frame S.

Energy and Momentum

44 • The approximate total energy of a particle of mass m moving at speed $u \ll c$ is

(a) mc^2.
(b) $\frac{1}{2}mu^2$.
(c) cmu.
(d) $\frac{1}{2}mc^2$.
(e) $\frac{1}{2}cmu$.

45 • Find the ratio of the total energy to the rest energy of a particle of rest mass m_0 moving with speed (a) $0.1c$, (b) $0.5c$, (c) $0.8c$, and (d) $0.99c$.

46 • A proton (rest energy 938 MeV) has a total energy of 1400 MeV. (a) What is its speed? (b) What is its momentum?

47 • How much energy would be required to accelerate a particle of mass m_0 from rest to (a) $0.5c$, (b) $0.9c$, and (c) $0.99c$? Express your answers as multiples of the rest energy.

48 • If the kinetic energy of a particle equals its rest energy, what error is made by using $p = m_0 u$ for its momentum?

49 • What is the energy of a proton whose momentum is $3m_0 c$?

50 •• A particle with momentum of 6 MeV/c has total energy of 8 MeV. (a) Determine the rest mass of the particle. (b) What is the energy of the particle in a reference frame in which its momentum is 4 MeV/c? (c) What are the relative velocities of the two reference frames?

51 •• Show that

$$d\left(\frac{m_0 u}{\sqrt{1 - u^2/c^2}}\right) = m_0\left(1 - \frac{u^2}{c^2}\right)^{-3/2} du$$

52 •• Use Equations 39-21 and 39-25 to derive the equation $E^2 = p^2 c^2 + (m_0 c^2)^2$.

53 •• Use the binomial expansion (Equation 39-27) and Equation 39-28 to show that when $pc \ll m_0 c^2$, the total energy is given approximately by

$$E \approx m_0 c^2 + \frac{p^2}{2m_0}$$

54 •• (a) Show that the speed u of a particle of mass m_0 and total energy E is given by

$$\frac{u}{c} = \left[1 - \frac{(m_0 c^2)^2}{E^2}\right]^{1/2}$$

and that when E is much greater than $m_0 c^2$, this can be approximated by

$$\frac{u}{c} \approx 1 - \frac{(m_0 c^2)^2}{2E^2}$$

Find the speed of an electron with kinetic energy of (b) 0.51 MeV and (c) 10 MeV.

55 •• The rest energy of a proton is about 938 MeV. If its kinetic energy is also 938 MeV, find (a) its momentum and (b) its speed.

56 •• What percentage error is made in using $\frac{1}{2}m_0 u^2$ for the kinetic energy of a particle if its speed is (a) $0.1c$ and (b) $0.9c$?

57 •• The K^0 particle has a rest mass of 497.7 MeV/c^2. It decays into a π^- and π^+, each with rest mass 139.6 MeV/c^2. Following the decay of a K^0, one of the pions is at rest in the laboratory. Determine the kinetic energy of the other pion and of the K^0 prior to the decay.

58 •• The sun radiates energy at the rate of about 4×10^{26} W. Assume that this energy is produced by a reaction whose net result is the fusion of 4 H nuclei to form 1 He nucleus, with the release of 25 MeV for each He nucleus formed. Calculate the sun's loss of rest mass per day.

59 •• Two protons approach each other head on at $0.5c$ relative to reference frame S'. (a) Calculate the total kinetic energy of the two protons as seen in frame S'. (b) Calculate the total kinetic energy of the protons as seen in reference frame S, which is moving with speed $0.5c$ relative to S' such that one of the protons is at rest.

60 •• An antiproton $\bar{p}$ has the same rest energy as a proton. It is created in the reaction $p + p \rightarrow p + p + p + \bar{p}$. In an experiment, protons at rest in the laboratory are bombarded with protons of kinetic energy K_L, which must be great enough so that kinetic energy equal to $2m_0 c^2$ can be converted into the rest energy of the two particles. In the frame of the laboratory, the total kinetic energy cannot be converted into rest energy because of conservation of momentum. However, in the zero-momentum reference frame in which the two initial protons are moving toward each other with equal speed u, the total kinetic energy can be converted into rest energy. (a) Find the speed of each proton u such that the total kinetic energy in the zero-momentum frame is $2m_0 c^2$. (b) Transform to the laboratory's frame in which one proton is at rest, and find the speed u' of the other proton. (c) Show that the kinetic energy of the moving proton in the laboratory's frame is $K_L = 6m_0 c^2$.

61 ••• A particle of rest mass 1 MeV/c^2 and kinetic energy 2 MeV collides with a stationary particle of rest mass 2 MeV/c^2. After the collision, the particles stick together. Find (a) the speed of the first particle before the collision, (b) the total energy of the first particle before the collision, (c) the initial total momentum of the system, (d) the total kinetic energy after the collision, and (e) the rest mass of the system after the collision.

Exploring . . . General Relativity

62 • A set of twins work in an office building. One works on the top floor and the other works in the basement. Considering general relativity, which one will age more quickly?

(a) They will age at the same rate.

(b) The twin who works on the top floor will age more quickly.

(c) The twin who works in the basement will age more quickly.

(d) It depends on the speed of the office building.

(e) None of these is correct.

63 ••• A horizontal turntable rotates with angular speed ω. There is a clock at the center of the turntable and one at a distance r from the center. In an inertial reference frame, the clock at distance r is moving with speed $u = r\omega$. (a) Show that from time dilation according to special relativity, time intervals Δt_0 for the clock at rest and Δt_r for the moving clock are related by

$$\frac{\Delta t_r - \Delta t_0}{\Delta t_0} \approx -\frac{r^2\omega^2}{2c^2} \quad \text{if } r\omega \ll c$$

(b) In a reference frame rotating with the table, both clocks are at rest. Show that the clock at distance r experiences a pseudoforce $F_r = mr\omega^2$ in this accelerated frame and that this is equivalent to a difference in gravitational potential between r and the origin of $\phi_r - \phi_0 = \frac{1}{2}r^2\omega^2$. Use this potential difference in Equation 2 to show that in this frame the difference in time intervals is the same as in the inertial frame.

General Problems

64 • True or false:

(a) The speed of light is the same in all reference frames.

(b) Proper time is the shortest time interval between two events.

(c) Absolute motion can be determined by means of length contraction.

(d) The light-year is a unit of distance.

(e) Simultaneous events must occur at the same place.

(f) If two events are not simultaneous in one frame, they cannot be simultaneous in any other frame.

(g) If two particles are tightly bound together by strong attractive forces, the rest mass of the system is less than the sum of the masses of the individual particles when separated.

65 • An observer sees a system consisting of a mass oscillating on the end of a spring moving past at a speed u and notes that the period of the system is T. Another observer, who is moving with the mass–spring system, also measures its period. The second observer will find a period that is

(a) equal to T.

(b) less than T.

(c) greater than T.

(d) either (a) or (b) depending on whether the system was approaching or receding from the first observer.

(e) There is not sufficient information to answer the question.

66 • The Lorentz transformation for y and z is the same as the classical result: $y = y'$ and $z = z'$. Yet the relativistic velocity transformation does not give the classical result $u_y = u_y'$ and $u_z = u_z'$. Explain.

67 • A spaceship departs from earth for the star Alpha Centauri, which is 4 light-years away. The spaceship travels at 0.75c. How long does it take to get there (a) as measured on earth and (b) as measured by a passenger on the spaceship?

68 • The total energy of a particle is twice its rest energy. (a) Find u/c for the particle. (b) Show that its momentum is given by $p = \sqrt{3}\, m_0 c$.

69 • How fast must a muon travel so that its mean lifetime is 46 μs if its mean lifetime at rest is 2 μs?

70 • A distant galaxy is moving away from the earth with a speed that results in each wavelength received on earth being shifted such that $\lambda' = 2\lambda_0$. Find the speed of the galaxy relative to the earth.

71 • How fast must a meterstick travel relative to you in the direction parallel to the stick so that its length as measured by you is 50 cm?

72 • Show that if V is much less than c, the Doppler shift is given approximately by $\Delta f/f \approx \pm V/c$.

73 •• If a plane flies at a speed of 2000 km/h, for how long must it fly before its clock loses 1 s because of time dilation?

74 • The radius of the orbit of a charged particle in a magnetic field is related to the momentum of the particle by

$$p = BqR \qquad\qquad 39\text{-}41$$

This equation holds classically for $p = mu$ and relativistically for $p = m_0 u/\sqrt{1 - u^2/c^2}$. An electron with kinetic energy of 1.50 MeV moves in a circular orbit perpendicular to a uniform magnetic field $B = 5 \times 10^{-3}$ T. (a) Find the radius of the orbit. (b) What result would you obtain if you used the classical relations $p = mu$ and $K = p^2/2m$?

75 •• Oblivious to economics and politics, Professor Spenditt proposes building a circular accelerator around the earth's circumference using bending magnets that provide a magnetic field of magnitude 1.5 T. (a) What would be the kinetic energy of protons orbiting in this field in a circle of radius R_E? (See Problem 74.) (b) What would be the period of rotation of these protons?

76 •• Frames S and S' are moving relative to each other along the x and x' axis. Observers in the two frames set their clocks to $t = 0$ when the origins coincide. In frame S, event 1 occurs at $x_1 = 1.0\ c\cdot y$ and $t_1 = 1$ y and event 2 occurs at $x_2 = 2.0\ c\cdot y$ and $t_2 = 0.5$ y. These events occur simultaneously in frame S'. (a) Find the magnitude and direction of the velocity of S' relative to S. (b) At what time do both these events occur as measured in S'?

77 •• An interstellar spaceship travels from the earth to a distant star system 12 light-years away (as measured in the earth's frame). The trip takes 15 y as measured on the ship. (a) What is the speed of the ship relative to the earth? (b) When the ship arrives, it sends a signal to the earth. How long after the ship leaves the earth will it be before the earth receives the signal?

78 •• The neutral pion π^0 has a rest mass of 135 MeV/c^2. This particle can be created in a proton–proton collision:

$$p + p \rightarrow p + p + \pi^0$$

Determine the threshold kinetic energy for the creation of a π^0 in a collision of a moving and stationary proton. (See Problem 60.)

79 •• A rocket with a proper length of 1000 m moves in the $+x$ direction at $0.6c$ with respect to an observer on the ground. An astronaut stands at the rear of the rocket and fires a bullet toward the front of the rocket at $0.8c$ relative to the rocket. How long does it take the bullet to reach the front of the rocket (a) as measured in the frame of the rocket, (b) as measured in the frame of the ground, and (c) as measured in the frame of the bullet?

80 ••• In a simple thought experiment, Einstein showed that there is mass associated with electromagnetic radiation. Consider a box of length L and mass M resting on a frictionless surface. At the left wall of the box is a light source that emits radiation of energy E, which is absorbed at the right wall of the box. According to classical electromagnetic theory, this radiation carries momentum of magnitude $p = E/c$ (Equation 32-13). (a) Find the recoil velocity of the box such that momentum is conserved when the light is emitted. (Since p is small and M is large, you may use classical mechanics.) (b) When the light is absorbed at the right wall of the box, the box stops, so the total momentum remains zero. If we neglect the very small velocity of the box, the time it takes for the radiation to travel across the box is $\Delta t = L/c$. Find the distance moved by the box in this time. (c) Show that if the center of mass of the system is to remain at the same place, the radiation must carry mass $m = E/c^2$.

81 ••• A rocket with a proper length of 700 m is moving to the right at a speed of $0.9c$. It has two clocks, one in the nose and one in the tail, that have been synchronized in the frame of the rocket. A clock on the ground and the nose clock on the rocket both read $t = 0$ as they pass. (a) At $t = 0$, what does the tail clock on the rocket read as seen by an observer on the ground? When the tail clock on the rocket passes the ground clock, (b) what does the tail clock read as seen by an observer on the ground, (c) what does the nose clock read as seen by an observer on the ground, and (d) what does the nose clock read as seen by an observer on the rocket? (e) At $t = 1$ h, as measured on the rocket, a light signal is sent from the nose of the rocket to an observer standing by the ground clock. What does the ground clock read when the observer receives this signal? (f) When the observer on the ground receives the signal, he sends a return signal to the nose of the rocket. When is this signal received at the nose of the rocket as seen on the rocket?

82 ••• An observer in frame S standing at the origin observes two flashes of colored light separated spatially by $\Delta x = 2400$ m. A blue flash occurs first, followed by a red flash 5 μs later. An observer in S' moving along the x axis at speed V relative to S also observes the flashes 5 μs apart and with a separation of 2400 m, but the red flash is observed first. Find the magnitude and direction of V.

83 ••• Reference frame S' is moving along the x' axis at $0.6c$ relative to frame S. A particle that is originally at $x' = 10$ m at $t_1' = 0$ is suddenly accelerated and then moves at a constant speed of $c/3$ in the $-x'$ direction until time $t_2' = 60$ m/c, when it is suddenly brought to rest. As observed in frame S, find (a) the speed of the particle, (b) the distance and direction the particle traveled from t_1' to t_2', and (c) the time the particle traveled.

84 ••• In reference frame S the acceleration of a particle is $\vec{a} = a_x\hat{i} + a_y\hat{j} + a_z\hat{k}$. Derive expressions for the acceleration components a_x', a_y', and a_z' of the particle in reference frame S' that is moving relative to S in the x direction with velocity V.

85 ••• When a projectile particle with kinetic energy greater than the threshold kinetic energy K_{th} strikes a stationary target particle, one or more particles may be created in the inelastic collision. Show that the threshold kinetic energy of the projectile is given by

$$K_{th} = \frac{(\Sigma\, m_{in} + \Sigma\, m_{fin})(\Sigma\, m_{fin} - \Sigma\, m_{in})c^2}{2m_{target}}$$

Here Σm_{in} is the sum of the rest masses of the projectile and target particles, $\Sigma\, m_{fin}$ is the sum of the rest masses of the final particles, and m_{target} is the rest mass of the target particle. Use this expression to determine the threshold kinetic energy of protons incident on a stationary proton target for the production of a proton–antiproton pair; compare your result with that of Problem 60.

86 ••• A particle of rest mass M_0 decays into two identical particles of rest mass m_0, where $m_0 = 0.3M_0$. Prior to the decay, the particle of rest mass M_0 has an energy of $4M_0c^2$ in the laboratory. The velocities of the decay products are along the direction of motion of M_0. Find the velocities of the decay products in the laboratory.

87 ••• A stick of proper length L_p makes an angle θ with the x axis in frame S. Show that the angle θ' made with the x' axis in frame S', which is moving along the $+x$ axis with speed V, is given by $\tan\theta' = \gamma\tan\theta$ and that the length of the stick in S' is

$$L' = L_p\left(\frac{1}{\gamma^2}\cos^2\theta + \sin^2\theta\right)^{1/2}$$

88 ••• Show that if a particle moves at an angle θ with the x axis with speed u in frame S, it moves at an angle θ' with the x' axis in S' given by

$$\tan\theta' = \frac{\sin\theta}{\gamma(\cos\theta - V/u)}$$

89 ••• For the special case of a particle moving with speed u along the y axis in frame S, show that its momentum and energy in frame S' are related to its momentum and energy in S by the transformation equations

$$P_x' = \gamma\left(p_x - \frac{VE}{c^2}\right), \quad p_y' = p_y, \quad p_z' = p_z$$

$$\frac{E'}{c} = \gamma\left(\frac{E}{c} - \frac{Vp_x}{c^2}\right)$$

Compare these equations with the Lorentz transformation for x', y', z', and t'. These equations show that the quantities p_x, p_y, p_z, and E/c transform in the same way as do x, y, z, and ct.

90 ••• The equation for the spherical wavefront of a light pulse that begins at the origin at time $t = 0$ is $x^2 + y^2 + z^2 -$

$(ct)^2 = 0$. Using the Lorentz transformation, show that such a light pulse also has a spherical wavefront in frame S' by showing that $x'^2 + y'^2 + z'^2 - (ct')^2 = 0$ in S'.

91 ••• In Problem 90, you showed that the quantity $x^2 + y^2 + z^2 - (ct)^2$ has the same value (0) in both S and S'. Such a quantity is called an *invariant*. From the results of Problem 89, the quantity $p_x^2 + p_y^2 + p_z^2 - (E/c)^2$ must also be an invariant. Show that this quantity has the value $-m_0 c^2$ in both the S and S' reference frames.

92 ••• Two identical particles of rest mass m_0 are each moving toward the other with speed u in frame S. The particles collide inelastically with a spring that locks shut (Figure 39-14) and come to rest in S, and their initial kinetic energy is transformed into potential energy. In this problem you are going to show that the conservation of momentum in reference frame S', in which one of the particles is initially at rest, requires that the total rest mass of the system after the collision be $2m_0/\sqrt{1 - u^2/c^2}$. (a) Show that the speed of the particle not at rest in frame S' is $u' = 2u/(1 + u^2/c^2)$ and use this result to show that

$$\sqrt{1 - \frac{u'^2}{c^2}} = \frac{1 - u^2/c^2}{1 + u^2/c^2}$$

(b) Show that the initial momentum in frame S' is $p' = 2m_0 u/(1 - u^2/c^2)$. (c) After the collision, the composite particle moves with speed u in S' (since it is at rest in S). Write the total momentum after the collision in terms of the final rest mass M_0, and show that the conservation of momentum implies that $M_0 = 2m_0/\sqrt{1 - u^2/c^2}$. (d) Show that the total energy is conserved in each reference frame.

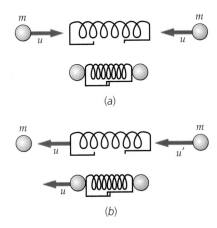

(a)

(b)

Figure 39-14 An inelastic collision between two identical objects (a) in the zero-momentum reference frame S and (b) in frame S', which is moving to the right with speed $V = u$ relative to frame S such that one of the particles is initially at rest. The spring, which is assumed to be massless, is merely a device for visualizing the storage of potential energy.

CHAPTER 40

Nuclear Physics

A nuclear power plant in Germany. The fission reactor core is housed in a hemispherical containment structure (center). Two large cooling towers are to its left.

To the chemist, the atomic nucleus is essentially a point charge that contains most of the mass of the atom. It plays a negligible role in the structure of atoms and molecules. When examined closely, the nucleus is found to contain protons and neutrons, whose interactions have played important roles in our everyday life as well as in the history and structure of the universe. The fission of very heavy nuclei such as uranium is a major source of power today, while the fusion of very light nuclei is the energy source that powers the stars, including our sun, and may hold the key to our energy needs of the future.

40-1   Properties of Nuclei

The nucleus of an atom contains just two kinds of particles: protons and neutrons,* which have approximately the same mass (the neutron is about 0.2% more massive). The proton has a charge of $+e$ and the neutron is uncharged. The number of protons, Z, is the atomic number of the atom, which also equals the number of electrons in the atom. The number of neutrons, N, is ap-

---

* The normal hydrogen nucleus contains a single proton.

proximately equal to Z for light nuclei, and for heavier nuclei is increasingly greater than Z. The total number of nucleons* $A = N + Z$ is called the mass number of the nucleus. A particular nuclear species is called a **nuclide.** Two or more nuclides with the same atomic number Z but different N and A numbers are called **isotopes.** A particular nuclide is designated by its atomic symbol (H for hydrogen, He for helium, etc.) with the mass number A as a pre-superscript. The lightest element, hydrogen, has three isotopes: ordinary hydrogen, ^{1}H, whose nucleus is just a single proton; deuterium, ^{2}H, whose nucleus contains one proton and one neutron; and tritium, ^{3}H, whose nucleus contains one proton and two neutrons. Although the mass of the deuterium atom is about twice that of the hydrogen atom and that of the tritium atom about three times that of hydrogen, these three atoms have nearly identical chemical properties because they each have one electron. On the average, there are about three stable isotopes for each atom, although some atoms have only one while others have five or six. The most common isotope of the second lightest atom, helium, is ^{4}He. The ^{4}He nucleus is also known as an α particle. Another isotope of helium is ^{3}He.

Inside the nucleus, the nucleons exert a strong attractive force on their nearby neighbors. This force, called the **strong nuclear force** or the **hadronic force,** is much stronger than the electrostatic force of repulsion between the protons and is very much stronger than the gravitational forces between the nucleons. (Gravity is so weak that it can always be neglected in nuclear physics.) The strong nuclear force is roughly the same between two neutrons, two protons, or a neutron and a proton. Two protons, of course, also exert a repulsive electrostatic force on each other due to their charges, which tends to weaken the attraction between them somewhat. The strong nuclear force decreases rapidly with distance, and it is negligible when two nucleons are more than a few femtometers apart.

Size and Shape

The size and shape of the nucleus can be determined by bombarding it with high-energy particles and observing the scattering. The results depend somewhat on the kind of experiment. For example, a scattering experiment using electrons measures the charge distribution of the nucleus, whereas one using neutrons determines the region of influence of the strong nuclear force. Despite these differences, a wide variety of experiments suggest that most nuclei are approximately spherical, with radii given approximately by

$$R = R_0 A^{1/3}$$

40-1

Nuclear radius

where R_0 is about 1.5 fm. The fact that the radius of a spherical nucleus is proportional to $A^{1/3}$ implies that the volume of the nucleus is proportional to A. Since the mass of the nucleus is also approximately proportional to A, the densities of all nuclei are approximately the same. This is analogous to a drop of liquid, which also has constant density independent of its size. The **liquid-drop model** of the nucleus has proved quite successful in explaining nuclear behavior, especially the fission of heavy nuclei.

N and Z Numbers

For light nuclei, the greatest stability is achieved when the numbers of protons and neutrons are approximately equal, $N \approx Z$. For heavier nuclei, insta-

---

* The word nucleon refers to either a neutron or a proton.

bility caused by the electrostatic repulsion between the protons is minimized when there are more neutrons than protons. We can see this by looking at the N and Z numbers for the most abundant isotopes of some representative elements: for $^{16}_{8}O$, $N = 8$ and $Z = 8$; for $^{40}_{20}Ca$, $N = 20$ and $Z = 20$; for $^{56}_{26}Fe$, $N = 30$ and $Z = 26$; for $^{207}_{82}Pb$, $N = 125$ and $Z = 82$; and for $^{238}_{92}U$, $N = 146$ and $Z = 92$. (The atomic number Z has been included here as a presubscript of the atomic symbol for emphasis. It is not actually needed because the atomic number is implied by the atomic symbol.)

Figure 40-1 shows a plot of N versus Z for the known stable nuclei. The curve follows the straight line $N = Z$ for small values of N and Z. We can understand this tendency for N and Z to be equal by considering the total energy of A particles in a one-dimensional box. Figure 40-2 shows the energy levels for eight neutrons and for four neutrons and four protons. Because of the exclusion principle, only two identical particles (with opposite spins) can be in the same space state. Since protons and neutrons are not identical, we can put two each in a state as in Figure 40-2b. Thus, the total energy for four protons and four neutrons is less than that for eight neutrons (or eight protons) as in Figure 40-2a. When the Coulomb energy of repulsion, which is proportional to Z^2, is included, this result changes somewhat. For large values of A and Z, the total energy may be increased less by adding two neutrons than by adding one neutron and one proton because of the electrostatic repulsion involved in the latter case. This explains why $N > Z$ for the heavier nuclei.

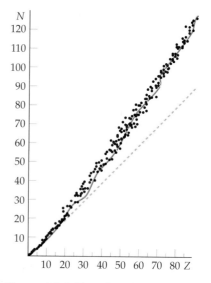

Figure 40-1 Plot of number of neutrons N versus number of protons Z for the stable nuclides. The dashed line is $N = Z$.

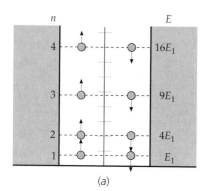

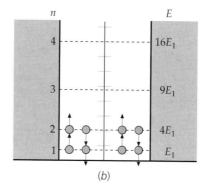

(a)   (b)

Figure 40-2 (a) Eight neutrons in a one-dimensional box. In accordance with the exclusion principle, only two neutrons (with opposite spins) can be in a given energy level. (b) Four neutrons and four protons in a one-dimensional box. Because protons and neutrons are not identical particles, two of each can be in each energy level. The total energy is much less for this case than for that in (a).

Mass and Binding Energy

The mass of a nucleus is less than the mass of its parts by $\Delta E/c^2$, where ΔE is the binding energy and c is the speed of light. When two or more nucleons fuse together to form a nucleus, the total mass decreases and energy is given off. Conversely, to break up a nucleus into its parts, energy must be put into the system to produce the increase in rest mass.

Atomic and nuclear masses are often given in unified mass units (u), defined as one-twelfth the mass of the neutral ^{12}C atom. The rest energy of one unified mass unit is

$$(1 \text{ u})c^2 = 931.5 \text{ MeV} \qquad\qquad 40\text{-}2$$

Consider ^4He, for example, which consists of two protons and two neutrons. The mass of an atom can be accurately measured in a mass spectrometer. The mass of the ^4He atom is 4.002 603 u. This includes the masses of the two electrons in the atom. The mass of the ^{1}H atom is 1.007 825 u, and that of the neutron is 1.008 665 u. The sum of the masses of two ^{1}H atoms plus two neutrons is $2(1.007\,825\ \text{u}) + 2(1.008\,665\ \text{u}) = 4.032\,98\ \text{u}$, which is greater than the mass of the ^{4}He atom by 0.030 377 u.* We can find the binding energy of the ^{4}He nucleus from this mass difference of 0.030 377 u by using the mass conversion factor $(1\ \text{u})c^2 = 931.5\ \text{MeV}$ from Equation 40-2. Then

$$(0.030\,377\ \text{u})c^2 = (0.030\,377\ \text{u})c^2 \times \frac{931.5\ \text{MeV}/c^2}{1\ \text{u}}$$

$$= 28.30\ \text{MeV}$$

The total binding energy of ^{4}He is thus 28.3 MeV. In general, the binding energy of a nucleus of an atom of atomic mass M_A containing Z protons and N neutrons is found by calculating the difference between the mass of the parts and the mass of the nucleus and then multiplying by c^2:

$$E_b = (ZM_H + Nm_n - M_A)c^2 \qquad \text{40-3}$$

Total nuclear binding energy

where M_H is the mass of the ^{1}H atom and m_n that of the neutron. (Note that the mass of the Z electrons in the term ZM_H is canceled by the mass of the Z electrons in the term M_A.) The atomic masses of the neutron and of some selected isotopes are listed in Table 40-1.

---

* Note that by using the masses of two ^{1}H atoms rather than two protons, the masses of the electrons in the atom are accounted for. We do this because it is atomic masses, not nuclear masses, that are measured directly and listed in mass tables.

Table 40-1

Atomic Masses of the Neutron and Selected Isotopes

Element	Symbol	Z	Atomic mass, u
Neutron	n	0	1.008 665
Hydrogen	^{1}H	1	1.007 825
Deuterium	^{2}H or D	1	2.014 102
Tritium	^{3}H or T	1	3.016 050
Helium	^{3}He	2	3.016 030
	^{4}He	2	4.002 603
Lithium	^{6}Li	3	6.015 125
	^{7}Li	3	7.016 004
Boron	^{10}B	5	10.012 939
Carbon	^{12}C	6	12.000 000
	^{13}C	6	13.003 354
	^{14}C	6	14.003 242
Nitrogen	^{13}N	7	13.005 738
	^{14}N	7	14.003 074
Oxygen	^{16}O	8	15.994 915
Sodium	^{23}Na	11	22.989 771
Potassium	^{39}K	19	38.963 710
Iron	^{56}Fe	26	55.939 395
Copper	^{63}Cu	29	62.929 592
Silver	^{107}Ag	47	106.905 094
Gold	^{197}Au	79	196.966 541
Lead	^{208}Pb	82	207.976 650
Polonium	^{212}Po	84	211.989 629
Radon	^{222}Rn	86	222.017 531
Radium	^{226}Ra	88	226.025 360
Uranium	^{238}U	92	238.048 608
Plutonium	^{242}Pu	94	242.058 725

Example 40-1

Find the binding energy of the last neutron in ^{4}He.

Picture the Problem The binding energy is c^2 times the difference in mass of ^{3}He plus a neutron and ^{4}He. We find these masses from Table 40-1 and convert to energy using Equation 40-3.

1. Add the mass of the neutron to that of ^{3}He:

$$m_{^3\text{He}} + m_n = 3.016\,030\ \text{u} + 1.008\,665\ \text{u} = 4.024\,695\ \text{u}$$

2. Subtract the mass of ^{4}He from the result:

$$(m_{^3\text{He}} + m_n) - m_{^4\text{He}} = 4.024\,695\ \text{u} - 4.002\,603\ \text{u}$$

$$= 0.022\,092\ \text{u}$$

3. Multiply this mass difference by c^2 and convert to MeV:

$$E_b = (\Delta m)c^2 = (0.022\,092\ \text{u})c^2 \times \frac{931.5\ \text{MeV}/c^2}{1\ \text{u}}$$

$$= 20.58\ \text{MeV}$$

Figure 40-3 shows the binding energy per nucleon E_b/A versus A. The mean value is about 8.3 MeV. The flatness of this curve for $A > 50$ shows that E_b is approximately proportional to A. This indicates that there is saturation of nuclear forces in the nucleus as would be the case if each nucleon were attracted only to its nearest neighbors. Such a situation also leads to a constant nuclear density consistent with the measurements of the radius. If, for example, there were no saturation and each nucleon bonded to each other nucleon, there would be $A - 1$ bonds for each nucleon and a total of $A(A - 1)$ bonds altogether. The total binding energy, which is a measure of the energy needed to break all these bonds, would then be proportional to $A(A - 1)$, and E_b/A would not be approximately constant. The steep rise in the curve for low A is due to the increase in the number of nearest neighbors and therefore to the increased number of bonds per nucleon. The gradual decrease at high A is due to the Coulomb repulsion of the protons, which increases as Z^2 and decreases the binding energy. Eventually, for very large A this Coulomb repulsion becomes so great that a nucleus with A greater than about 300 is unstable and undergoes spontaneous fission.

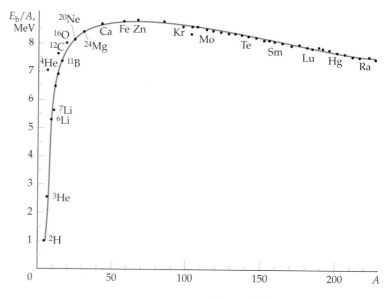

Figure 40-3 The binding energy per nucleon versus the mass number A. For nuclei with values of A greater than about 50, the curve is approximately flat, indicating that the total binding energy is approximately proportional to A.

40-2   Radioactivity

Many nuclei are radioactive; that is, they decay into other nuclei by the emission of particles, such as photons, electrons, neutrons, or α particles. The terms α decay, β decay, and γ decay were used before it was known that α particles are ^{4}He nuclei, β particles are either electrons (β^-) or positrons* (β^+), and γ rays are photons. The rate of decay is not constant over time, but decreases exponentially. *This exponential time dependence is characteristic of all radioactivity and indicates that radioactive decay is a statistical process.* Because each nucleus is well shielded from others by the atomic electrons, pressure and temperature changes have little or no effect on the rate of radioactive decay or other nuclear properties.

Let N be the number of radioactive nuclei at some time t. If the decay of an individual nucleus is a random event, we expect the number of nuclei that decay in some time interval dt to be proportional to N and to dt. Because of these decays, the number N will decrease. The change in N is given by

$$dN = -\lambda N \, dt \qquad\qquad 40\text{-}4$$

where λ is a constant of proportionality called the **decay constant.** The rate of change of N, dN/dt, is proportional to N. This is characteristic of exponential decay. To solve Equation 40-4 for N, we first divide each side by N, thus separating the variables N and t:

$$\frac{dN}{N} = -\lambda \, dt$$

---

* The positron is identical to an electron except it has a charge of $+e$.

Integrating, we obtain

$$\ln N = -\lambda t + C \qquad\qquad \textbf{40-5}$$

where C is some constant of integration. Taking the exponential of each side we obtain

$$N = e^{-\lambda t + C} = e^{C}e^{-\lambda t}$$

or

$$N = N_0 e^{-\lambda t} \qquad\qquad \textbf{40-6}$$

where $N_0 = e^C$ is the number of nuclei at $t = 0$. The number of radioactive decays per second is called the decay rate R:

$$R = -\frac{dN}{dt} = \lambda N = \lambda N_0 e^{-\lambda t} = R_0 e^{-\lambda t} \qquad\qquad \textbf{40-7}$$

Decay rate

where

$$R_0 = \lambda N_0 \qquad\qquad \textbf{40-8}$$

is the rate of decay at time $t = 0$. The decay rate R is the quantity that is determined experimentally.

The average or **mean lifetime** τ is the reciprocal of the decay constant:

$$\tau = \frac{1}{\lambda} \qquad\qquad \textbf{40-9}$$

(See Problem 33.) The mean lifetime is analogous to the time constant in the exponential decrease in the charge on a capacitor in an RC circuit that we discussed in Section 26-6. After a time equal to the mean lifetime, the number of radioactive nuclei and the decay rate have each decreased to 37% of their original values. The **half-life** $t_{1/2}$ is defined as the time it takes for the number of nuclei and the decay rate to decrease by half. Setting $t = t_{1/2}$ and $N = N_0/2$ in Equation 40-6 gives

$$\frac{N_0}{2} = N_0 e^{-\lambda t_{1/2}} \qquad\qquad \textbf{40-10}$$

or

$$e^{+\lambda t_{1/2}} = 2$$

Solving for $t_{1/2}$ gives

$$t_{1/2} = \frac{\ln 2}{\lambda} = \frac{0.693}{\lambda} = 0.693\tau \qquad\qquad \textbf{40-11}$$

Figure 40-4 shows a plot of N versus t. If we multiply the numbers on the N axis by λ, this graph becomes a plot of R versus t. After each time interval of one half-life, the number of nuclei left and the decay rate have decreased to half of their previous values. For example, if the decay rate is R_0 initially, it will be $\frac{1}{2}R_0$ after one half-life, $(\frac{1}{2})(\frac{1}{2})R_0$ after two half-lives, and so forth. After n half-lives, the decay rate will be

$$R = (\tfrac{1}{2})^n R_0 \qquad\qquad \textbf{40-12}$$

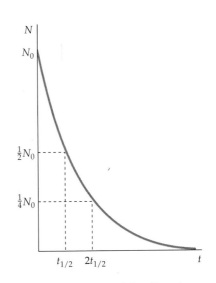

Figure 40-4 Exponential radioactive decay. After each half-life $t_{1/2}$, the number of nuclei remaining has decreased by one-half. The decay rate $R = \lambda N$ has the same time dependence.

The half-lives of radioactive nuclei vary from very small times (less than 1 μs) to very large times (up to 10^{16} y).

Example 40-2

A radioactive source has a half-life of 1 min. At time $t = 0$, it is placed near a detector, and the counting rate (the number of decay particles detected per unit time) is observed to be 2000 counts/s. Find the counting rate at times $t = 1$ min, 2 min, 3 min, and 10 min.

Picture the Problem The counting rate decreases by a factor of 2 each minute.

1. Since the half-life is 1 min, the counting rate will be half as great at $t = 1$ min as at $t = 0$:

$$r_1 = \tfrac{1}{2} r_0 = \tfrac{1}{2}(2000 \text{ counts/s}) = 1000 \text{ counts/s at 1 min}$$

2. At $t = 2$ min, the rate is half that at 1 min. It decreases by one-half each minute:

$$r_2 = \tfrac{1}{2} r_1 = \tfrac{1}{2}(1000 \text{ counts/s}) = 500 \text{ counts/s at 2 min}$$
$$r_3 = \tfrac{1}{2} r_2 = \tfrac{1}{2}(500 \text{ counts/s}) = 250 \text{ counts/s at 3 min}$$

3. At $t = 10$ min, the rate will be $\left(\tfrac{1}{2}\right)^{10}$ times the initial rate:

$$r_{10} = \left(\tfrac{1}{2}\right)^{10} r_0 = \left(\tfrac{1}{2}\right)^{10} (2000 \text{ counts/s})$$
$$= 1.95 \text{ counts/s} \approx 2 \text{ counts/s at 10 min}$$

Example 40-3

If the detection efficiency in Example 40-2 is 20%, (a) how many radioactive nuclei are there at time $t = 0$? (b) At time $t = 1$ min? (c) How many nuclei decay in the first minute?

Picture the Problem The detection efficiency depends on the probability that a radioactive decay particle will enter the detector and the probability that upon entering the detector it will produce a count. If the efficiency is 20%, the decay rate must be 5 times the counting rate.

(a)1. The number of radioactive nuclei is related to the decay rate R, and the decay constant λ:

$$R = \lambda N$$

2. The decay constant is related to the half-life:

$$\lambda = \frac{0.693}{t_{1/2}} = \frac{0.693}{1 \text{ min}}$$

3. Calculate the decay rate from the counting rate:

$$R_0 = 5 \times 2000 \text{ counts/s} = 10^4 \text{ s}^{-1}$$

4. Substitute to calculate N_0 at $t = 0$:

$$N_0 = \frac{R_0}{\lambda} = \frac{10{,}000 \text{ s}^{-1}}{0.693 \text{ min}^{-1}} \times \frac{60 \text{ s}}{1 \text{ min}} = 8.66 \times 10^5$$

(b) At time $t = 1$ min $= t_{1/2}$, there are half as many radioactive nuclei as at $t = 0$:

$$N_1 = \tfrac{1}{2}(8.66 \times 10^5) = 4.33 \times 10^5$$

(c) The number of nuclei that decay in the first minute is $N_0 - N_1$:

$$\Delta N = N_0 - N_1 = 8.66 \times 10^5 - 4.33 \times 10^5 = 4.33 \times 10^5$$

The SI unit of radioactive decay is the **becquerel** (Bq), which is defined as one decay per second:

$$1 \text{ Bq} = 1 \text{ decay/s} \qquad\qquad 40\text{-}13$$

A historical unit that applies to all types of radioactivity is the **curie** (Ci), which is defined as

$$1 \text{ Ci} = 3.7 \times 10^{10} \text{ decays/s} = 3.7 \times 10^{10} \text{ Bq} \qquad\qquad 40\text{-}14$$

The curie is the rate at which radiation is emitted by 1 g of radium. Since this is a very large unit, the millicurie (mCi) or microcurie (μCi) are often used.

Beta Decay

Beta decay occurs in nuclei that have too many or too few neutrons for stability. In β decay, A remains the same while Z either increases by 1 (β^- decay) or decreases by 1 (β^+ decay).

The simplest example of β decay is the decay of the free neutron into a proton plus an electron. (The half-life of a free neutron is about 10.8 min.) The energy of decay is 0.782 MeV, which is the difference between the rest energy of the neutron and that of the proton plus electron. More generally, in β^- decay, a nucleus of mass number A and atomic number Z decays into a nucleus, referred to as the **daughter nucleus,** of mass number A and atomic number $Z' = Z + 1$ with the emission of an electron. If the decay energy were shared by only the daughter nucleus and the emitted electron, the energy of the electron would be uniquely determined by the conservation of energy and momentum. Experimentally, however, the energies of the electrons emitted in the β^- decay of a nucleus are observed to vary from zero to the maximum energy available. A typical energy spectrum for these electrons is shown in Figure 40-5.

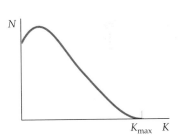

Figure 40-5 Number of electrons emitted in β^- decay versus kinetic energy. The fact that all the electrons do not have the same energy K_{max} suggests that another particle that shares the energy available for decay is emitted.

To explain the apparent nonconservation of energy in β decay, W. Pauli in 1930 suggested that a third particle, which he called the **neutrino,** is also emitted. Because the measured maximum energy of the emitted electrons is equal to the total available for the decay, the rest energy and therefore the mass of the neutrino was assumed to be zero. (It is now believed that the mass of the neutrino is very small but not zero.) In 1948, measurements of the momenta of the emitted electron and the recoiling nucleus showed that the neutrino was also needed for the conservation of linear momentum in β decay. The neutrino was first observed experimentally in 1957. It is now known that there are at least three kinds of neutrinos, one (ν_e) associated with electrons, one (ν_μ) associated with muons, and one (ν_τ) not yet observed experimentally, associated with the newly discovered tau particle, τ. Moreover, each neutrino has an antiparticle, written $\bar\nu_e$, $\bar\nu_\mu$, and $\bar\nu_\tau$. It is the electron antineutrino that is emitted in the decay of a neutron, which is written*

$$n \rightarrow p + \beta^- + \bar\nu_e \qquad\qquad 40\text{-}15$$

In β^+ decay, a proton changes into a neutron with the emission of a positron (and a neutrino). A free proton cannot decay by positron emission because of conservation of energy (the rest mass of the neutron plus the positron is greater than that of the proton), but because of binding-energy effects, a proton inside a nucleus can decay. A typical β^+ decay is

$$^{13}_{7}\text{N} \rightarrow {}^{13}_{6}\text{C} + \beta^+ + \nu_e \qquad\qquad 40\text{-}16$$

The electrons or positrons emitted in β decay do not exist inside the nucleus. They are created in the process of decay, just as photons are created when an atom makes a transition from a higher to a lower energy state.

---

* This reaction is also written $n \rightarrow p + e^- + \bar\nu_e$.

An important example of β decay is that of ^{14}C, which is used in **radioactive carbon dating:**

$$^{14}C \rightarrow {}^{14}N + \beta^- + \bar{\nu}_e \qquad\qquad 40\text{-}17$$

The half-life for this decay is 5730 y. The radioactive isotope ^{14}C is produced in the upper atmosphere in nuclear reactions caused by cosmic rays. The chemical behavior of carbon atoms with ^{14}C nuclei is the same as those with ordinary ^{12}C nuclei. For example, atoms with these nuclei combine with oxygen to form CO_2 molecules. Since living organisms continually exchange CO_2 with the atmosphere, the ratio of ^{14}C to ^{12}C in a living organism is the same as the equilibrium ratio in the atmosphere, which is about 1.3×10^{-12}. After an organism dies, it no longer absorbs ^{14}C from the atmosphere, so the ratio of ^{14}C to ^{12}C continually decreases due to the radioactive decay of ^{14}C. The number of ^{14}C decays per minute per gram of carbon in a living organism can be calculated from the known half-life of ^{14}C and the number of ^{14}C nuclei in a gram of carbon. The result is that there are about 15.0 decays per minute per gram of carbon in a living organism. Using this result and the measured number of decays per minute per gram of carbon in a nonliving sample of bone, wood, or other object containing carbon, we can determine the age of the sample. For example, if the measured rate were 7.5 decays per minute per gram, the sample would be one half-life = 5730 years old.

Example 40-4

A bone containing 200 g of carbon has a β-decay rate of 400 decays/min. How old is the bone?

Picture the Problem We first obtain a rough estimate of the age of the bone. If the bone were from a living organism, we would expect the decay rate to be (15 decays/min·g)(200 g) = 3000 decays/min. Since 400/3000 is roughly 1/8 (actually 1/7.5), the sample must be about three half-lives old, which is about 3(5730) y = 17,190 y. To find the age of the bone more accurately, we note that after n half-lives, the decay rate will have decreased by a factor of $(\frac{1}{2})^n$.

1. Write the decay rate after n half-lives in terms of the initial decay rate:

$$R_n = (\tfrac{1}{2})^n R_0$$

2. Calculate the initial decay rate for 200 g:

$$R_0 = (15 \text{ decays/min·g})(200 \text{ g}) = 3000 \text{ decays/min}$$

3. Substitute the measured decay rate and simplify:

$$R_n = (\tfrac{1}{2})^n \, 3000 \text{ decays/min} = 400 \text{ decays/min}$$

$$(\tfrac{1}{2})^n = \frac{400}{3000}$$

$$2^n = \frac{3000}{400} = 7.5$$

4. We solve for n by taking the logarithm of each side:

$$n \ln 2 = \ln 7.5$$

$$n = \frac{\ln 7.5}{\ln 2} = 2.91$$

5. The age of the bone is $n t_{1/2}$:

$$t = n t_{1/2} = 2.91(5730 \text{ y}) = 1.67 \times 10^4 \text{ y}$$

Gamma Decay

In γ decay a nucleus in an excited state decays to a lower-energy state by the emission of a photon. This is the nuclear counterpart of spontaneous emission of photons by atoms and molecules. Unlike β or α decay, the radioactive nucleus remains the same nucleus after γ decay. Since the spacing of the nuclear energy levels is of the order of 1 MeV (as compared with spacing of the order of 1 eV in atoms), the wavelengths of the emitted photons are of the order of 1 pm (1 pm = 10^{-12} m):

$$\lambda = \frac{hc}{E} \approx \frac{1240 \text{ eV·nm}}{1 \text{ MeV}} = 0.00124 \text{ nm} = 1.24 \text{ pm}$$

The mean lifetime for γ decay is often very short. Usually it is observed only because it follows either α or β decay. For example, if a radioactive parent nucleus decays by β decay to an excited state of the daughter nucleus, the daughter nucleus then decays to its ground state by γ emission. Direct measurements of mean lifetimes as short as about 10^{-11} s are possible. Measurements of mean lifetimes shorter than 10^{-11} s are difficult, but they can sometimes be made by indirect methods.

A few γ emitters have very long lifetimes, of the order of hours. Nuclear energy states that have such long lifetimes are called **metastable states.**

Alpha Decay

All very heavy nuclei ($Z > 83$) are theoretically unstable to α decay because the mass of the original radioactive nucleus is greater than the sum of the masses of the decay products—an α particle and the daughter nucleus. Consider the decay of ^{232}Th ($Z = 90$) into ^{228}Ra ($Z = 88$) plus an α particle. This is written as

$$^{232}\text{Th} \rightarrow {}^{228}\text{Ra} + \alpha = {}^{228}\text{Ra} + {}^{4}\text{He} \qquad\qquad 40\text{-}18$$

The mass of the ^{232}Th atom is 232.038 124 u. The mass of the daughter atom ^{228}Ra is 228.031 139 u. Adding 4.002 603 u to this for the mass of ^{4}He, we get 232.033 742 u for the total mass of the decay products. This is less than the mass of ^{232}Th by 0.004 382 u, which multiplied by 931.5 MeV/c^2 gives 4.08 MeV/c^2 for the excess rest mass of ^{232}Th over that of the decay products. The isotope ^{232}Th is therefore theoretically unstable to α decay. This decay does in fact occur in nature with the emission of an α particle of kinetic energy 4.08 MeV. (The kinetic energy of the α particle is actually somewhat less than 4.08 MeV because some of the decay energy is shared by the recoiling ^{228}Ra nucleus.)

In general, when a nucleus emits an α particle, both N and Z decrease by 2 and A decreases by 4. The daughter of a radioactive nucleus is often itself radioactive and decays by either α or β decay or both. If the original nucleus has a mass number A that is 4 times an integer, the daughter nucleus and all those in the chain will also have mass numbers equal to 4 times an integer. Similarly, if the mass number of the original nucleus is $4n + 1$, where n is an integer, all the nuclei in the decay chain will have mass numbers given by $4n + 1$, with n decreasing by one at each decay. We can see, therefore, that there are four possible α-decay chains, depending on whether A equals $4n$, $4n + 1$, $4n + 2$, or $4n + 3$, where n is an integer. All but one of these decay chains are found on Earth. The $4n + 1$ series is not found because its longest lived member (other than the stable end product ^{209}Bi) is ^{237}Np, which has a half-life of only 2×10^6 y. Because this is much less than the age of the earth, this series has disappeared.

Figure 40-6 shows the thorium series, for which $A = 4n$. It begins with an α decay from ^{232}Th to ^{228}Ra. The daughter nuclide of an α decay is on the left or neutron-rich side of the stability curve (the dashed line in the figure), so it often decays by β^- decay. In the thorium series, ^{228}Ra decays by β^- decay to ^{228}Ac, which in turn decays by β^- decay to ^{228}Th. There are then four α decays to ^{212}Pb, which decays by β^- decay to ^{212}Bi. The series branches at ^{212}Bi, which decays either by α decay to ^{208}Tl or by β^- decay to ^{212}Po. The branches meet at the stable lead isotope ^{208}Pb.

The energies of α particles from natural radioactive sources range from about 4 to 7 MeV, and the half-lives of the sources range from about 10^{-5} s to 10^{10} y. In general, the smaller the energy of the emitted α particle, the longer the half-life. As we discussed in Section 36-4, the enormous variation in half-lives was explained by George Gamow in 1928. He considered α decay to be a process in which an α particle is first formed inside a nucleus and then tunnels through the Coulomb barrier (Figure 40-7). A slight increase in the energy of the α particle reduces the relative height $U - E$ of the barrier and also the thickness. Because the probability of penetration is so sensitive to the relative height and thickness of the barrier, a small increase in E leads to a large increase in the probability of barrier penetration and therefore to a shorter lifetime. Gamow was able to derive an expression for the half-life as a function of E that is in excellent agreement with experimental results.

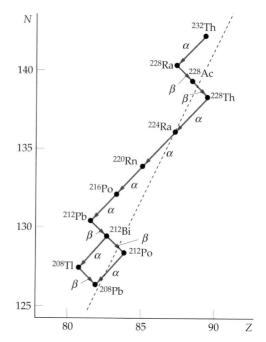

Figure 40-6 The thorium (4n) α decay series. The dashed line is the curve of stability.

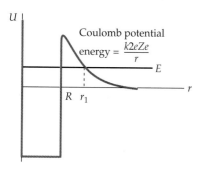

Figure 40-7 A model of the potential energy for an α particle and a nucleus. The strong attractive nuclear force that exists for values of r less than the nuclear radius R is indicated by the potential well. Outside the nucleus, the nuclear force is negligible, and the potential energy is given by Coulomb's law $U = +k2eZe/r$, where Ze is the nuclear charge and $2e$ is the charge of the α particle. The energy E is the kinetic energy of the α particle when it is far away from the nucleus. A small increase in E reduces the relative height of the barrier and also its thickness, leading to a much greater chance of penetration. An increase in the energy of the emitted α particles by a factor of 2 results in a reduction of the half-life by a factor of more than 10^{20}.

40-3   **Nuclear Reactions**

Information about nuclei is typically obtained by bombarding them with various particles and observing the results. Although the first experiments of this type were limited by the need to use naturally occurring radiation, they produced many important discoveries. In 1932 J. D. Cockcroft and E. T. S. Walton succeeded in producing the reaction

$$p + {}^7\text{Li} \rightarrow {}^8\text{Be} \rightarrow {}^4\text{He} + {}^4\text{He}$$

using artificially accelerated protons. At about the same time, the Van de Graaff electrostatic generator was built (by R. Van de Graaff in 1931), as was the first cyclotron (by E. O. Lawrence and M. S. Livingston in 1932). Since then, enormous advances in the technology for accelerating and detecting particles have been made, and many nuclear reactions have been studied.

When a particle is incident on a nucleus, several different things can happen. The incident particle may be scattered elastically or inelastically, or the incident particle may be absorbed by the nucleus, and another particle or particles may be emitted. In inelastic scattering the nucleus is left in an excited state and subsequently decays by emitting photons (or other particles).

The amount of energy released or absorbed in a reaction (in the center of mass reference frame) is called the **Q value** of the reaction. The Q value equals c^2 times this mass difference. When energy is released by a nuclear reaction, the reaction is said to be an **exothermic reaction.** In an exothermic reaction, the total mass of the incoming particles is greater than that of the outgoing particles, and the Q value is positive. If the total mass of the incoming particles is less than that of the outgoing particles, energy is required for the reaction to take place, and the reaction is said to be an **endothermic reaction.** The Q value of an endothermic reaction is negative. In general, if Δm is the *increase* in mass, the Q value is

$$Q = -(\Delta m)c^2$$

40-19

Q value

An endothermic reaction cannot take place below a certain threshold energy. In the laboratory reference frame in which stationary particles are bombarded by incoming particles, the threshold energy is somewhat greater than $|Q|$ because the outgoing particles must have some kinetic energy to conserve momentum.

A measure of the effective size of a nucleus for a particular nuclear reaction is the **cross section** σ. If I is the number of incident particles per unit time per unit area (the incident intensity) and R is the number of reactions per unit time per nucleus, the cross section is

$$\sigma = \frac{R}{I}$$

40-20

The cross section σ has the dimensions of area. Since nuclear cross sections are of the order of the square of the nuclear radius, a convenient unit for them is the **barn,** which is defined as

$$1 \text{ barn} = 10^{-28} \text{ m}^2$$

40-21

The cross section for a particular reaction is a function of energy. For an endothermic reaction, it is zero for energies below the threshold energy.

Example 40-5

Find the Q value of the reaction p + ^{7}Li $\rightarrow$ ^{4}He + ^{4}He and state whether the reaction is exothermic or endothermic.

Picture the Problem We find the masses of the atoms from Table 40-1 and calculate the difference in the total mass of the outgoing particles and incomming particles. The Q value is $-(\Delta m)c^2$. If we use the mass of hydrogen rather than the mass of the proton, there will be 4 electrons on each side of the reaction so the electron masses will cancel.

1. Find the mass of each atom from Table 40-1:

^{1}H	1.007 825 u
^{7}Li	7.016 004 u
^{4}He	4.002 603 u

2. Calculate the initial mass m_i of the incoming particles:

$m_i = 1.007\ 825\ u + 7.016\ 004\ u = 8.023\ 829\ u$

3. Calculate the final mass m_f:

$m_f = 2(4.002\ 603\ u) = 8.005\ 206\ u$

4. Calculate the increase in mass:

$\Delta m = m_f - m_i = 8.005\ 206\ u - 8.023\ 829\ u$

$= -0.018\ 623\ u$

5. Calculate the Q value:

$Q = -(\Delta m)c^2 = (+0.018\ 623\ u)c^2(931.5\ MeV/u\ c^2)$

$= 17.35\ MeV$

Q is positive so the reaction is exothermic.

Remarks Since the initial mass is greater than the final mass, mass is converted into energy and the reaction is exothermic, yeilding 17.35 MeV.

Reactions With Neutrons

Nuclear reactions involving neutrons are important for understanding nuclear reactors. The most likely reaction between a nucleus and a neutron having an energy of more than about 1 MeV is scattering. However, even if the scattering is elastic, the neutron loses some energy to the nucleus because the nucleus recoils. If a neutron is scattered many times in a material, its energy decreases until it is of the order of the energy of thermal motion kT, where k is Boltzmann's constant and T is the absolute temperature. (At ordinary room temperatures, kT is about 0.025 eV.) The neutron is then equally likely to gain or lose energy from a nucleus when it is elastically scattered. A neutron with energy of the order of kT is called a **thermal neutron.**

At low energies, a neutron is likely to be captured, with the emission of a γ ray from the excited nucleus. Figure 40-8 shows the neutron-capture cross section for silver as a function of the energy of the neutron. The large peak in this curve is called a **resonance.** Except for the resonance, the cross section varies fairly smoothly with energy, decreasing with increasing energy roughly as $1/v$, where v is the speed of the neutron. We can understand this energy dependence as follows:

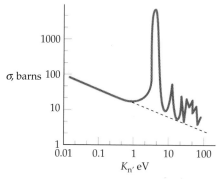

Figure 40-8 Neutron-capture cross section for silver versus energy of the neutron. The straight line indicates the $1/v$ dependence of the cross section, which is proportional to the time spent by the neutron near the silver nucleus. Superimposed on this dependence are a large resonance and several smaller resonances.

Consider a neutron moving with speed v near a nucleus of diameter $2R$. The time it takes the neutron to pass the nucleus is $2R/v$. Thus, the neutron-capture cross section is proportional to the time spent by the neutron in the vicinity of the nucleus. The dashed line in Figure 40-8 indicates this $1/v$ dependence. At the maximum of the resonance, the value of the cross section is very large ($\sigma > 5000$ barns) compared with a value of only about 10 barns just past the resonance. Many elements show similar resonances in their neutron-capture cross sections. For example, the maximum cross section for ^{113}Cd is about 57,000 barns. This material is thus very useful for shielding against low-energy neutrons.

An important nuclear reaction that involves neutrons is fission, which is discussed in the next section.

(a)

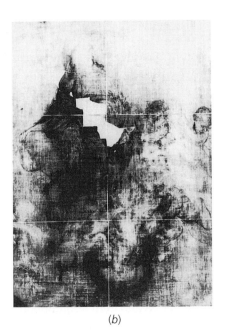

(b)

(c)

Hidden layers in paintings are analyzed by bombarding the painting with neutrons and observing the radiative emissions from nuclei that have captured a neutron. Different elements used in the painting have different half-lives. (a) Van Dyck's painting "Saint Rosalie Interceding for the Plague-Stricken of Palermo." The black-and-white images in (b) and (c) were

formed using a special film sensitive to electrons emitted by the radioactively decaying elements. Image (b), taken a few hours after the neutron irradiation, reveals the presence of manganese, found in umber, a dark earth-pigment used for the painting's base layer. (Blank areas show where modern repairs, free of manganese, have been made.) The image in (c) was

taken four days later, after the umber emissions had died away and when phosphorus, found in charcoal and boneblack, was the main radiating element. Upside down is revealed a sketch of Van Dyck himself. The self-portrait, executed in charcoal, had been overpainted by the artist.

40-4   **Fission and Fusion**

Figure 40-9 shows a plot of the nuclear mass difference per nucleon $(M - Zm_p - Nm_n)/A$ in units of MeV/c^2 versus A. This is just the negative of the binding-energy curve shown in Figure 40-3. From Figure 40-9, we can see that the rest mass per nucleon for both very heavy ($A \approx 200$) and very light ($A \lesssim 20$) nuclides is more than that for nuclides of intermediate mass. Thus, energy is released when a very heavy nucleus, such as ^{235}U, breaks up into two lighter nuclei—a process called **fission**—or when two

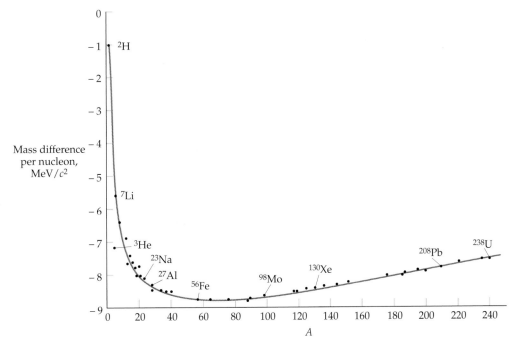

Figure 40-9 Plot of mass difference per nucleon $(M - Zm_p - Nm_n)/A$ in units of MeV/c^2 versus A. The rest mass per nucleon is less for nuclei of intermediate mass than for either very light or very heavy nuclei.

very light nuclei, such as ^{2}H and ^{3}H, fuse together to form a nucleus of greater mass—a process called **fusion.**

The application of both fission and fusion to the development of nuclear weapons has had a profound effect on our lives during the past 50 years. The peaceful application of these reactions to the development of energy resources may have an even greater effect in the future. We will look at some of the features of fission and fusion that are important for their application in reactors to generate power.

Fission

Very heavy nuclei ($Z > 92$) are subject to spontaneous fission. They break apart into two nuclei even if left to themselves with no outside disturbance. We can understand this by considering the analogy of a charged liquid drop. If the drop is not too large, surface tension can overcome the repulsive forces of the charges and hold the drop together. There is, however, a certain maximum size beyond which the drop will be unstable and will spontaneously break apart. Spontaneous fission puts an upper limit on the size of a nucleus and therefore on the number of elements that are possible.

Some heavy nuclei—uranium and plutonium, in particular—can be induced to fission by the capture of a neutron. In the fission of ^{235}U, for example, the uranium nucleus is excited by the capture of a neutron, causing it to split into two nuclei and emit several neutrons. The Coulomb force of repulsion drives the fission fragments apart, with the energy eventually showing up as thermal energy. Consider, for example, the fission of a nucleus of mass number $A = 200$ into two nuclei of mass number $A = 100$. Since the rest energy for $A = 200$ is about 1 MeV per nucleon greater than that for $A = 100$, about 200 MeV per nucleus is released in such a fission. This is a large amount of energy. By contrast, in the chemical reaction of combustion, only about 4 eV of energy is released per molecule of oxygen consumed.

Example 40-6

Calculate the total energy in kilowatt-hours released in the fission of 1 g of ^{235}U, assuming that 200 MeV is released per fission.

Picture the Problem We need to find the number of uranium nuclei in one gram of ^{235}U, which we do using the fact that there are Avogadro's number ($N_A = 6.02 \times 10^{23}$) of nuclei in 235 grams.

1. The total energy is the number of nuclei times the energy per nucleus:

$$E = NE_{\text{nucleus}} = N(200 \text{ MeV/nucleus})$$

2. Calculate N:

$$N = \frac{6.02 \times 10^{23} \text{ nuclei/mol}}{235 \text{ g/mol}} \times 1 \text{ g} = 2.56 \times 10^{21} \text{ nuclei}$$

3. Calculate the energy per gram in eV and convert to kW·h:

$$E = \frac{200 \times 10^6 \text{ eV}}{\text{nucleus}} \times 2.56 \times 10^{21} \text{ nuclei}$$

$$\times \frac{1.6 \times 10^{-19} \text{ J}}{1 \text{ eV}} \times \frac{1 \text{ h}}{3600 \text{ s}} \times \frac{1 \text{ kW}}{1000 \text{ J/s}}$$

$$= 2.28 \times 10^4 \text{ kW·h}$$

The fission of uranium was discovered in 1939 by Hahn and Strassmann, who found, by careful chemical analysis, that medium-mass elements (such as barium and lanthanum) were produced in the bombardment of uranium with neutrons. The discovery that several neutrons are emitted in the fission process led to speculation concerning the possibility of using these neutrons to cause further fissions, thereby producing a chain reaction. When ^{235}U captures a neutron, the resulting ^{236}U nucleus emits γ rays as it deexcites to the ground state about 15% of the time and undergoes fission about 85% of the time. The fission process is somewhat analogous to the oscillation of a liquid drop, as shown in Figure 40-10. If the oscillations are violent enough, the drop splits in two. Using the liquid-drop model, Bohr and Wheeler calculated the critical energy E_c needed by the ^{236}U nucleus to undergo fission. (^{236}U is the nucleus formed momentarily by the capture of a neutron by ^{235}U.) For this nucleus, the critical energy is 5.3 MeV, which is less than the 6.4 MeV of excitation energy produced when ^{235}U captures a neutron. The capture of a neutron by ^{235}U therefore produces an excited state of the ^{236}U nucleus that has more than enough energy to break apart. On the other hand, the critical energy for fission of the ^{239}U nucleus is 5.9 MeV. The capture of a neutron by a ^{238}U nucleus produces an excitation energy of only 5.2 MeV. Therefore, when a neutron is captured by ^{238}U to form ^{239}U, the excitation energy is not great enough for fission to occur. In this case, the excited ^{239}U nucleus deexcites by γ emission and then decays to ^{239}N$_p$ by β decay, and then again to ^{239}Pu by β decay.

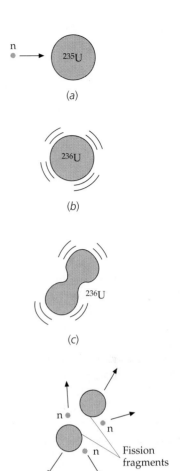

(a)

(b)

(c)

(d)

Figure 40-10 Schematic illustration of nuclear fission. (*a*) The absorption of a neutron by ^{235}U leads to (*b*) ^{236}U in an excited state. In (*c*), the oscillation of ^{236}U has become unstable. (*d*) The nucleus splits apart into two nuclei of medium mass and emits several neutrons that can produce fission in other nuclei.

A fissioning nucleus can break into two medium-mass fragments in many different ways, as shown in Figure 40-11. Depending on the particular reaction, 1, 2, or 3 neutrons may be emitted. The average number of neutrons emitted in the fission of ^{235}U is about 2.5. A typical fission reaction is

$$n + {}^{235}U \rightarrow {}^{141}Ba + {}^{92}Kr + 3n$$

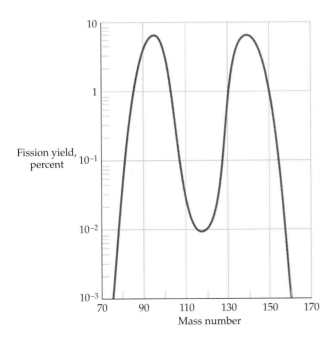

Fission yield, percent

Mass number

Figure 40-11 Distribution of the possible fission fragments of ^{235}U. The splitting of ^{235}U into two fragments of unequal mass is more likely than its splitting into fragments of equal mass.

Nuclear Fission Reactors

To sustain a chain reaction in a fission reactor, one of the neutrons (on the average) emitted in the fission of ^{235}U must be captured by another ^{235}U nucleus and cause it to fission. The **reproduction constant** k of a reactor is defined as the average number of neutrons from each fission that cause a subsequent fission. The maximum possible value of k is 2.5, but it is normally less than this for two important reasons: (1) Some of the neutrons may escape from the region containing fissionable nuclei, and (2) some of the neutrons may be captured by nonfissioning nuclei in the reactor. If k is exactly 1, the reaction will be self-sustaining. If it is less than 1, the reaction will die out. If k is significantly greater than 1, the reaction rate will increase rapidly and "run away." In the design of nuclear bombs, such a runaway reaction is desired. In power reactors, the value of k must be kept very nearly equal to 1.

The inside of a nuclear power plant in Kent, England. A technician is standing on the reactor charge transfer plate, into which uranium fuel rods fit.

Since the neutrons emitted in fission have energies of the order of 1 MeV, whereas the chance for neutron capture leading to fission in ^{235}U is largest at small energies, the chain reaction can be sustained only if the neutrons are slowed down before they escape from the reactor. At high energies (1 to 2 MeV), neutrons lose energy rapidly by inelastic scattering from ^{238}U, the principal constituent of natural uranium. (Natural uranium contains 99.3% ^{238}U and only 0.7% fissionable ^{235}U.) Once the neutron energy is below the excitation energies of the nuclei in the reactor (about 1 MeV), the main process of energy loss is by elastic scattering, in which a fast neutron collides with a nucleus at rest and transfers some of its kinetic energy to that nucleus. Such energy transfers are efficient only if the masses of the two bodies are comparable. A neutron will not transfer much energy in an elastic collision with a heavy uranium nucleus. Such a collision is like one between a marble and a billiard ball. The marble will be deflected by the much more massive billiard ball, and very little of its kinetic energy will be transferred to the billiard ball. A **moderator** consisting of material such as water or carbon that contains light nuclei is therefore placed around the fissionable material in the core of the reactor to slow down the neutrons. The neutrons are slowed down by elastic collisions with the nuclei of the moderator until they are in thermal equilibrium with the moderator. Because of the relatively large neutron-capture cross section of the hydrogen nucleus in water, reactors using ordinary water as a moderator cannot easily achieve $k \approx 1$ unless they use enriched uranium, in which the ^{235}U content has been increased from 0.7% to between 1 and 4%. Natural uranium can be used if heavy water (D_2O) is used instead of ordinary (light) water (H_2O) as the moderator. Although heavy water is expensive, most Canadian reactors use it for a moderator to avoid the cost of constructing uranium-enrichment facilities.

Figure 40-12 shows some of the features of a pressurized-water reactor commonly used in the United States to generate electricity. Fission in the core heats the water to a high temperature in the primary loop, which is closed. This water, which also serves as the moderator, is under high pressure to prevent it from boiling. The hot water is pumped to a heat exchanger, where it heats the water in the secondary loop and converts it to steam, which is then used to drive the turbines that produce electrical power. Note that the water in the secondary loop is isolated from that in the primary loop to prevent its contamination by the radioactive nuclei in the reactor core.

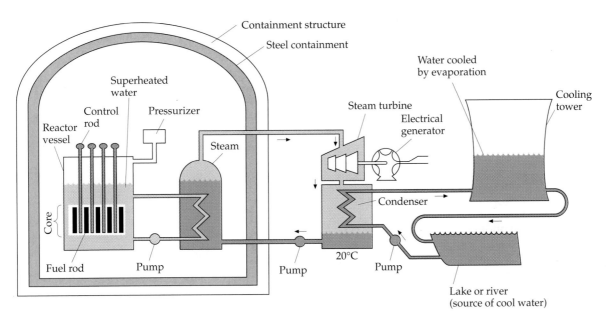

Figure 40-12 Simplified drawing of a pressurized-water reactor. The water in contact with the reactor core serves as both the moderator and the heat-transfer material. It is isolated from the water used to produce the steam that drives the turbines. Many features, such as the backup cooling mechanisms, are not shown here.

The ability to control the reproduction factor k precisely is important if a power reactor is to be operated safely. Both natural negative-feedback mechanisms and mechanical methods of control are used. If k is greater than 1 and the reaction rate increases, the temperature of the reactor increases. If water is used as a moderator, its density decreases with increasing temperature and it becomes a less effective moderator. A second important control method is the use of control rods made of a material, such as cadmium, that has a very large neutron-capture cross section. When the reactor is started up, the control rods are inserted so that k is less than 1. As the rods are gradually withdrawn from the reactor, fewer neutrons are captured by them and k increases to 1. If k becomes greater than 1, the rods are inserted again.

Mechanical control of the reaction rate of a nuclear reactor using control rods is possible only because some of the neutrons emitted in the fission process are **delayed neutrons.** The time needed for a neutron to slow down from 1 or 2 MeV to the thermal-energy level and then be captured is only of the order of a millisecond. If all the neutrons emitted in fission were prompt neutrons, that is, emitted immediately in the fission process, mechanical control would not be possible because the reactor would run away before the rods could be inserted. However, about 0.65% of the neutrons emitted are delayed by an average time of about 14 s. These neutrons are emitted not in the fission process itself but in the decay of the fission fragments. The effect of the delayed neutrons can be seen in the following examples.

Example 40-7

If the average time between fission generations (the time it takes for a neutron emitted in one fission to cause another) is $t_1 = 1 \text{ ms} = 0.001 \text{ s}$ and the reproduction constant is 1.001, how long will it take for the reaction rate to double?

Picture the Problem The time to double is the number of generations N needed to double times the generation time. If $k = 1.001$, the reaction rate after N generations is 1.001^N. We find the number of generations by setting 1.001^N equal to 2 and solving for N.

1. Set 1.001^N equal to 2 and solve for N:

$$(1.001)^N = 2$$

$$N \ln 1.001 = \ln 2$$

$$N = \frac{\ln 2}{\ln 1.001} = 693$$

2. Multiply the number of generations by the generation time: $t = Nt_1 = 693(0.001 \text{ s}) = 0.693 \text{ s}$

Remark The doubling time of about 0.7 s is not enough time for insertion of control rods.

Example 40-8   *try it yourself*

Assuming that 0.65% of the neutrons emitted are delayed by 14 s, find the average generation time and the doubling time if $k = 1.001$.

Picture the Problem The doubling time is Nt_{av}, where t_{av} is the average time between generations. Since 99.35% of the generation times are 0.001 s and 0.65% are 14 s, the average generation time is $0.9935(0.001 \text{ s}) + 0.0065(14 \text{ s})$.

Cover the column to the right and try these on your own before looking at the answers.

Steps	Answers
1. Compute the average generation time.	1. $t_{av} = 0.092 \text{ s}$
2. Use your result to find the time for 693 generations.	2. $t = 63.8 \text{ s}$

Remarks Even though the number of delayed neutrons is less than 1%, they have a large effect on the doubling time. Here they increase the generation time by a factor of 92, resulting in a doubling time of about 64 s, which is plenty of time for mechanical insertion of control rods.

Because of the limited supply of natural uranium, the small fraction of ^{235}U in natural uranium, and the limited capacity of enrichment facilities, reactors based on the fission of ^{235}U cannot meet our energy needs for very long. A promising alternative is the **breeder reactor.** When the relatively plentiful but nonfissionable ^{238}U nucleus captures a neutron, it decays by β decay (with a half-life of 20 min) to ^{239}Np, which in turn decays by β decay (with a half-life of 2.35 days) to the fissionable nuclide ^{239}Pu. Since ^{239}Pu fissions with fast neutrons, no moderator is needed. A reactor initially fueled with a mixture of ^{238}U and ^{239}Pu will breed as much fuel as it uses or more if one or more of the neutrons emitted in the fission of ^{239}Pu is captured by ^{238}U. Practical studies indicate that a typical breeder reactor can be expected to double its fuel supply in 7 to 10 years.

There are two major safety problems inherent with breeder reactors. The fraction of delayed neutrons is only 0.3% for the fission of ^{239}Pu, so the time between generations is much less than that for ordinary reactors. Mechanical control is therefore much more difficult. Also, since the operating temperature of a breeder reactor is relatively high and a moderator is not desired, a heat-transfer material such as liquid sodium metal is used rather than water (which is the moderator as well as the heat-transfer material in an ordinary reactor). If the temperature of the reactor increases, the resulting decrease in the density of the heat-transfer material leads to positive feedback, since it will absorb fewer neutrons than before. Because of these safety considerations, breeder reactors are not yet in commercial use in the United States. There are, however, several in operation in France, Great Britain, and the Soviet Union.

Fusion

In fusion, two light nuclei such as deuterium (^{2}H) and tritium (^{3}H) fuse together to form a heavier nucleus. A typical fusion reaction is

$$^2\text{H} + {}^3\text{H} \rightarrow {}^4\text{He} + \text{n} + 17.6\,\text{MeV}$$

The energy released in fusion depends on the particular reaction. For the ^{2}H + ^{3}H reaction, it is 17.6 MeV. Although this is less than the energy released in a fission reaction, it is a greater amount of energy per unit mass. The energy released in this fusion reaction is (17.6 MeV)/(5 nucleons) = 3.52 MeV per nucleon. This is about 3.5 times as great as the 1 MeV per nucleon released in fission.

The production of power from the fusion of light nuclei holds great promise because of the relative abundance of the fuel and the absence of some of the dangers inherent in fission reactors. Unfortunately, the technology necessary to make fusion a practical source of energy has not yet been developed. We will consider the ^{2}H + ^{3}H reaction; other reactions present similar problems.

Because of the Coulomb repulsion between the ^{2}H and ^{3}H nuclei, very large kinetic energies, of the order of 1 MeV, are needed to get the nuclei close enough together for the attractive nuclear forces to become effective and cause fusion. Such energies can be obtained in an accelerator, but since the scattering of one nucleus by the other is much more probable than fusion, the bombardment of one nucleus by another in an accelerator requires the input of more energy than is recovered. To obtain energy from fusion, the particles must be heated to a temperature great enough for the fusion reaction to occur as the result of random thermal collisions. Because a significant number of particles have kinetic energies greater than the mean kinetic energy, $\frac{3}{2}kT$, and because some particles can tunnel through the Coulomb barrier, a temperature T corresponding to $kT \approx 10$ keV is adequate to ensure that a reasonable number of fusion reactions will occur if the density of particles is sufficiently high. The temperature corresponding to $kT = 10$ keV is of the order of 10^8 K. Such temperatures occur in the interiors of stars, where such reactions are common. At these temperatures, a gas consists of positive ions and negative electrons and is called a **plasma.** One of the problems arising in attempts to produce controlled fusion reactions is that of confining the plasma long enough for the reactions to take place. In the interior of the sun the plasma is confined by the enormous gravitational field of the sun. In a laboratory on earth, confinement is a difficult problem.

The energy required to heat a plasma is proportional to the density of its ions, n, whereas the collision rate is proportional to n^2, the square of the density. If τ is the confinement time, the output energy is proportional to $n^2\tau$. If the output energy is to exceed the input energy, we must have

$$C_1 n^2 \tau > C_2 n$$

where C_1 and C_2 are constants. In 1957, the British physicist J. D. Lawson evaluated these constants from estimates of the efficiencies of various hypothetical fusion reactors and derived the following relation between density and confinement time, known as **Lawson's criterion:**

$$n\tau > 10^{20} \text{ s·particles/m}^3 \qquad\qquad \text{40-22}$$

Lawson's criterion

If Lawson's criterion is met and the thermal energy of the ions is great enough ($kT \sim 10$ keV), the energy released by a fusion reactor will just equal the energy input; that is, the reactor will just break even. For the reactor to be practical, much more energy must be released.

Two schemes for achieving Lawson's criterion are currently under investigation. In one scheme, **magnetic confinement,** a magnetic field is used to confine the plasma (see Section 28-2). In the most common arrangement, first developed in the USSR and called the Tokamak, the plasma is confined in a large toroid. The magnetic field is a combination of the doughnut-shaped magnetic field due to the windings of the toroid and the self-field due to the current of the circulating plasma. The break-even point has been achieved recently using magnetic confinement, but we are still a long way from building a practical fusion reactor.

In a second scheme, called **inertial confinement,** a pellet of solid deuterium and tritium is bombarded from all sides by intense pulsed laser beams of energies of the order of 10^4 J lasting about 10^{-8} s. (Intense beams of ions are also used.) Computer simulation studies indicate that the pellet should be compressed to about 10^4 times its normal density and heated to a temperature greater than 10^8 K. This should produce about 10^6 J of fusion energy in 10^{-11} s, which is so brief that confinement is achieved by inertia alone.

Because the break-even point is just barely being achieved in magnetic-confinement fusion, and because the building of a fusion reactor involves many practical problems that have not yet been solved, the availability of fusion to meet our energy needs is not expected for at least several decades. However, fusion holds great promise as an energy source for the future.

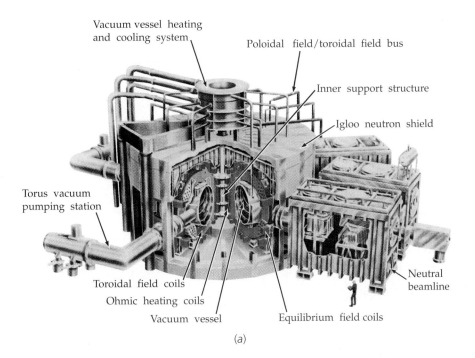

Vacuum vessel heating and cooling system

Poloidal field/toroidal field bus

Inner support structure

Igloo neutron shield

Torus vacuum pumping station

Neutral beamline

Toroidal field coils

Ohmic heating coils

Vacuum vessel

Equilibrium field coils

(a)

(b)

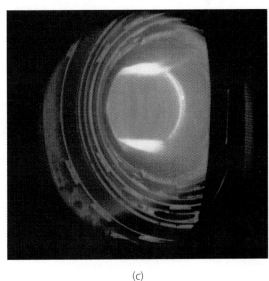

(c)

(a) Schematic of the Tokamak Fusion Test Reactor (TFTR). The toroidal coils, surrounding the doughnut-shaped vacuum vessel, are designed to conduct current for 3-s pulses, separated by waiting times of 5 min. Pulses peak at 73,000 A, producing a magnetic field of 5.2 T. This field is the principal means of confining the deuterium–tritium plasma that circulates within the vacuum vessel. Current for the pulses is delivered by converting the rotational energy of two 600-ton flywheels. Sets of poloidal coils, perpendicular to the toroidal coils, carry an oscillating current that generates a current through the confined plasma itself, heating it ohmically. Additional poloidal fields help stabilize the confined plasma. Between four and six neutral-beam injection systems (only one of which is shown in the schematic) are used to inject high-energy deuterium atoms into the deuterium–tritium plasma, heating beyond what could be obtained ohmically—ultimately to the point of fusion. (b) The TFTR itself. The diameter of the vacuum vessel is 7.7 m (see also photo on page 900). (c) An 800-kA plasma, lasting 1.6 s, as it discharges within the vacuum vessel.

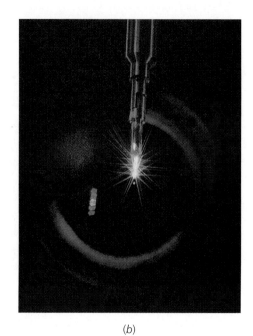

(a)   (b)

(a) The Nova target chamber, an aluminum sphere approximately 5 m in diameter, inside which 10 beams from the world's most powerful laser converge onto a hydrogen-containing pellet 0.5 mm in diameter. The resulting fusion reaction is visible as a tiny star (b), lasting 10^{-10} s, releasing 10^{13} neutrons.

Summary

Topic	Relevant Equations and Remarks
1. Nuclear Properties	Nuclei have N neutrons, Z protons, and a mass number $A = N + Z$. For light nuclei, N and Z are approximately equal, whereas for heavy nuclei, N is greater than Z.
Isotopes	Isotopes consist of two or more nuclei having the same value of Z but different values of N and A.
Size and shape	Most nuclei are approximately spherical in shape and have a volume that is proportional to A. Because the mass is proportional to A, nuclear density is independent of A.
Radius	$$R = R_0 A^{1/3} \approx (1.5 \text{ fm}) A^{1/3} \qquad \textbf{40-1}$$
Mass and binding energy	The mass of a stable nucleus is less than the sum of the masses of its nucleons. The mass difference times c^2 equals the binding energy of the nucleus. The binding energy is approximately proportional to the mass number A.
2. Radioactivity	Unstable nuclei are radioactive and decay by emitting α particles (^{4}He nuclei), β particles (electrons or positrons), or γ rays (photons). All radioactivity is statistical in nature and follows an exponential decay law:
Decay law	$$N = N_0 e^{-\lambda t} \qquad \textbf{40-6}$$
Decay rate	$$R = \lambda N = R_0 e^{-\lambda t} \qquad \textbf{40-7}$$
Mean life	$$\tau = \frac{1}{\lambda} \qquad \textbf{40-9}$$

| Half-life | $t_{1/2} = \dfrac{0.693}{\lambda}$ | **40-11** |

Half-lives of α decay range from a fraction of a second to millions of years. For β decay they range up to hours or days and for γ decay half-lives are usually less than a microsecond.

| Curie | The number of decays per second of 1 g of radium is the curie, which equals 3.7×10^{10} decays/s $= 3.7 \times 10^{10}$ Bq. |

3. Nuclear Reactions

| Q value | The Q value equals c^2 times the difference in the total mass of the incoming particles and the total mass of the outgoing particles. If the net mass increase is Δm, the Q value is |

$$Q = -(\Delta m)c^2 \qquad \textbf{40-19}$$

| Exothermic reaction | The mass decreases, Q is positive and measures the energy released. |

| Endothermic reaction | The mass increases, Q is negative. Then $|Q|$ is the threshold energy for the reaction in the center of mass reference frame. |

3. Fission

| | Fission occurs when some heavy elements, such as ^{235}U or ^{239}Pu, capture a neutron and split apart into two medium-mass nuclei. The two nuclei then fly apart because of electrostatic repulsion, releasing a large amount of energy. A chain reaction is possible because several neutrons are emitted by a nucleus when it undergoes fission. A chain reaction can be sustained in a reactor if, on the average, one of the emitted neutrons is slowed down by scattering in the reactor and is then captured by another fissionable nucleus. Very heavy nuclei ($Z > 92$) are subject to spontaneous fission. |

4. Fusion

| | A large amount of energy is released when two light nuclei, such as ^2H and ^3H, fuse together. Fusion takes place spontaneously inside the sun and other stars, where the temperature is great enough (about 10^8 K) for thermal motion to bring the charged hydrogen ions close enough together to fuse. Although controlled fusion holds great promise as a future energy source, practical difficulties have thus far prevented its development. |

| Lawson criterion | The minimum product of particle density n and confinement time τ to get more energy out of a fusion reactor than is put in is $n\tau > 10^{20}$ s·particles/m³. |

Problem-Solving Guide

Summary of Worked Examples

Type of Calculation	Procedure and Relevant Examples

1. Nuclear Properties

| Find the binding energy of a nucleon in a nucleus. | Compute the difference in mass between the final nucleus plus nucleon and the original nucleus and multiply Δm by c^2.   **Example 40-1** |

2.   Radioactivity

Find the counting rate of a radioactive source at various times.	The rate decreases by a factor of 2 for each half-life.	**Examples 40-2, 40-3**
Find the number of nuclei that decay during a given time interval.	Find the number of nuclei at the beginning and end of the interval and subtract.	**Example 40-3**
Find the age of an organic sample.	Compare the actual decay rate of carbon in the sample with the original rate of 15 decays per gram per minute. You can compute the number of half-lives from $R = (0.5)^n R_0$ and then find t from $t_{1/2} = 5730$ y.	**Example 40-4**

3.   Find the Q value of a reaction.

	Calculate the difference in the total mass of the incoming particles and outgoing particles and multiply by c^2.	**Example 40-5**

4.   Nuclear Fission

Calculate the total energy released in fission given the mass decrease.	Use 200 MeV per nuclei and find the number of nuclei in the sample.	**Example 40-6**
Find the time for the reaction rate in a reactor to double given the reproduction constant.	Multiply the generation time by the number of generations. If there are delayed neutrons, you need to compute the average generation time.	**Examples 40-7, 40-8**

Problems

Conceptual Problems

Problems from Optional and Exploring sections

In a few problems, you are given more data than you actually need; in a few other problems, you are required to supply data from your general knowledge, outside sources, or informed estimates.

•   Single-concept, single-step, relatively easy
••   Intermediate-level, may require synthesis of concepts
••• Challenging, for advanced students

Properties of Nuclei

1   •   Give the symbols for two other isotopes of (a) ^{14}N, (b) ^{56}Fe, and (c) ^{118}Sn.

2   •   Calculate the binding energy and the binding energy per nucleon from the masses given in Table 40-1 for (a) ^{12}C, (b) ^{56}Fe, and (c) ^{238}U.

3   •   Repeat Problem 2 for (a) ^{6}Li, (b) ^{39}K, and (c) ^{208}Pb.

4   •   Use Equation 40-1 to compute the radii of the following nuclei: (a) ^{16}O, (b) ^{56}Fe, and (c) ^{197}Au.

5   •   (a) Given that the mass of a nucleus of mass number A is approximately $m = CA$, where C is a constant, find an expression for the nuclear density in terms of C and the constant R_0 in Equation 40-1. (b) Compute the value of this nuclear density in grams per cubic centimeter using the fact that C has the approximate value of 1 g per Avogadro's number of nucleons.

6   •   Derive Equation 40-2; that is, show that the rest energy of one unified mass unit is 931.5 MeV.

7   •   Use Equation 40-1 for the radius of a spherical nucleus and the approximation that the mass of a nucleus of mass number A is A u to calculate the density of nuclear matter in grams per cubic centimeter.

8   ••   The electrostatic potential energy of two charges q_1 and q_2 separated by a distance r is $U = kq_1q_2/r$, where k is the Coulomb constant. (a) Use Equation 40-1 to calculate the radii of ^{2}H and ^{3}H. (b) Find the electrostatic potential energy when these two nuclei are just touching, that is, when their centers are separated by the sum of their radii.

9   ••   (a) Calculate the radii of $^{141}_{56}$Ba and $^{92}_{36}$Kr from Equation 40-1. (b) Assume that after the fission of ^{235}U into ^{141}Ba and ^{92}Kr, the two nuclei are momentarily separated by a distance r equal to the sum of the radii found in (a), and calculate the electrostatic potential energy for these two nuclei at this separation. (See Problem 8.) Compare your result with the measured fission energy of 175 MeV.

Radioactivity

10   •   Why is the decay series $A = 4n + 1$ not found in nature?

11   •   A decay by α emission is often followed by β decay. When this occurs, it is by β^- and not β^+ decay. Why?

12 • The half-life of ^{14}C is much less than the age of the universe, yet ^{14}C is found in nature. Why?

13 • What effect would a long-term variation in cosmic-ray activity have on the accuracy of ^{14}C dating?

14 • Homer enters the visitors' chambers, and his geiger-beeper goes off. He shuts off the beep, removes the device from his shoulder patch and holds it near the only new object in the room: an orb which is to be presented as a gift from the visiting Cartesians. Pushing a button marked "monitor," Homer reads that the orb is a radioactive source with a counting rate of 4000 counts/s. After 10 min, the counting rate has dropped to 1000 counts/s. The source's half-life appears on the geiger-beeper display. (a) What is the half-life? (b) What will the counting rate be 20 min after the monitoring device was switched on?

15 • A certain source gives 2000 counts/s at time $t = 0$. Its half-life is 2 min. (a) What is the counting rate after 4 min? (b) After 6 min? (c) After 8 min?

16 • The counting rate from a radioactive source is 8000 counts/s at time $t = 0$, and 10 min later the rate is 1000 counts/s. (a) What is the half-life? (b) What is the decay constant? (c) What is the counting rate after 20 min?

17 • The half-life of radium is 1620 y. Calculate the number of disintegrations per second of 1 g of radium, and show that the disintegration rate is approximately 1 Ci.

18 • A radioactive silver foil ($t_{1/2} = 2.4$ min) is placed near a Geiger counter and 1000 counts/s are observed at time $t = 0$. (a) What is the counting rate at $t = 2.4$ min and at $t = 4.8$ min? (b) If the counting efficiency is 20%, how many radioactive nuclei are there at time $t = 0$? At time $t = 2.4$ min? (c) At what time will the counting rate be about 30 counts/s?

19 • Use Table 40-1 to calculate the energy in MeV for the α decay of (a) ^{226}Ra and (b) ^{242}Pu.

20 • Suppose that two billion years ago 10% of the mass of the earth was ^{14}C. Approximately what percentage of the mass of the earth today would be ^{14}C, neglecting formation of ^{14}C in the atmosphere?

21 • At the scene of the crime, in the museum's west wing, Angela found some wood chips, so she slipped them into her purse for future analysis. They were allegedly from an old wooden mask, which the guard said he threw at the would-be thief. Later, in the lab, she determined the age of the chips, using a sample that contained 10 g of carbon and showed a ^{14}C decay rate of 100 counts/min. How old are they?

22 • The thief in Problem 21 had been after a valuable carving made from a 10,000-year-old bone. The guard said that he chased the thief away, but Angela suspects that the guard is an accomplice, and that the bone in the display case is, in fact, a fake. If a sample of the bone containing 15 g of carbon were to be analyzed, what should the decay rate of ^{14}C be if it is a 10,000-year-old bone?

23 • Through a friend in security at the museum, Angela got a sample having 175 g of carbon. The decay rate of ^{14}C was 8.1 Bq. (a) How old is it? (b) Is it from the carving described in Problem 22?

24 • A sample of a radioactive isotope is found to have an activity of 115.0 Bq immediately after it is pulled from the reactor that formed it. Its activity 2 h 15 min later is measured to be 85.2 Bq. (a) Calculate the decay constant and the half-life of the sample. (b) How many radioactive nuclei were there in the sample initially?

25 •• Derive the result that the activity of 1 g of natural carbon due to the β decay of ^{14}C is 15 decays/min = 0.25 Bq.

26 •• Measurements of the activity of a radioactive sample have yielded the following results. Plot the activity as a function of time, using semilogarithmic paper, and determine the decay constant and half-life of the radioisotope.

Time, min	Activity	Time, min	Activity
0	4287	20	880
5	2800	30	412
10	1960	40	188
15	1326	60	42

27 •• (a) Show that if the decay rate is R_0 at time $t = 0$ and R_1 at some later time t_1, the decay constant is given by $\lambda = t_1^{-1} \ln(R_0/R_1)$ and the half-life is given by $t_{1/2} = 0.693 t_1 / \ln(R_0/R_1)$. (b) Use these results to find the decay constant and the half-life if the decay rate is 1200 Bq at $t = 0$ and 800 Bq at $t_1 = 60$ s.

28 •• A wooden casket is thought to be 18,000 years old. How much carbon would have to be recovered from this object to yield a ^{14}C counting rate of no less than 5 counts/min?

29 •• A 1.00-mg sample of substance of atomic mass 59.934 u emits β particles with an activity of 1.131 Ci. Find the decay constant for this substance in s^{-1} and its half-life in years.

30 •• The counting rate from a radioactive source is measured every minute. The resulting counts per second are 1000, 820, 673, 552, 453, 371, 305, 250. Plot the counting rate versus time on semilog graph paper, and use your graph to find the half-life of the source.

31 •• A sample of radioactive material is initially found to have an activity of 115.0 decays/min. After 4 d 5 h, its activity is measured to be 73.5 decays/min. (a) Calculate the half-life of the material. (b) How long (from the initial time) will it take for the sample to reach an activity level of 10.0 decays/min?

32 •• The rubidium isotope ^{87}Rb is a β emitter with a half-life of 4.9×10^{10} y that decays into ^{87}Sr. It is used to determine the age of rocks and fossils. Rocks containing the fossils of early animals contain a ratio of ^{87}Sr to ^{87}Rb of 0.0100. Assuming that there was no ^{87}Sr present when the rocks were formed, calculate the age of these fossils.

33 ••• If there are N_0 radioactive nuclei at time $= 0$, the number that decay in some time interval dt at time t is $-dN = \lambda N_0 e^{-\lambda t} \, dt$. If we multiply this number by the lifetime t of these nuclei, sum over all the possible lifetimes from $t = 0$ to $t = \infty$, and divide by the total number of nuclei, we get the mean lifetime τ:

$$\tau = \frac{1}{N_0} \int_0^\infty t |dN| = \int_0^\infty t \lambda e^{-\lambda t}$$

Show that $\tau = 1/\lambda$.

1310 CHAPTER 40 Nuclear Physics

Nuclear Reactions

34 • Using Table 40-1, find the Q values for the following reactions: (a) $^1\text{H} + {}^3\text{H} \rightarrow {}^3\text{He} + n + Q$ and (b) $^2\text{H} + {}^2\text{H} \rightarrow {}^3\text{He} + n + Q$.

35 • Using Table 40-1, find the Q values for the following reactions: (a) $^2\text{H} + {}^2\text{H} \rightarrow {}^3\text{H} + {}^1\text{H} + Q$, (b) $^2\text{H} + {}^3\text{He} \rightarrow {}^4\text{He} + {}^1\text{H} + Q$, and (c) $^6\text{Li} + n \rightarrow {}^3\text{H} + {}^4\text{He} + Q$.

36 •• (a) Use the atomic masses $m = 14.00324$ u for $^{14}_6\text{C}$ and $m = 14.00307$ u for $^{14}_7\text{N}$ to calculate the Q value (in MeV) for the β decay

$$^{14}_6\text{C} \rightarrow {}^{14}_7\text{N} + \beta^- + \bar{\nu}_e$$

(b) Explain why you do not need to add the mass of the β^- to that of atomic $^{14}_7\text{N}$ for this calculation.

37 •• (a) Use the atomic masses $m = 13.00574$ u for $^{13}_7\text{N}$ and $m = 13.003354$ u for $^{13}_6\text{C}$ to calculate the Q value (in MeV) for the β decay

$$^{13}_7\text{N} \rightarrow {}^{13}_6\text{C} + \beta^+ + \nu_e$$

(b) Explain why you need to add two electron masses to the mass of $^{13}_6\text{C}$ in the calculation of the Q value for this reaction.

Fission and Fusion

38 • Why isn't there an element with $Z = 130$?

39 • Why is a moderator needed in an ordinary nuclear fission reactor?

40 • Explain why water is more effective than lead in slowing down fast neutrons.

41 • What happens to the neutrons produced in fission that do not produce another fission?

42 • What is the advantage of a breeder reactor over an ordinary one? What are the disadvantages?

43 • Assuming an average energy of 200 MeV per fission, calculate the number of fissions per second needed for a 500-MW reactor.

44 • If the reproduction factor in a reactor is $k = 1.1$, find the number of generations needed for the power level to (a) double, (b) increase by a factor of 10, and (c) increase by a factor of 100. Find the time needed in each case if (d) there are no delayed neutrons, so the time between generations is 1 ms, and (e) there are delayed neutrons that make the average time between generations 100 ms.

45 • Compute the temperature T for which $kT = 10$ keV, where k is Boltzmann's constant.

46 •• In 1989, researchers claimed to have achieved fusion in an electrochemical cell at room temperature. They claimed a power output of 4 W from deuterium fusion reactions in the palladium electrode of their apparatus. If the two most likely reactions are

$$^2\text{H} + {}^2\text{H} \rightarrow {}^3\text{He} + n + 3.27 \text{ MeV}$$

and

$$^2\text{H} + {}^2\text{H} \rightarrow {}^3\text{H} + {}^1\text{H} + 4.03 \text{ MeV}$$

with 50% of the reactions going by each branch, how many neutrons per second would we expect to be emitted in the generation of 4 W of power?

47 •• A fusion reactor using only deuterium for fuel would have the two reactions in Problem 46 taking place in it. The ^3H produced in the second reaction reacts immediately with another ^2H to produce

$$^3\text{H} + {}^2\text{H} \rightarrow {}^4\text{He} + n + 17.6 \text{ MeV}$$

The ratio of ^2H to ^1H atoms in naturally occurring hydrogen is 1.5×10^{-4}. How much energy would be produced from 4 L of water if all of the ^2H nuclei undergo fusion?

48 ••• The fusion reaction between ^2H and ^3H is

$$^3\text{H} + {}^2\text{H} \rightarrow {}^4\text{He} + n + 17.6 \text{ MeV}$$

Using the conservation of momentum and the given Q value, find the final energies of both the ^4He nucleus and the neutron, assuming that the initial momentum of the system is zero.

49 ••• Energy is generated in the sun and other stars by fusion. One of the fusion cycles, the proton–proton cycle, consists of the following reactions:

$$^1\text{H} + {}^1\text{H} \rightarrow {}^2\text{H} + \beta^+ + \nu_e$$
$$^1\text{H} + {}^2\text{H} \rightarrow {}^3\text{He} + \gamma$$

followed by

$$^1\text{H} + {}^3\text{He} \rightarrow {}^4\text{He} + \beta^+ + \nu_e$$

(a) Show that the net effect of these reactions is

$$4{}^1\text{H} \rightarrow {}^4\text{He} + 2\beta^+ + 2\nu_e + \gamma$$

(b) Show that rest energy of 24.7 MeV is released in this cycle (not counting the energy of 1.02 MeV released when each positron meets an electron and the two annihilate). (c) The sun radiates energy at the rate of about 4×10^{26} W. Assuming that this is due to the conversion of four protons into helium plus γ rays and neutrinos, which releases 26.7 MeV, what is the rate of proton consumption in the sun? How long will the sun last if it continues to radiate at its present level? (Assume that protons constitute about half of the total mass $[2 \times 10^{30}$ kg] of the sun.)

General Problems

50 • True or false:

(a) The atomic nucleus contains protons, neutrons, and electrons.
(b) The mass of ^2H is less than the mass of a proton plus a neutron.
(c) After two half-lives, all the radioactive nuclei in a given sample have decayed.
(d) In a breeder reactor, fuel can be produced as fast as it is consumed.

51 • Why do extreme changes in the temperature or pressure of a radioactive sample have little or no effect on the radioactivity?

52 • The stable isotope of sodium is ^{23}Na. What kind of radioactivity would you expect of (a) ^{22}Na and (b) ^{24}Na?

53 • Why does fusion occur spontaneously in the sun but not on earth?

54 • (a) Show that $ke^2 = 1.44$ MeV·fm, where k is the Coulomb constant and e is the electron charge. (b) Show that $hc = 1240$ MeV·fm.

55 • The counting rate from a radioactive source is 6400 counts/s. The half-life of the source is 10 s. Make a plot of the counting rate as a function of time for times up to 1 min. What is the decay constant for this source?

56 • Find the energy needed to remove a neutron from (a) ^{4}He and (b) ^{7}Li.

57 • The isotope ^{14}C decays according to ^{14}C $\rightarrow$ ^{14}N + e^- + $\bar{\nu}_e$. The atomic mass of ^{14}N is 14.003074 u. Determine the maximum kinetic energy of the electron. (Neglect recoil of the nitrogen atom.)

58 • A neutron star is an object of nuclear density. If our sun were to collapse to a neutron star, what would be the radius of that object?

59 • Nucleus A has a half-life that is twice that of nucleus B. At $t = 0$ the number of B nuclei in a sample is twice that of A nuclei. If the half-life of A is 1 h, will there ever be an instant when the number of A and B nuclei are equal? If so, when will this moment occur?

60 • Calculate the nuclear radii of ^{19}F, ^{145}La, and ^{246}Cm.

61 • The relative abundance of ^{40}K (molecular mass 40.0 g/mol) is 1.2×10^{-4}. The isotope ^{40}K is radioactive with a half-life of 1.3×10^9 y. Potassium is an essential element of every living cell. In the human body the mass of potassium constitutes approximately 0.36% of the total mass. Determine the activity of this radioactive source in a student whose mass is 60 kg.

62 •• A 0.05394-kg sample of ^{144}Nd (atomic mass 143.91 u) emits an average of 2.36 α particles each second. Find the decay constant in s^{-1} and the half-life in years.

63 •• The isotope ^{24}Na is a β emitter with a half-life of 15 h. A saline solution containing this radioactive isotope with an activity of 600 kBq is injected into the bloodstream of a patient. Ten hours later, the activity of 1 mL of blood from this individual yields a counting rate of 60 Bq. Determine the volume of blood in this patient.

64 •• (a) Determine the closest distance of approach of an 8-MeV α particle in a head-on collision with a nucleus of ^{197}Au and a nucleus of ^{10}B, neglecting the recoil of the struck nuclei. (b) Repeat the calculation taking into account the recoil of the struck nuclei.

65 •• Twelve nucleons are in a one-dimensional infinite square well of length $L = 3$ fm. (a) Using the approximation that the mass of a nucleon is 1 u, find the lowest energy of a nucleon in the well. Express your answer in MeV. What is the ground-state energy of the system of 12 nucleons in the well if (b) all the nucleons are neutrons so that there can be only 2 in each state and (c) 6 of the nucleons are neutrons and 6 are protons so that there can be 4 nucleons in each state? (Neglect the energy of Coulomb repulsion of the protons.)

66 •• The helium nucleus or α particle is a very tightly bound system. Nuclei with $N = Z = 2n$, where n is an integer, such as ^{12}C, ^{16}O, ^{20}Ne, and ^{24}Mg, may be thought of as agglomerates of α particles. (a) Use this model to estimate the binding energy of a pair of α particles from the atomic masses of ^{4}He and ^{16}O. Assume that the four α particles in ^{16}O form a regular tetrahedron with one α particle at each vertex. (b) From the result obtained in part (a) determine, on the basis of this model, the binding energy of ^{12}C and compare your result with that obtained from the atomic mass of ^{12}C.

67 •• Radioactive nuclei with a decay constant of λ are produced in an accelerator at a constant rate R_p. The number of radioactive nuclei N then obeys the equation $dN/dt = R_p - \lambda N$. (a) If N is zero at $t = 0$, sketch N versus t for this situation. (b) The isotope ^{62}Cu is produced at a rate of 100 per second by placing ordinary copper (^{63}Cu) in a beam of high-energy photons. The reaction is

$$\lambda + {}^{63}\text{Cu} \rightarrow {}^{62}\text{Cu} + \text{n}$$

^{62}Cu decays by β decay with a half-life of 10 min. After a time long enough so that $dN/dt \approx 0$, how many ^{62}Cu nuclei are there?

68 •• The total energy consumed in the United States in 1 y is about 7.0×10^{19} J. How many kilograms of ^{235}U would be needed to provide this amount of energy if we assume that 200 MeV of energy is released by each fissioning uranium nucleus, that all of the uranium atoms undergo fission, and that all of the energy-conversion mechanisms used are 100% efficient?

69 •• (a) Find the wavelength of a particle in the ground state of a one-dimensional infinite square well of length $L = 2$ fm. (b) Find the momentum in units of MeV/c for a particle with this wavelength. (c) Show that the total energy of an electron with this wavelength is approximately $E \approx pc$. (d) What is the kinetic energy of an electron in the ground state of this well? This calculation shows that if an electron were confined in a region of space as small as a nucleus, it would have a very large kinetic energy.

70 •• (a) How many α decays and how many β decays must a ^{222}Rn nucleus undergo before it becomes a ^{210}Pb nucleus? (b) Calculate the total energy released in the decay of one ^{222}Rn nucleus to ^{210}Pb. (The mass of ^{210}Pb is 209.984187 u.)

71 ••• Assume that a neutron decays into a proton plus an electron without the emission of a neutrino. The energy shared by the proton and electron is then 0.782 MeV. In the rest frame of the neutron, the total momentum is zero, so the momentum of the proton must be equal and opposite that of the electron. This determines the relative energies of the two particles, but because the electron is relativistic, the exact calculation of these relative energies is somewhat difficult. (a) Assume that the kinetic energy of the electron is 0.782 MeV and calculate the momentum p of the electron in units of MeV/c. (Hint: Use Equation 39-28.) (b) From your result for (a), calculate the kinetic energy $p^2/2m_p$ of the proton. (c) Since the total energy of the electron plus proton is 0.782 MeV, the calculation in (b) gives a correction to the assumption that the energy of the electron is 0.782 MeV. What percentage of 0.782 MeV is this correction?

72 ••• Consider a neutron of mass m moving with speed v_L and making an elastic head-on collision with a nucleus of mass M that is at rest in the laboratory frame of reference. (a) Show that the speed of the center of mass in the lab frame is $V = mv_L/(m + M)$. (b) What is the speed of the nucleus in the center-of-mass frame before the collision? After the collision? (c) What is the speed of the nucleus in the lab frame after the collision? (d) Show that the energy of the nucleus after the collision in the lab frame is

$$\frac{1}{2}M(2V)^2 = \frac{4mM}{(m + M)^2}\left(\frac{1}{2}mv_L^2\right)$$

(e) Show that the fraction of the energy lost by the neutron in this elastic collision is

$$\frac{-\Delta E}{E} = \frac{4mM}{(m + M)^2} = \frac{4(m/M)}{(1 + m/M)^2} \qquad \text{40-23}$$

73 ••• (a) Use the result of part (e) of Problem 72 (Equation 40-23) to show that after N head-on collisions of a neutron with carbon nuclei at rest, the energy of the neutron is approximately $(0.714)^N E_0$, where E_0 is its original energy. (b) How many head-on collisions are required to reduce the energy of the neutron from 2 MeV to 0.02 eV, assuming stationary carbon nuclei?

74 ••• On the average, a neutron loses 63% of its energy in a collision with a hydrogen atom and 11% of its energy in a collision with a carbon atom. Calculate the number of collisions needed to reduce the energy of a neutron from 2 MeV to 0.02 eV if the neutron collides with (a) hydrogen atoms and (b) carbon atoms. (See Problem 73.)

75 ••• Frequently, the "daughter" of a radioactive "parent" is itself radioactive. Suppose the parent, designated by A, has a decay constant λ_A, while the daughter, designated B, has a decay constant λ_B. The number of nuclei of B are then given by the solution to the differential equation

$$dN_B/dt = \lambda_A N_A - \lambda_B N_B$$

(a) Justify this differential equation. (b) Show that the solution for this equation is

$$N_B(t) = \frac{N_{A0}\lambda_A}{\lambda_B - \lambda_A}(e^{-\lambda_A t} - e^{-\lambda_B t})$$

where N_{A0} is the number of A nuclei present at $t = 0$ when there are no B nuclei. (c) Show that $N_B(t) > 0$ whether $\lambda_A > \lambda_B$ or $\lambda_B > \lambda_A$. (d) Make a plot of $N_A(t)$ and $N_B(t)$ as a function of time when $\tau_B = 3\tau_A$.

76 ••• Suppose isotope A decays to isotope B with a decay constant λ_A, and isotope B in turn decays with a decay constant λ_B. Suppose a sample contains, at $t = 0$, only isotope A. Derive an expression for the time at which the number of isotope B nuclei will be a maximum. (See Problem 75.)

77 ••• An example of the situation discussed in Problem 75 is the radioactive isotope ^{229}Th, an α emitter with a half-life of 7300 y. Its daughter, ^{225}Ra, is a β emitter with a half-life of 14.8 d. In this, as in many instances, the half-life of the parent is much longer than that of the daughter. Using the expression given in Problem 75 (b), show that, starting with a sample of pure ^{229}Th containing N_{A0} nuclei, the number, N_B, of ^{225}Ra nuclei will, after several years, be a constant, given by

$$N_B = \frac{\lambda_A}{\lambda_B}N_A$$

The number of daughter nuclei are said to be in "secular equilibrium."

CHAPTER 41

Elementary Particles and the Beginning of the Universe

Tracks in a bubble chamber produced by an incoming high-energy proton (yellow) incident from the left, colliding with a proton at rest. The small green spiral is an electron knocked out of an atom. It curves to the left because of an external magnetic field in the chamber. The collision produces seven negative particles (π^-)(blue); a neutral particle Λ^0 that leaves no track; and nine positive particles (red) including seven π^+, a K^+, and a proton. The Λ^0 travels in the original direction of the incoming proton before decaying into a proton (yellow) and a π^- (purple).

In Dalton's atomic theory of matter (1808), the atom was considered to be the smallest indivisible constituent of matter, that is, an elementary particle. Then, with the discovery of the electron by Thomson (1897), the Bohr theory of the nuclear atom (1913), and the discovery of the neutron (1932), it became clear that atoms and even nuclei have considerable structure. For a time, it was thought that there were just four "elementary" particles: proton, neutron, electron, and photon. However, the positron or antielectron was discovered in 1932, and shortly thereafter the muon, the pion, and many other particles were predicted and discovered.

Since the 1950s, enormous sums of money have been spent constructing particle accelerators of greater and greater energies in hopes of finding particles predicted by various theories. At present, we know of several hundred particles that at one time or another have been considered to be elementary, and research teams at the giant accelerator laboratories around the world are searching for and finding new particles. Some of these have such short lifetimes (of the order of 10^{-23} s) that they can be detected only indirectly. Many are observed only in nuclear reactions with high-energy accelerators. In addition to the usual particle properties of mass, charge, and spin, new properties have been found and given whimsical names such as strangeness, charm, color, topness, and bottomness.

In this chapter, we will first look at the various ways of classifying the multitude of particles that have been found. We will then describe the current theory of elementary particles, called the *standard model,* in which all matter in nature—from the exotic particles produced in the giant accelerator laboratories to ordinary grains of sand—is considered to be constructed from just two families of elementary particles, leptons and quarks. In the final section we will use our knowledge of elementary particles to discuss the Big Bang theory of the origin of the universe.

41-1   Hadrons and Leptons

All the different forces observed in nature, from ordinary friction to the tremendous forces involved in supernova explosions, can be understood in terms of the four basic interactions: (1) the strong nuclear interaction (also called the hadronic interaction), (2) the electromagnetic interaction, (3) the weak (nuclear) interaction, and (4) the gravitational interaction. The four basic interactions provide a convenient structure for the classification of particles. Some particles participate in all four interactions, whereas others participate in only some of them. For example, all particles participate in gravity, the weakest of the interactions. All particles that carry electric charge participate in the electromagnetic interaction.

Particles that interact via the strong interaction are called **hadrons.** There are two kinds of hadrons: **baryons,** which have spin $\frac{1}{2}$ (or $\frac{3}{2}$, $\frac{5}{2}$, and so on), and **mesons,** which have zero or integral spin. Baryons, which include nucleons, are the most massive of the elementary particles. Mesons have intermediate masses between the mass of the electron and the mass of the proton. Particles that decay via the strong interaction have very short lifetimes of the order of 10^{-23} s, which is about the time it takes light to travel a distance equal to the diameter of a nucleus. On the other hand, particles that decay via the weak interaction have much longer lifetimes of the order of 10^{-10} s. Table 41-1 lists some of the properties of those hadrons that are stable against decay via the strong interaction.

Hadrons are rather complicated entities with complex structures. If we use the term "elementary particle" to mean a point particle without structure that is not constructed from some more elementary entities, hadrons do not fit the bill. It is now believed that all hadrons are composed of more fundamental entities called *quarks,* which are truly elementary particles.

Particles that participate in the weak interaction but not in the strong interaction are called **leptons.** These include electrons, muons, and neutrinos, which are all less massive than the lightest hadron. The word *lepton,* meaning "light particle," was chosen to reflect the relatively small mass of these particles. However, the most recently discovered lepton, the *tau,* found by Perl in 1975, has a mass of 1784 MeV/c^2, nearly twice that of the proton (938 MeV/c^2), so we now have a "heavy lepton." As far as we know, leptons are point particles with no structure and can be considered to be truly elementary in the sense that they are not composed of other particles.

There are six leptons, each of which has an antiparticle. They are the electron, the muon, and the tau, and a distinct neutrino associated with each of these three particles. (The neutrino associated with the tau has not yet been observed experimentally.) The masses of these particles are quite different. The mass of the electron is 0.511 MeV/c^2, the mass of the muon is 106 MeV/c^2, and that of the tau is 1784 MeV/c^2. The neutrinos were originally thought to be massless, but there is now strong evidence that their mass,

The Super-Kamiokande detector, built in Japan in 1996 as a joint Japanese–American experiment, is essentially a water tank the size of a large cathedral installed in a deep zinc mine one mile inside a mountain. When neutrinos pass through the tank, one of them occasionally collides with an atom, sending blue light through the water to an array of detectors. This is a picture of the detector wall and top with about 9000 photomultiplier tubes which help detect the neutrinos. Experimental results reported in June 1998 indicate that the mass of the neutrino cannnot be zero.

Table 41-1

Hadrons That Are Stable Against Decay via the Strong Nuclear Interaction

Name	Symbol	Mass MeV/c^2	Spin, $\hbar$	Charge, e	Antiparticle	Mean Lifetime, s	Typical Decay Products[a]
Baryons							
Nucleon	p (proton)	938.3	$\frac{1}{2}$	+1	$\bar{p}^-$	Infinite	
	n (neutron)	939.6	$\frac{1}{2}$	0	$\bar{n}$	930	$p + e^- + \bar{\nu}_e$
Lambda	Λ^0	1116	$\frac{1}{2}$	0	$\bar{\Lambda}^0$	2.5×10^{-10}	$p + \pi^-$
Sigma[b]	Σ^+	1189	$\frac{1}{2}$	+1	$\bar{\Sigma}^-$	0.8×10^{-10}	$n + \pi^+$
	Σ^0	1193	$\frac{1}{2}$	0	$\bar{\Sigma}^0$	10^{-20}	$\Lambda^0 + \gamma$
	Σ^-	1197	$\frac{1}{2}$	−1	$\bar{\Sigma}^+$	1.7×10^{-10}	$n + \pi^-$
Xi	Ξ^0	1315	$\frac{1}{2}$	0	$\bar{\Xi}^0$	3.0×10^{-10}	$\Lambda^0 + \pi^0$
	Ξ^-	1321	$\frac{1}{2}$	−1	$\bar{\Xi}^+$	1.7×10^{-10}	$\Lambda^0 + \pi^-$
Omega	Ω^-	1672	$\frac{3}{2}$	−1	$\bar{\Omega}^+$	1.3×10^{-10}	$\Xi^0 + \pi^-$
Mesons							
Pion	π^+	139.6	0	+1	π^-	2.6×10^{-8}	$\mu^+ + \nu_\mu$
	π^0	135	0	0	π^0	0.8×10^{-16}	$\gamma + \gamma$
	π^-	139.6	0	−1	π^+	2.6×10^{-8}	$\mu^- + \bar{\nu}_\mu$
Kaon[c]	K^+	493.7	0	+1	K^-	1.24×10^{-8}	$\pi^+ + \pi^0$
	K^0	497.7	0	0	$\bar{K}^0$	0.88×10^{-10} and 5.2×10^{-8}	$\pi^+ + \pi^-$ $\pi^+ + e^- + \nu_e$
Eta	η^0	549	0	0		2×10^{-19}	$\gamma + \gamma$

[a]Other decay modes also occur for most particles.
[b]The Σ^0 is included here for completeness even though it does decay via the strong interaction.
[c]The K^0 has two distinct lifetimes, sometimes referred to as K^0_{short} and K^0_{long}. All other particles have a unique lifetime.

though very small, is not zero. There is no present measurement of the neutrino mass, but it is expected that it is of the order of a few eV/c^2. Experiments designed to detect neutrinos emitted from the sun have found a much smaller number than expected, which could be explained if the mass of the neutrino were not zero. In addition, a mass even as small as 40 eV/c^2 for the neutrino would have great cosmological significance. The answer to the question of whether the universe will continue to expand indefinitely or will reach a maximum size and begin to contract depends on the total mass in the universe. Thus, the answer could depend on whether the rest mass of the neutrino is merely small rather than zero since the cosmic density of each species of neutrino is about 100 cm^3. The observation of electron neutrinos from the supernova 1987A puts an upper limit on the mass of these neutrinos. Since the velocity of a particle with mass depends on its energy, the arrival time of a burst of neutrinos with mass from a supernova would be spread out in time. The fact that the electron neutrinos from the 1987 supernova all arrived at the earth within 13 s of one another results in an upper limit of about 16 eV/c^2 for their mass. Note that an upper limit does not imply that the mass is not zero. Recent measurements of the relative number of muon neutrinos and electron neutrinos entering a huge underground detector in Japan called Super-Kamiokande suggest that at least one type of neutrino can oscillate between types (for example, between a mu neutrino and a tau neutrino). Such oscillation is only possible if the neutrino has mass.

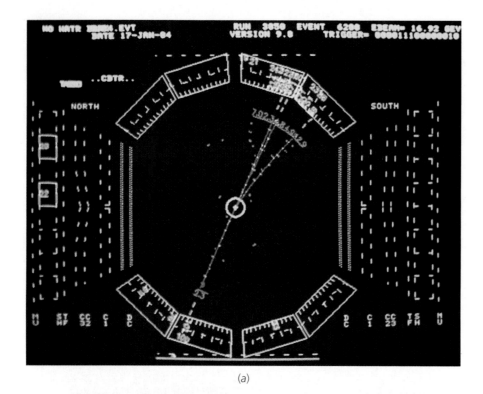

(a)

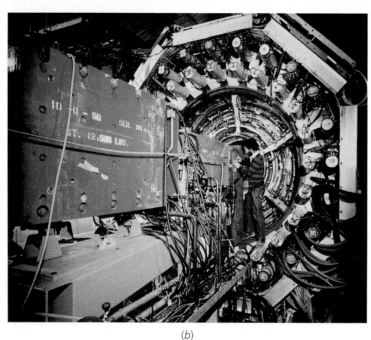

(b)

(*a*) A computer display of the production and decay of a τ^+ and τ^- pair. An electron and positron annihilate at the center marked by the yellow cross, producing a τ^+ and τ^- pair, which travel in opposite directions, but quickly decay while still inside the beam pipe (yellow circle). The τ^+ decays into two invisible neutrinos and a μ^+, which travels toward the bottom left. Its track in the drift chamber is calculated by a computer and indicated in red. It penetrates the lead–argon counters outlined in purple and is detected at the blue dot near the bottom blue line that marks the end of a muon detector. The τ^- decays into three charged pions (red tracks moving upward) plus invisible neutrinos. (*b*) The Mark I detector, built by a team from the Stanford Linear Accelerator Center (SLAC) and the Lawrence Berkeley Laboratory, became famous for many discoveries, including the ψ/J meson and the τ lepton. Tracks of particles are recorded by wire spark chambers wrapped in concentric cylinders around the beam pipe extending out to the ring where physicist Carl Friedberg has his right foot. Beyond this are two rings of protruding tubes, housing photomultipliers that view various scintillation counters. The rectangular magnets at the left guide the counterrotating beams that collide in the center of the detector.

41-2   Spin and Antiparticles

One important characteristic of a particle is its intrinsic spin angular momentum. We have already discussed the fact that the electron has a quantum number m_s that corresponds to the z component of its intrinsic spin characterized by the quantum number $s = \frac{1}{2}$. Protons, neutrons, neutrinos, and the various other particles that also have an intrinsic spin characterized by the quantum number $s = \frac{1}{2}$ are called **spin-$\frac{1}{2}$ particles.** Particles that have spin $\frac{1}{2}$

(or $\frac{3}{2}, \frac{5}{2}, \ldots$) are called fermions and obey the Pauli exclusion principle. Particles such as pions and other mesons have zero spin or integral spin ($s = 0, 1, 2, \ldots$). These particles are called bosons and do not obey the Pauli exclusion principle. Any number of these particles can be in the same quantum state.

Spin-$\frac{1}{2}$ particles are described by the Dirac equation, an extension of the Schrödinger equation that includes special relativity. One feature of Dirac's theory, proposed in 1927, is the prediction of the existence of antiparticles. In special relativity, the energy of a particle is related to the mass and momentum of the particle by $E = \pm\sqrt{p^2c^2 + m^2c^4}$ (Equation 39-28). We usually choose the positive solution and dismiss the negative-energy solution with a physical argument. However, the Dirac equation requires the existence of wave functions that correspond to the negative-energy states. Dirac got around this difficulty by postulating that all the negative-energy states were filled and would therefore not

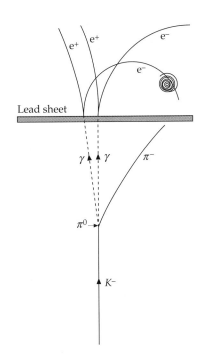

A negative kaon (K^-) enters a bubble chamber from the bottom and decays into a π^-, which moves off to the right, and a π^0, which immediately decays into two photons whose paths are indicated by the dashed lines in the drawing. Each photon interacts in the lead sheet, producing an electron–positron pair. The spiral at the right is another electron that has been knocked out of an atom in the chamber. (Other extraneous tracks have been removed from the photograph.)

be observable. Only holes in the "infinite sea" of negative-energy states would be observed. For example, a hole in the negative sea of electron energy states would appear as a particle identical to the electron except with positive charge. When such a particle came in the vicinity of an electron the two particles would annihilate, releasing energy of $2m_ec^2$. This interpretation received little attention until the positron with just these properties was discovered in 1932 by Carl Anderson.

Antiparticles are never created alone but always in particle–antiparticle pairs. In the creation of an electron–positron pair by a photon, the energy of the photon must be greater than the rest energy of the electron plus that of the positron, which is $2m_ec^2 \approx 1.02$ MeV, where m_e is the mass of the electron. Although the positron is stable, it has only a short-term existence in our universe because of the large supply of electrons in matter. The fate of a positron is annihilation according to the reaction

$$e^+ + e^- \rightarrow \gamma + \gamma \qquad\qquad 41\text{-}1$$

The probability of this reaction is large only if the positron is at rest or nearly at rest. Two photons moving in opposite directions are needed to conserve linear momentum.

The fact that we call electrons *particles* and positrons *antiparticles* does not imply that positrons are less fundamental than electrons. It merely reflects the nature of our part of the universe. If our matter were made up of negative protons and positive electrons, then positive protons and negative electrons would suffer quick annihilation and would be called antiparticles.

The antiproton (p^-) was discovered in 1955 by E. Segrè and O. Chamberlain using a beam of protons in the Bevatron at Berkeley to produce the reaction*

$$p^+ + p^+ \rightarrow p^+ + p^+ + p^+ + p^- \qquad\qquad 41\text{-}2$$

*The antiproton is sometimes denoted by $\bar{p}$ rather than p^-. For neutral particles, such as the neutron, the bar must be used to denote the antiparticle. Thus the antineutron is denoted by $\bar{n}$. The normal electron and proton are often denoted by e and p without the minus or plus superscripts.

The creation of a proton–antiproton pair (Figure 41-1) requires kinetic energy of at least $2m_pc^2 = 1877$ MeV $= 1.877$ GeV in the zero-momentum reference frame in which the two protons approach each other with equal and opposite momenta. In the laboratory frame in which one of the protons is initially at rest, the kinetic energy of the incoming proton must be at least $6m_pc^2 = 5.63$ GeV (see Problem 60 of Chapter 39). This energy was not available in laboratories before the development of high-energy accelerators in the 1950s. Antiprotons annihilate with protons to produce two gamma rays in a reaction similar to that in Equation 41-1.

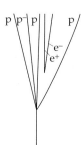
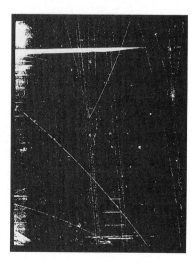

Figure 41-1 Bubble-chamber tracks showing the creation of a proton–antiproton pair in the collision of an incident 25-GeV proton with a stationary proton in liquid hydrogen.

Air view of the European Laboratory for Particle Physics (CERN) just outside of Geneva, Switzerland. The large circle shows the Large Electron–Positron collider (LEP) tunnel, which is 27 km in circumference. The irregular dashed line is the border between France and Switzerland.

The tunnel of the proton–antiproton collider at CERN. The same bending magnets and focusing magnets can be used for protons or antiprotons moving in opposite directions. The rectangular box in the foreground is a focusing magnet; the next four boxes are bending magnets.

Example 41-1

A proton and an antiproton at rest annihilate according to the reaction $p^+ + p^- \rightarrow \gamma + \gamma$. Find the energies and wavelengths of the photons.

Picture the Problem Because the proton and the antiproton are at rest, conservation of momentum requires that the two photons created in their annihilation have equal and opposite momenta and therefore equal energies. Conservation of energy implies that the total energy of the photons equals the rest energy of the proton plus that of the antiproton (approximately 938 MeV each).

1. Set the total energy of the two photons, $2E_\gamma$, equal to the rest energy of the proton plus antiproton and solve for E_γ:

$$2E_\gamma = 2m_p c^2$$
$$E_\gamma = m_p c^2 = 938 \text{ MeV}$$

2. Set the energy of the photon equal to $hf = hc/\lambda$ and solve for the wavelength λ:

$$E_\gamma = hf = \frac{hc}{\lambda}$$

$$\lambda = \frac{hc}{E_\gamma} = \frac{1240 \text{ eV·nm}}{938 \text{ MeV}} = 1.32 \times 10^{-6} \text{ nm}$$

$$= 1.32 \text{ fm}$$

41-3   The Conservation Laws

One of the maxims of nature is "anything that can happen does." If a conceivable decay or reaction does not occur, there must be a reason. The reason is usually expressed in terms of a conservation law. The conservation of energy rules out the decay of any particle for which the total rest mass of the decay products would be greater than the initial rest mass of the particle before decay. The conservation of linear momentum requires that when an electron and positron at rest annihilate, two photons must be emitted. Angular momentum must also be conserved in a reaction or decay. A fourth conservation law that restricts the possible particle decays and reactions is that of electric charge. The net electric charge before a decay or reaction must equal the net charge after the decay or reaction.

There are two additional conservation laws that are important in the reactions and decays of elementary particles: the conservation of baryon number and the conservation of lepton number. Consider the possible decay

$$p \rightarrow \pi^0 + e^+$$

This decay would conserve charge, energy, angular momentum, and linear momentum, but it does not occur. It does not conserve either lepton number or baryon number. The conservation of lepton number and baryon number implies that whenever a lepton or baryon is created, an antiparticle of the same type is also created. We assign the **lepton number** $L = +1$ to all leptons, $L = -1$ to all antileptons, and $L = 0$ to all other particles. Similarly, the **baryon number** $B = +1$ is assigned to all baryons, $B = -1$ to all antibaryons, and $B = 0$ to all other particles. The baryon and lepton numbers cannot change in a reaction or decay. The conservation of baryon number along with the conservation of energy implies that the least massive baryon, the proton, must be stable.

The conservation of lepton number implies that the neutrino emitted in the β decay of the free neutron is an antineutrino:

$$n \rightarrow p^+ + e^- + \bar{\nu}_e \qquad\qquad 41\text{-}3$$

The fact that neutrinos and antineutrinos are different is illustrated by an experiment in which ^{37}Cl is bombarded with an intense antineutrino beam from the decay of reactor neutrons. If neutrinos and antineutrinos were the same, we would expect the following reaction:

$$^{37}\text{Cl} + \bar{\nu}_e \rightarrow {}^{37}\text{Ar} + e^- \qquad\qquad 41\text{-}4$$

This reaction is not observed. However, if *protons* are bombarded with antineutrinos, the reaction

$$p + \bar{\nu}_e \rightarrow n + e^+ \qquad\qquad 41\text{-}5$$

is observed. Note that the lepton number is -1 on the left side of reaction 41-4 and $+1$ on the right side. But the lepton number is -1 on both sides of reaction 41-5.

Not only are neutrinos and antineutrinos distinct particles, but the neutrinos associated with electrons are distinct from the neutrinos associated with muons. Electron-like leptons (e and ν_e), muon-like leptons (μ and ν_μ), and tau-like leptons (τ and ν_τ) are each separately conserved, so we assign separate lepton numbers L_e, L_μ, and L_τ to the particles. For e and ν_e, $L_e = +1$; for their antiparticles, $L_e = -1$; and for all other particles, $L_e = 0$. The lepton numbers L_μ and L_τ are similarly assigned.

Example 41-2

What conservation laws (if any) are violated by the following decays?
(a) $n \rightarrow p + \pi^-$ (b) $\Lambda^0 \rightarrow p^- + \pi^+$ (c) $\mu^- \rightarrow e^- + \gamma$

(a) There are no leptons in this decay, so there is no problem with the conservation of lepton number. The net charge is zero before and after the decay, so charge is conserved. Also, the baryon number is +1 before and after the decay. However, the rest energy of the proton (938.3 MeV) plus that of the pion (139.6 MeV) is greater than the rest energy of the neutron (939.6 MeV). Thus, this decay violates the conservation of energy.

(b) Again, there are no leptons involved, and the net charge is zero before and after the decay. Also, the rest energy of the Λ^0 (1116 MeV) is greater than the rest energy of the antiproton (938.3 MeV) plus that of the pion (139.6 MeV), so energy is conserved, with the loss in rest energy equaling the gain in kinetic energy of the decay products. However, this decay does not conserve baryon number, which is +1 for the Λ^0 and −1 for the antiproton.

(c) This reaction does not conserve muon lepton number or electron lepton number. The muon does decay via

$$\mu^- \to e^- + \bar{\nu}_e + \nu_\mu$$

which does conserve both muon and electron lepton numbers.

There are some conservation laws that are not universal but apply only to certain kinds of interactions. In particular, there are quantities that are conserved in decays and reactions that occur via the strong interaction but not in decays or reactions that occur via the weak interaction. One of these quantities that is particularly important is **strangeness,** introduced by M. Gell-Mann and K. Nishijima in 1952 to explain the strange behavior of some of the heavy baryons and mesons. Consider the reaction

$$p + \pi^- \to \Lambda^0 + K^0 \qquad\qquad 41\text{-}6$$

The proton and pion interact via the strong interaction. Both the Λ^0 and K^0 decay into hadrons

$$\Lambda^0 \to p + \pi^- \qquad\qquad 41\text{-}7$$

and

$$K^0 \to \pi^+ + \pi^- \qquad\qquad 41\text{-}8$$

However, the decay times for both the Λ^0 and K^0 are of the order of 10^{-10} s, which is characteristic of the weak interaction, rather than 10^{-23} s, which would be expected for the strong interaction. Other particles showing similar behavior were called **strange particles.** These particles are always produced in pairs and never singly, even when all other conservation laws are met. This behavior is described by assigning a new property called strangeness to these particles. In reactions and decays that occur via the strong interaction, strangeness is conserved. In those that occur via the weak interaction the strangeness can change by ±1. The strangeness of the ordinary hadrons—the nucleons and pions—was arbitrarily taken to be zero. The strangeness of the K^0 was arbitrarily chosen to be +1. The strangeness of the Λ^0 particle must then be −1 so that strangeness is

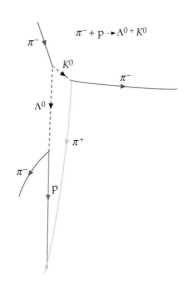

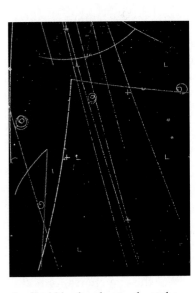

An early photograph of bubble-chamber tracks at the Lawrence Berkeley Laboratory, showing the production and decay of two strange particles, the K^0 and the Λ^0. These neutral particles are identified by the tracks of their decay particles. The lambda particle was named because of the similarity of the tracks of its decay particles to the Greek letter Λ. (The blue tracks are particles not involved in the reaction of Equation 41-6.)

conserved in reaction 41-6. The strangeness of other particles could then be assigned by looking at their various reactions and decays. In those that occur via the weak interaction, the strangeness can change by ±1.

Figure 41-2 shows the masses of the baryons and mesons that are stable against decay via the strong interaction versus strangeness. We can see from this figure that these particles cluster in multiplets of one, two, or three particles of approximately equal mass, and that the strangeness of a multiplet of particles is related to the "center of charge" of the multiplet.

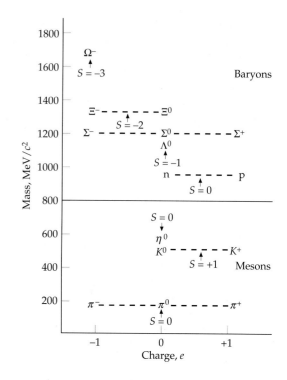

Figure 41-2 The strangeness of hadrons shown on a plot of rest mass versus charge. The strangeness of a baryon-charge multiplet is related to the number of places the center of charge of the multiplet is displaced from that of the nucleon doublet. For each displacement of $\frac{1}{2}e$, the strangeness changes by ±1. For mesons, the strangeness is related to the number of places the center of charge is displaced from that of the pion triplet. Because of the unfortunate original assignment of +1 for the strangeness of kaons, all of the baryons that are stable against decay via the strong interaction have negative or zero strangeness.

Example 41-3

State whether the following decays can occur via the strong interaction, via the weak interaction, or not at all:
(a) $\Sigma^+ \rightarrow p + \pi^0$ (b) $\Sigma^0 \rightarrow \Lambda^0 + \gamma$ (c) $\Xi^0 \rightarrow n + \pi^0$

Picture the Problem We first note that the mass of each decaying particle is greater than that of the decay products, so there is no problem with energy conservation in any of the decays. In addition, there are no leptons involved in any of the decays, and charge and baryon number are both conserved in all the decays. The decay will occur via the strong interaction if strangeness is conserved. If $\Delta S = \pm 1$, it will occur via the weak interaction. If S changes by more than 1, the decay will not occur.

(a) From Figure 41-2, we can see that the strangeness of the Σ^+ is -1, whereas the strangeness of both the proton and the pion is zero. This decay is possible via the weak interaction but not the strong interaction. It is, in fact, one of the decay modes of the Σ^+ particle with a lifetime of the order of 10^{-10} s.

(b) Since the strangeness of both the Σ^0 and Λ^0 is -1, this decay can proceed via the strong interaction. It is, in fact, the dominant mode of decay of the Σ^0 particle with a lifetime of about 10^{-20} s.

(c) The strangeness of the Ξ^0 is -2, whereas the strangeness of both the neutron and pion is zero. Since strangeness cannot change by 2 in a decay or reaction, this decay cannot occur.

41-4   Quarks

Leptons appear to be truly elementary particles in that they do not break down into smaller entities and they seem to have no measurable size or structure. Hadrons, on the other hand, are complex particles with size and structure, and they decay into other hadrons. Furthermore, at the present time, there are only six known leptons, whereas there are many more hadrons. Except for the Σ^0 particle, Table 41-1 includes only hadrons that are stable against decay via the strong interaction. Hundreds of other hadrons have been discovered, and their properties, such as charge, spin, mass, strangeness, and decay schemes, have been measured.

The most important advance in our understanding of elementary particles was the quark model proposed by M. Gell-Mann and G. Zweig in 1963 in which all hadrons consist of combinations of two or three truly elementary particles called **quarks.*** In the original model, quarks came in three types, called **flavors,** labeled u, d, and s (for *up*, *down*, and *strange*). An unusual property of quarks is that they carry fractional electron charges. The charge of the u quark is $+\frac{2}{3}e$ and that of the d and s quarks is $\frac{1}{3}e$. Each quark has spin $\frac{1}{2}$ and a baryon number of $\frac{1}{3}$. The strangeness of the u and d quark is 0 and that of the s quark is -1. Each quark has an antiquark with the opposite electric charge, baryon number, and strangeness. Baryons consist of three quarks (or three antiquarks for antiparticles), whereas mesons consist of a quark and an antiquark, giving them a baryon number $B = 0$, as required. The proton consists of the combination uud and the neutron, udd. Baryons with a strangeness $S = -1$ contain one s quark. All the particles listed in Table 41-1 can be constructed from these three quarks and three antiquarks.[†] The great strength of the quark model is that all the allowed combinations of three quarks or quark–antiquark pairs result in known hadrons. Strong evidence for the existence of quarks inside a nucleon is provided by high-energy scattering experiments called *deep inelastic scattering.* In these experiments, a nucleon is bombarded with electrons, muons, or neutrinos of energies from 15 to 200 GeV. Analyses of particles scattered at large angles indicate that inside the nucleon are 3 spin-$\frac{1}{2}$ particles of sizes much smaller than that of the nucleon. These experiments are analogous to Rutherford's scattering of α particles by atoms in which the presence of a tiny nucleus in the atom was inferred from the large-angle scattering of the α particles.

*The name *quark* was chosen by Gell-Mann from a quotation from *Finnegans Wake* by James Joyce.

[†]The correct quark combinations of hadrons are not always obvious, because of the symmetry requirements on the total wave function. For example, the π^0 meson is represented by a linear combination of $u\bar{u}$ and $d\bar{d}$.

Example 41-4

What are the properties of the particles made up of the following quarks: (a) $u\bar{d}$, (b) $\bar{u}d$, (c) dds, and (d) uss?

Picture the Problem Baryons are made up of 3 quarks, whereas mesons consist of a quark and an antiquark. We add the electric charges of the quarks to find the total charge of the hadron. We also find the strangeness of the hadron by adding the strangeness of the quarks.

(a) Since $u\bar{d}$ is a quark–antiquark combination, it has baryon number 0 and is therefore a meson. There is no strange quark here, so the strangeness of the meson is zero. The charge of the up quark is $+\frac{2}{3}e$ and that of the antidown quark is $+\frac{1}{3}e$, so the charge of the meson is $+ 1e$. This is the quark combination of the π^+ meson.

(b) The particle $\bar{u}d$ is also a meson with zero strangeness. Its electric charge is $-\frac{2}{3}e + (-\frac{1}{3}e) = -1e$. This is the quark combination of the π^- meson.

(c) The particle dds is a baryon with strangeness -1 since it contains one strange quark. Its electric charge is $-\frac{1}{3}e - \frac{1}{3}e - \frac{1}{3}e = -1e$. This is the quark combination for the Σ^- particle.

(d) The particle uss is a baryon with strangeness -2. Its electric charge is $+\frac{2}{3}e - \frac{1}{3}e - \frac{1}{3}e = 0$. This is the quark combination for the Ξ^0 particle.

In 1967, a fourth quark was proposed to explain some discrepancies between experimental determinations of certain decay rates and calculations based on the quark model. The fourth quark is labeled c for a new property called **charm.** Like strangeness, charm is conserved in strong interactions but changes by ± 1 in weak interactions. In 1975, a new heavy meson called the **ψ/J particle** (or simply the **ψ particle**) was discovered that has the properties expected of a $c\bar{c}$ combination. Since then other mesons with combinations such as $c\bar{d}$ and $\bar{c}d$, as well as baryons containing the charmed quark, have been discovered. Two more quarks labeled t and b (for top and bottom) were proposed in the 1970s. In 1977, a massive new meson called the **Y meson** or **bottomonium,** which is considered to have the quark combination $b\bar{b}$, was discovered. The top quark was observed in 1995. The properties of the six quarks are listed in Table 41-2.

Table 41-2

Properties of Quarks and Antiquarks

Flavor	Spin	Charge	Baryon Number	Strangeness	Charm	Topness	Bottomness
Quarks							
u (up)	$\frac{1}{2}\hbar$	$+\frac{2}{3}e$	$+\frac{1}{3}$	0	0	0	0
d (down)	$\frac{1}{2}\hbar$	$-\frac{1}{3}e$	$+\frac{1}{3}$	0	0	0	0
s (strange)	$\frac{1}{2}\hbar$	$-\frac{1}{3}e$	$+\frac{1}{3}$	-1	0	0	0
c (charmed)	$\frac{1}{2}\hbar$	$+\frac{2}{3}e$	$+\frac{1}{3}$	0	$+1$	0	0
t (top)	$\frac{1}{2}\hbar$	$+\frac{2}{3}e$	$+\frac{1}{3}$	0	0	$+1$	0
b (bottom)	$\frac{1}{2}\hbar$	$-\frac{1}{3}e$	$+\frac{1}{3}$	0	0	0	$+1$
Antiquarks							
$\bar{u}$	$\frac{1}{2}\hbar$	$-\frac{2}{3}e$	$-\frac{1}{3}$	0	0	0	0
$\bar{d}$	$\frac{1}{2}\hbar$	$+\frac{1}{3}e$	$-\frac{1}{3}$	0	0	0	0
$\bar{s}$	$\frac{1}{2}\hbar$	$+\frac{1}{3}e$	$-\frac{1}{3}$	$+1$	0	0	0
$\bar{c}$	$\frac{1}{2}\hbar$	$-\frac{2}{3}e$	$-\frac{1}{3}$	0	-1	0	0
$\bar{t}$	$\frac{1}{2}\hbar$	$-\frac{2}{3}e$	$-\frac{1}{3}$	0	0	-1	0
$\bar{b}$	$\frac{1}{2}\hbar$	$+\frac{1}{3}e$	$-\frac{1}{3}$	0	0	0	-1

The six quarks and six leptons (and their antiparticles) are thought to be the fundamental, elementary particles of which all matter is composed. Table 41-3 lists the masses of the fundamental particles. In this table, the masses given for neutrinos are upper limits. The masses given for quarks are educated guesses. There is experimental evidence for the existence of each of these particles.

Table 41-3

Masses of Fundamental Particles

Particle	Mass
Quarks	
u (up)	$336 \ \mathrm{MeV}/c^2$
d (down)	$338 \ \mathrm{MeV}/c^2$
s (strange)	$540 \ \mathrm{MeV}/c^2$
c (charmed)	$1{,}500 \ \mathrm{MeV}/c^2$
t (top)	$174{,}000 \ \mathrm{MeV}/c^2$
b (bottom)	$500 \ \mathrm{MeV}/c^2$
Leptons	
e^- (electron)	$0.511 \ \mathrm{MeV}/c^2$
ν_e (electron neutrino)	$< 7 \ \mathrm{eV}/c^2$
μ^- (muon)	$105.659 \ \mathrm{MeV}/c^2$
ν_μ (muon neutrino)	$< 0.27 \ \mathrm{MeV}/c^2$
τ^- (tau)	$1{,}784 \ \mathrm{MeV}/c^2$
ν_τ (tau neutrino)	$< 31 \ \mathrm{MeV}/c^2$

Quark Confinement

Despite considerable experimental effort, no isolated quark has ever been observed. It is now believed that it is impossible to obtain an isolated quark. Although the force between quarks is not known, it is believed that the potential energy of two quarks increases with increasing separation distance so that an infinite amount of energy would be needed to separate the quarks completely. This would be true, for example, if the force of attraction between two quarks remains constant or increases with separation distance, rather than decreasing with increasing separation distance as is the case for other fundamental forces, such as the electric force between two charges, the gravitational force between two masses, and the strong nuclear force between two hadrons.

When a large amount of energy is added to a quark system such as a nucleon, a quark–antiquark pair is created and the original quarks remain confined within the original system. Because quarks cannot be isolated, but are always bound in a baryon or meson, the mass of a quark cannot be accurately known, which is why the masses listed in Table 41-3 are merely educated guesses.

41-5   Field Particles

In addition to the six fundamental leptons and six fundamental quarks, there are other particles, called *field particles* or *field quanta*, that are associated with the forces exerted by one elementary particle on another. In **quantum electrodynamics** the electromagnetic field of a single charged particle is described by **virtual photons** that are continuously being emitted and reabsorbed by the particle. If we put energy into the system by accelerating the charge, some of these virtual photons can be "shaken off" and become real, observable photons. The photon is said to mediate the electromagnetic interaction. Each of the four basic interactions can be described in this way.

The field quantum associated with the gravitational interaction, called the **graviton,** has not yet been observed. The gravitational "charge" analogous to electric charge is mass.

The weak interaction is thought to be mediated by three field quanta called **vector bosons:** W^+, W^-, and Z^0. These particles were predicted by S. Glashow, A. Salam, and S. Weinberg in a theory called the *electroweak theory,* which we discuss in the next section. The W and Z particles were first observed in 1983 by a group of over a hundred scientists led by C. Rubbia using the high-energy accelerator at CERN in Geneva, Switzerland. The masses of the $W^\pm$ particles (about 80 GeV/c^2) and the Z particle (about 91 GeV/c^2) measured in this experiment were in excellent agreement with those predicted by the electroweak theory. (The W^- particle is the antiparticle of the W^+ particle, so they must have identical masses.)

The field quanta associated with the strong force between quarks are called **gluons.** Isolated gluons have not been observed experimentally. The "charge" responsible for the strong interactions comes in three varieties, labeled *red, green,* and *blue* (analogous with the three primary colors), and the strong charge is called the **color charge.** The field theory for strong interactions, analogous to quantum electrodynamics for electromagnetic interactions, is called **quantum chromodynamics (QCD).**

Table 41-4 lists the bosons responsible for mediating the basic interactions.

Table 41-4

Bosons That Mediate the Basic Interactions

Interaction	Boson	Spin	Mass	Electric Charge
Strong	g (gluon)^a	1	0	0
Weak	$W^\pm$	1	80.22 GeV/c^2	$\pm 1e$
	Z^0	1	91.19 GeV/c^2	0
Electromagnetic	γ (photon)	1	0	0
Gravitational	Graviton^a	2	0	0

^{a} Not yet observed.

41-6   The Electroweak Theory

In the **electroweak theory,** the electromagnetic and weak interactions are considered to be two different manifestations of a more fundamental electroweak interaction. At very high energies ($\gg$100 GeV), the electroweak interaction would be mediated by four bosons. From symmetry considerations, these would be a triplet consisting of W^+, W^0, and W^-, all of equal mass, and a singlet B^0 of some other mass. Neither the W^0 nor the B^0 would be observed directly, but one linear combination of the W^0 and the B^0 would be the Z^0 and another would be the photon. At ordinary energies, the symmetry is broken. This leads to the separation of the electromagnetic interaction mediated by the massless photon and the weak interaction mediated by the W^+, W^-, and Z^0 particles. The fact that the photon is massless and that the W and Z particles have masses of the order of 100 GeV/c^2 shows that the symmetry assumed in the electroweak theory does not exist at lower energies.

The symmetry-breaking mechanism is called a **Higgs field,** which requires a new boson, the **Higgs boson,** whose rest energy is expected to be of the order of 1 TeV (1 TeV $= 10^{12}$ eV). The Higgs boson has not yet been observed. Calculations show that Higgs bosons (if they exist) should be produced in head-on collisions between protons of energies of the order of 20 TeV. Such energies are not presently available.

41-7   The Standard Model

The combination of the quark model, electroweak theory, and quantum chromodynamics is called the **standard model.** In this model, the fundamental particles are the leptons and quarks, each of which comes in six flavors as shown in Table 41-3, and the force carriers are the photon, the $W^{\pm}$ and Z particles, and the gluons (of which there are eight types). The leptons and quarks are all spin-$\frac{1}{2}$ fermions, which obey the Pauli exclusion principle, and the force carriers are integral-spin bosons, which do not obey the Pauli exclusion principle. Every force in nature is due to one of the four basic interactions: strong, electromagnetic, weak, and gravitational. A particle experiences one of the basic interactions if it carries a charge associated with that interaction. Electric charge is the familiar charge that we have studied previously. Weak charge, also called flavor charge, is carried by leptons and quarks. The charge associated with the strong interaction is called color charge and is carried by quarks and gluons but not by leptons. The charge associated with the gravitational force is mass. It is important to note that the photon, which mediates the electromagnetic interaction, does not carry electric charge. Similarly, the $W^{\pm}$ and Z particles, which mediate the weak interaction, do not carry weak charge. However, the gluons, which mediate the strong interaction, do carry color charge. This fact is related to the confinement of quarks as discussed in Section 41-4.

All matter is made up of leptons or quarks. There are no known composite particles consisting of leptons bound together by the weak force. Leptons exist only as isolated particles. Hadrons (baryons and mesons) are composite particles consisting of quarks bound together by the color charge. A result of QCD theory is that only color-neutral combinations of quarks are allowed. Three quarks of different colors can combine to form color-neutral baryons, such as the neutron and proton. Mesons contain a quark and an antiquark and are also color-neutral. Excited states of hadrons are considered to be different particles. For example, the Δ^{+} particle is an excited state of the proton. Both are made up of the *uud* quarks, but the proton is in the ground state with spin $\frac{1}{2}$ and a rest energy of 938 MeV, whereas the Δ^{+} particle is in the first excited state with spin $\frac{3}{2}$ and a rest energy of 1232 MeV. The two *u* quarks can be in the same spin state in the Δ^{+} without violating the exclusion principle, because they have different color. All baryons eventually decay to the lightest baryon, the proton. The proton cannot decay because of conservation of energy and conservation of baryon number.

The strong interaction has two parts, the fundamental or color interaction and what is called the *residual strong interaction.* The fundamental interaction is responsible for the force exerted by one quark on another and is mediated by gluons. The residual strong interaction is responsible for the force between color-neutral nucleons, such as the neutron and proton. This force is due to the residual strong interactions between the color-charged quarks that make up the nucleons and can be viewed as being mediated by the exchange of mesons. The residual strong interaction between color-neutral nucleons

Table 41-5

Properties of the Basic Interactions

	Gravitational	Weak	Electromagnetic	Strong	
				Fundamental	Residual
Acts on	Mass	Flavor	Electric charge	Color charge	
Particles experiencing	All	Quarks, leptons	Electrically charged	Quarks, gluons	Hadrons
Particles mediating	Graviton	$W^\pm, Z$	γ	Gluons	Mesons
Strength for two quarks at 10^{-18} m^a	10^{-41}	0.8	1	25	(not applicable)
Strength for two protons in nucleus^a	10^{-36}	10^{-7}	1	(not applicable)	20

^{a} Strengths are relative to electromagnetic strength.

can be thought of as analogous to the residual electromagnetic interaction between neutral atoms that bind them together to form molecules. Table 41-5 lists some of the properties of the basic interactions.

For each particle there is an antiparticle. A particle and its antiparticle have identical mass and spin but opposite electric charge. For leptons, the lepton numbers L_e, L_μ, and L_τ of the antiparticles are the negatives of the corresponding numbers for the particles. For example, the lepton number for the electron is $L_e = +1$ and that for the positron is $L_e = -1$. For hadrons, the baryon number, strangeness, charm, topness, and bottomness are the sums of those quantities for the quarks that make up the hadron. The number of each antiparticle is the negative of the number for the corresponding particle. For example, the lambda particle Λ^0, which is made up of the uds quarks, has $B = 1$ and $S = -1$, whereas its antiparticle $\overline{\Lambda}^0$, which is made up of the $\overline{u}\overline{d}\overline{s}$ quarks, has $B = -1$ and $S = +1$. A particle such as the photon γ or the Z^0 particle that has zero electric charge, $B = 0$, $L = 0$, $S = 0$, and zero charm, topness, and bottomness, is its own antiparticle. Note that the K^0 meson ($d\overline{s}$) has a zero value for all of these quantities except strangeness, which is +1. Its antiparticle, the $\overline{K}^0$ meson ($\overline{d}s$), has strangeness -1, which makes it distinct from the K^0. The π^+ ($u\overline{d}$) and π^- ($\overline{u}d$) are somewhat special in that they have electric charge but zero values for L, B, and S. They are antiparticles of each other, but since there is no conservation law for mesons, it is impossible to say which is the particle and which is the antiparticle. Similarly, the W^+ and W^- are antiparticles of each other.

Grand Unification Theories

With the success of the electroweak theory, attempts have been made to combine the strong, electromagnetic, and weak interactions in various **grand unification theories** known as **GUTs**. In one of these theories, leptons and quarks are considered to be two aspects of a single class of particles. Under certain conditions, a quark could change into a lepton and vice versa, even though this would appear to violate the conservation of lepton number and baryon number. One of the exciting predictions of this theory is that the proton is not stable but merely has a very long lifetime of the order of 10^{31} y. Such a long lifetime makes proton decay difficult to observe. However, projects are ongoing in which detectors monitor very large numbers of protons in search of an event indicating the decay of a proton.

41-8 **Evolution of the Universe**

In the presently accepted model, the universe began with a singular cataclysmic event called the **Big Bang** and is expanding. The first evidence that the universe is expanding was the astronomer E. P. Hubble's discovery of the relation between the redshifts in the spectra of galaxies and their distances from us. This relation is illustrated in Figure 41-3 for a group of spiral galaxies used by astronomers for calibrating distances. Provided that the redshift is due to the Doppler effect, the recession velocity v of a galaxy is related to its distance r from us by **Hubble's law,**

$$v = Hr \qquad\qquad 41\text{-}9$$

where H is the **Hubble constant.** In principle, the value of H is easy to obtain since it relies on the direct calculation of v from redshift measurements. However, astronomical distances are very difficult to obtain and they have been computed for only a fraction of the 10^{10} or so galaxies in the observable universe. Thus, the value of H changes as distance calibration data are refined. The currently accepted value of the Hubble constant is

$$H = \frac{23 \text{ km/s}}{10^6\, c\cdot\text{y}} \qquad\qquad 41\text{-}10$$

Hubble's law tells us that the galaxies are all rushing away from us, with those the farthest away moving the fastest. However, there is no reason why our location should be special. An observer in any galaxy would make the same observations and compute the same Hubble constant. Thus, Hubble's law suggests that all of the galaxies are receding from each other at an average speed of 23 km/s per $10^6\, c\cdot\text{y}$ of separation. In other words, the universe is expanding. Notice that the basic dimension of H is reciprocal time. The quantity $1/H$ is called the **Hubble age** and equals about 1.3×10^{10} y. This would correspond to the age of the universe if the gravitational pull on the receding galaxies were ignored.

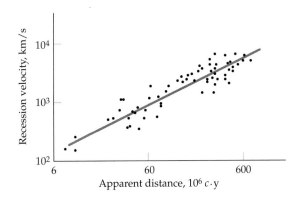

Figure 41-3 A plot of the recession velocities of individual galaxies versus apparent distance.

Example **41-5**

Redshift measurements of a galaxy in the constellation Virgo yield a recession velocity of 1200 km/s. How far is it to that galaxy?

Picture the Problem We calculate the distance from Hubble's law.

1. Use Hubble's law to find r: $\qquad\qquad r = \dfrac{v}{H} = (1200 \text{ km/s})\dfrac{10^6\, c\cdot\text{y}}{23 \text{ km/s}} = 52 \times 10^6\, c\cdot\text{y}$

Exercise Show that $1/H = 1.3 \times 10^{10}$ y.

The 2.7-K Background Radiation

In investigating ways of accounting for the cosmic abundance of elements heavier than hydrogen, cosmologists recognized that nucleosynthesis in stars could explain the abundance of elements heavier than helium but could not by itself explain that of helium. Helium must therefore have been formed during the Big Bang. To synthesize an amount of helium sufficient to account for its present abundance, the Big Bang would have to have occurred at an extremely high initial temperature to provide the necessary reaction rate before fusion was shut down by the decreasing density of the very rapid initial expansion. The high temperature implies a corresponding thermal (blackbody) radiation field that would cool as the expansion progressed. Theoretical analysis predicted that from the estimated time of the Big Bang to the present, the remnants of the radiation field should have cooled to a temperature of about 3 K, corresponding to a blackbody spectrum with peak wavelength λ_{max} in the microwave region. In 1965, the predicted cosmic background radiation was discovered by Arno Penzias and Robert Wilson at the Bell Labs. Since this landmark discovery, careful analysis has established that the temperature of the background field is 2.7 ± 0.1 K and has shown that it has an isotropic distribution in space.

The Big Bang

The singular event that initiated the expansion of the universe is thought to have been a huge explosion. Initially, the four forces of nature (strong, electromagnetic, weak, and gravity) were unified into a single force. Physicists have been successful in developing theoretical descriptions that unify the first three, but a theory of quantum gravity, needed for the extreme densities of the single-force period, does not yet exist. Consequently, until the cooling universe "froze" or "condensed out" the gravitational force at about 10^{-43} s after the Big Bang, when the temperature was still 10^{32} K, we have no means of describing what was occurring. At this point, the average energy of the particles created would have been about 10^{19} GeV. As the universe continued to cool below 10^{32} K, the three forces other than gravity remained unified and are described by the grand unification theories. Quarks and leptons were indistinguishable and particle quantum numbers were not conserved. It was during this period that a slight excess of quarks over antiquarks occurred, roughly 1 in 10^9, that ultimately resulted in the predominance of matter over antimatter that we now observe in the universe.

At 10^{-35} s, the universe had expanded sufficiently to cool to about 10^{27} K, at which point another phase transition occurred as the strong force condensed out of the GUTs group, leaving only the electromagnetic and weak forces still unified as the **electroweak force.** During this period, the previously free quarks in the dense mixture of roughly equal numbers of quarks, leptons, their antiparticles, and photons began to combine into hadrons and their antiparticles, including the nucleons. By the time the universe had cooled to about 10^{13} K, at about $t = 10^{-6}$ s, the hadrons had mostly disappeared. This is because 10^{13} K corresponds to $kT \sim 1$ GeV, which is the minimum energy needed to create nucleons and antinucleons from the photons present via the reactions

$$\gamma \rightarrow p^+ + p^- \qquad\qquad 41\text{-}11a$$

and

$$\gamma \rightarrow n^+ + \bar{n} \qquad\qquad 41\text{-}11b$$

The particle–antiparticle pairs annihilated and there was no new production to replace them. Only the slight earlier excess of quarks over antiquarks led to a slight excess of protons and neutrons over their antiparticles. The annihilations resulted in photons and leptons, and after about $t = 10^{-4}$ s, those particles in roughly equal numbers dominated the universe. This was the **lepton era.** At about $t = 10$ s, the temperature had fallen to 10^{10} K ($kT \sim 1$ MeV).

Further expansion and cooling dropped the average photon energy below that needed to form an electron–positron pair. Annihilation then removed all of the positrons as it had the antiprotons and antineutrons earlier, leaving only the small excess of electrons arising from charge conservation, and the **radiation era** began. The particles present were primarily photons and neutrinos.

Within a few more minutes, the temperature dropped sufficiently to enable fusing protons and neutrons to form nuclei that were not immediately photodisintegrated. The nuclei of deuterium, helium, and lithium were produced in this **nucleosynthesis period,** but the rapid expansion soon dropped the temperature too low for the fusion to continue and the formation of heavier elements had to await the birth of stars.

A long time later, when the temperature had dropped to about 3000 K as the universe grew to about $1/1000$ of its present size, kT dropped below typical atomic ionization energies and atoms were formed. By then the expansion had redshifted the radiation field so that the total radiation energy was about equal to the energy represented by the remaining mass. As expansion and cooling continued, the energy of the steadily redshifting radiation steadily declined until, at $t = 10^{10}$ y (now), matter came to dominate the universe, with its energy density exceeding that of the 2.7-K radiation remaining from the Big Bang by a factor of about 1000.

Summary

Topic	Remarks and Relevant Equations
1. Basic Interactions	There are four basic interactions: strong, electromagnetic, weak, and gravitational.
Gravitational	All particles with mass experience the force due to the gravitational interaction.
Electromagnetic	All particles with electric charge experience the force due to the electromagnetic interaction.
Weak	The "charge" associated with the weak interaction is called flavor. Quarks and leptons have flavor and experience the weak interaction. Decay times via weak interaction are typically 10^{-10} s.
Strong	The "charge" associated with the strong interaction is called color. Quarks and gluons have color and experience the strong interaction. Hadrons (baryons and mesons) experience a residual strong interaction resulting from the fundamental strong interaction between the quarks that make up the hadrons. Decay times via strong interaction are typically 10^{-23} s.
2. Fundamental Particles	There are two families of fundamental particles, leptons and quarks, each containing six members. It is thought that these particles have no size and no internal structure.
Leptons	Leptons are spin-$\frac{1}{2}$ fermions: the electron e and its neutrino ν_e, the muon μ and its neutrino ν_μ, and the tau τ and its neutrino ν_τ. The electron, muon, and tau have mass, electric charge, and flavor but not color, so they participate in the gravitational, electromagnetic, and weak interactions but not the strong interaction. The neutrinos have flavor but no electric charge and no color. They appear to have a very small mass.

Quarks	There are six quarks, called up u, down d, strange s, charmed c, top t, and bottom b. Each is a spin-$\frac{1}{2}$ fermion. The quarks participate in all of the basic interactions. Because they are always confined in mesons or baryons, their masses can only be estimated.
3. Hadrons	Hadrons are composite particles that are made up of quarks. There are two types, baryons and mesons. Baryons, which include the neutron and proton, are fermions of half-integral spin consisting of three quarks. Mesons, which include pions and kaons, have zero or integral spin. Hadrons interact with each other via the residual strong interaction.
4. Field Particles	In addition to the six fundamental leptons and six fundamental quarks, there are field particles that are associated with the basic interactions.

Interaction	*Field Particle*
Gravity	Graviton
Electromagnetic	Photon
Weak	W^+, W^-, Z^0
Strong	Gluons

5. Conservation Laws	Some quantities, such as energy, momentum, electric charge, angular momentum, baryon number, and each of the three lepton numbers, are strictly conserved in all reactions and decays. Others, such as strangeness and charm, are conserved in reactions and decays that proceed via the strong interaction but not in those that proceed via the weak interaction.
6. Particles and Antiparticles	Particles and their antiparticles have identical masses but opposite values for their other properties, such as charge, lepton number, baryon number, and strangeness. Particle–antiparticle pairs can be produced in various nuclear reactions if the energy available is greater than $2mc^2$, where m is the mass of the particle.
7. Hubble's Law	Hubble's law relates the recession velocity of a galaxy, determined from the redshift of its spectrum, to the distance of the galaxy from us:

$$v = Hr \qquad \textbf{41-9}$$

where the Hubble constant $H = 23$ km/s per million light-years. From Hubble's law, we conclude that the universe is expanding and that the expansion began approximately $1/H$ years ago.

8. The Big Bang	According to the model currently used to describe the evolution of the universe, the universe began with a Big Bang approximately 10^{10} years ago. The Big Bang model is supported by substantial experimental observations, including the isotropic, 2.7-K, background blackbody radiation spectrum.

Problem-Solving Guide

Summary of Worked Examples

Type of Calculation	**Procedure and Relevant Examples**
1. Spin and Antiparticles	
Find the energy of the photons emitted when a particle and an antiparticle annihilate.	When the particle and antiparticle annihilate from rest, two photons of equal energy and opposite momentum are emitted. The energy of each photon is mc^2, where m is the mass of the particle or antiparticle. **Example 41-1**

2.   Conservation Laws

Check a possible decay for violation of conservation laws.	Check the spin, electric charge, baryon number, and lepton number of each side of the equation. For energy conservation, check that the initial rest energy is greater than the final rest energy. **Example 41-2**
Determine if a decay occurs via the weak or strong interaction or not at all.	First check the rest masses for conservation of energy. Then compute the change in strangeness. If $\Delta S = 0$, the decay will be via the strong interaction. If $\Delta S = \pm 1$, the decay will be via the weak interaction. If $|\Delta S| \geq 2$ the decay will not occur. **Example 41-3**

3.   Quarks

Determine the properties of a particle given the quark combination.	Baryons contain 3 quarks; mesons contain a quark and an antiquark. Add the charges of the quarks to determine the charge of the particle, and add the strangenesses of the quarks to determine the strangeness of the particle. **Example 41-4**

4.   Hubble's Law

Given the recession velocity, find the distance to a galaxy.	Compute r from Hubble's law $v = Hr$. **Example 41-5**

Problems

In a few problems, you are given more data than you actually need; in a few other problems, you are required to supply data from your general knowledge, outside sources, or informed estimates.

Conceptual Problems

Problems from Optional and Exploring sections

• Single-concept, single-step, relatively easy
•• Intermediate-level, may require synthesis of concepts
••• Challenging, for advanced students

Hadrons and Leptons

1 • How are baryons and mesons similar? How are they different?

2 • The muon and the pion have nearly the same mass. How do these particles differ?

3 • How can you tell whether a decay proceeds via the strong interaction or the weak interaction?

4 • True or false:

(a) All baryons are hadrons.
(b) All hadrons are baryons.

Spin and Antiparticles

5 • True or false:

Mesons are spin-$\frac{1}{2}$ particles.

6 • Two pions at rest annihilate according to the reaction $\pi^+ + \pi^- \rightarrow \gamma + \gamma$. (a) Why must the energies of the two γ rays be equal? (b) Find the energy of each γ ray. (c) Find the wavelength of each γ ray.

7 • Find the minimum energy of the photon needed for the following pair-production reactions: (a) $\gamma \rightarrow \pi^+ + \pi^-$, (b) $\gamma \rightarrow p + p^-$, and (c) $\gamma \rightarrow \mu^- + \mu^+$.

The Conservation Laws

8 • State which of the decays or reactions that follow violate one or more of the conservation laws, and give the law or laws violated in each case: (a) $p^+ \rightarrow n + e^+ + \bar{\nu}_e$, (b) $n \rightarrow p^+ + \pi^-$, (c) $e^+ + e^- \rightarrow \gamma$, (d) $p + p^- \rightarrow \gamma + \gamma$, and (e) $\bar{\nu}_e + p \rightarrow n + e^+$.

9 • Determine the change in strangeness in each reaction that follows, and state whether the reaction can proceed via the strong interaction, the weak interaction, or not at all: (a) $\Omega^- \rightarrow \Xi^0 + \pi^-$, (b) $\Xi^0 \rightarrow p + \pi^- + \pi^0$, and (c) $\Lambda^0 \rightarrow p^+ + \pi^-$.

10 • Determine the change in strangeness for each decay, and state whether the decay can proceed via the strong interaction, the weak interaction, or not at all: (a) $\Omega^- \rightarrow \Lambda^0 + K^-$ and (b) $\Xi^0 \rightarrow p + \pi^-$.

11 • Determine the change in strangeness for each decay, and state whether the decay can proceed via the strong interaction, the weak interaction, or not at all: (*a*) $\Omega^- \rightarrow \Lambda^0 + \bar{\nu}_e + e^-$ and (*b*) $\Sigma^+ \rightarrow p + \pi^0$.

12 • (*a*) Which of the following decays of the τ particle is possible?

$$\tau \rightarrow \mu^- + \bar{\nu}_\pi + \nu_\tau$$

$$\tau \rightarrow \mu^- + \nu_\mu + \bar{\nu}_\tau$$

(*b*) Explain why the other is not possible. (*c*) Calculate the kinetic energy of the decay products for the decay that is possible.

13 •• Consider the following decay chain:

$$\Omega^- \rightarrow \Xi^0 + \pi^-$$

$$\Xi^0 \rightarrow \Sigma^+ + e^- + \bar{\nu}_e$$

$$\pi^- \rightarrow \mu^- + \bar{\nu}_\mu$$

$$\Sigma^+ \rightarrow n + \pi^+$$

$$\pi^+ \rightarrow \mu^+ + \nu_\mu$$

$$\mu^+ \rightarrow e^+ + \bar{\nu}_\mu + \nu_e$$

$$\mu^- \rightarrow e^- + \bar{\nu}_e + \nu_\mu$$

(*a*) Are all the final products shown stable? If not, finish the decay chain. (*b*) Write the overall decay reaction for Ω^- to the final products. (*c*) Check the overall decay reaction for the conservation of electric charge, baryon number, lepton number, and strangeness.

14 •• Test the following decays for violation of the conservation of energy, electric charge, baryon number, and lepton number: (*a*) n $\rightarrow \pi^+ + \pi^- + \mu^+ + \mu^-$; (*b*) $\pi^0 \rightarrow e^+ + e^- + \gamma$. Assume that linear and angular momentum are conserved. State which conservation laws (if any) are violated in each decay.

The Quark Model

15 • How can you tell whether a particle is a meson or a baryon by looking at its quark content?

16 • Are there any quark–antiquark combinations that result in a nonintegral electric charge?

17 • Find the baryon number, charge, and strangeness for the following quark combinations and identify the hadron: (*a*) *uud*, (*b*) *udd*, (*c*) *uus*, (*d*) *dds*, (*e*) *uss*, and (*f*) *dss*.

18 • Repeat Problem 17 for the following quark combinations: (*a*) $u\bar{d}$, (*b*) $\bar{u}d$, (*c*) $u\bar{s}$, and (*d*) $\bar{u}s$.

19 • The Δ^{++} particle is a baryon that decays via the strong interaction. Its strangeness, charm, topness, and bottomness are all zero. What combination of quarks gives a particle with these properties?

20 • Find a possible combination of quarks that gives the correct values for electric charge, baryon number, and strangeness for (*a*) K^+ and (*b*) K^0.

21 • The D^+ meson has no strangeness, but it has charm of +1. (*a*) What is a possible quark combination that will give the correct properties for this particle? (*b*) Repeat (*a*) for the D^- meson, which is the antiparticle of the D^+.

22 • Find a possible combination of quarks that gives the correct values for electric charge, baryon number, and strangeness for (*a*) K^- (the K^- is the antiparticle of the K^+) and (*b*) $\bar{K}^0$.

23 •• Find a possible quark combination for the following particles: (*a*) Λ^0, (*b*) p$^-$, and (*c*) Σ^-.

24 •• Find a possible quark combination for the following particles: (*a*) $\bar{n}$, (*b*) Ξ^0, and (*c*) Σ^+.

25 •• Find a possible quark combination for the following particles: (*a*) Ω^- and (*b*) Ξ^-.

26 •• State the properties of the particles made up of the following quarks: (*a*) *ddd*, (*b*) $u\bar{c}$, (*c*) $u\bar{b}$, and (*d*) $\bar{s}\bar{s}\bar{s}$.

General Problems

27 • True or false:

(*a*) Leptons consist of three quarks.
(*b*) The times for decays via the weak interaction are typically longer than those for decays via the strong interaction.
(*c*) Electrons interact with protons via the strong interaction.
(*d*) Strangeness is not conserved in weak interactions.
(*e*) Neutrons have no charm.

28 • (*a*) What conditions are necessary for a particle and its antiparticle to be the same? Find the antiparticle for (*b*) π^0 and (*c*) Ξ^0.

29 •• Consider the following decay chain:

$$\Xi^0 \rightarrow \Lambda^0 + \pi^0$$

$$\Lambda^0 \rightarrow p + \pi^-$$

$$\pi^0 \rightarrow \gamma + \gamma$$

$$\pi^- \rightarrow \mu^- + \bar{\nu}_\mu$$

$$\mu^- \rightarrow e^- + \bar{\nu}_e + \nu_\mu$$

(*a*) Are all the final products shown stable? If not, finish the decay chain. (*b*) Write the overall decay reaction for Ξ^0 to the final products. (*c*) Check the overall decay reaction for the conservation of electric charge, baryon number, lepton number, and strangeness. (*d*) In the first step of the chain, could the Λ^0 have been a Σ^0?

30 •• Test the following decays for violation of the conservation of energy, electric charge, baryon number, and lepton number: (*a*) $\Lambda^0 \rightarrow p + \pi^-$, (*b*) $\Sigma^- \rightarrow n + p^-$, (*c*) $\mu^- \rightarrow e^- + \bar{\nu}_e + \nu_\mu$. Assume that linear and angular momentum are conserved. State which conservation laws (if any) are violated in each decay.

31 ••• (*a*) Calculate the total kinetic energy of the decay products for the decay $\Lambda^0 \rightarrow p + \pi^-$. Assume the Λ^0 is initially at rest. (*b*) Find the ratio of the kinetic energy of the pion to the kinetic energy of the proton. (*c*) Find the kinetic energies of the proton and the pion for this decay.

32 ••• A Σ^0 particle at rest decays into a Λ^0 plus a photon. (*a*) What is the total energy of the decay products? (*b*) Assuming that the kinetic energy of the Λ^0 is negligible compared with the energy of the photon, calculate the approximate momentum of the photon. (c) Use your result for (b) to calculate the kinetic energy of the Λ^0. (d) Use your result for (c) to obtain a better estimate of the momentum and the energy of the photon.

33 ••• In this problem, you will calculate the difference in the time of arrival of two neutrinos of different energy from a supernova that is 170,000 light-years away. Let the energies of the neutrinos be $E_1 = 20$ MeV and $E_2 = 5$ MeV, and assume that the rest mass of a neutrino is 20 eV$/c^2$. Because their total energy is so much greater than their rest energy, the neutrinos have speeds that are very nearly equal to c and energies that are approximately $E \approx pc$. (a) If t_1 and t_2 are the times it takes for neutrinos of speeds u_1 and u_2 to travel a distance x, show that

$$\Delta t = t_2 - t_1 = x \frac{u_1 - u_2}{u_1 u_2} \approx \frac{x \, \Delta u}{c^2}$$

(b) The speed of a neutrino of rest mass m_0 and total energy E can be found from Equation 39-25. Show that when $E \gg m_0 c^2$, the speed u is given approximately by

$$\frac{u}{c} \approx 1 - \frac{1}{2} \left(\frac{m_0 c^2}{E} \right)^2$$

(c) Use the results for (b) to calculate $u_1 - u_2$ for the energies and rest mass given, and calculate Δt from the result for (a) for $x = 170,000 c \cdot$y. (d) Repeat the calculation in (c) using $m_0 c^2 = 40$ eV for the rest energy of a neutrino.

SI Units and Conversion Factors

Basic Units

Length	The *meter* (m) is the distance traveled by light in a vacuum in $1/299{,}792{,}458$ s.
Time	The *second* (s) is the duration of $9{,}192{,}631{,}770$ periods of the radiation corresponding to the transition between the two hyperfine levels of the ground state of the ^{133}Cs atom.
Mass	The *kilogram* (kg) is the mass of the international standard body preserved at Sèvres, France.
Current	The *ampere* (A) is that current in two very long parallel wires 1 m apart that gives rise to a magnetic force per unit length of 2×10^{-7} N/m.
Temperature	The *kelvin* (K) is $1/273.16$ of the thermodynamic temperature of the triple point of water.
Luminous intensity	The *candela* (cd) is the luminous intensity, in the perpendicular direction, of a surface of area $1/600{,}000$ m^2 of a blackbody at the temperature of freezing platinum at a pressure of 1 atm.

Derived Units

Force	newton (N)	$1\,\text{N} = 1\,\text{kg·m/s}^2$
Work, energy	joule (J)	$1\,\text{J} = 1\,\text{N·m}$
Power	watt (W)	$1\,\text{W} = 1\,\text{J/s}$
Frequency	hertz (Hz)	$1\,\text{Hz} = \text{s}^{-1}$
Charge	coulomb (C)	$1\,\text{C} = 1\,\text{A·s}$
Potential	volt (V)	$1\,\text{V} = 1\,\text{J/C}$
Resistance	ohm (Ω)	$1\,\Omega = 1\,\text{V/A}$
Capacitance	farad (F)	$1\,\text{F} = 1\,\text{C/V}$
Magnetic field	tesla (T)	$1\,\text{T} = 1\,\text{N/A·m}$
Magnetic flux	weber (Wb)	$1\,\text{Wb} = 1\,\text{T·m}^2$
Inductance	henry (H)	$1\,\text{H} = 1\,\text{J/A}^2$

Conversion Factors

Conversion factors are written as equations for simplicity;
relations marked with an asterisk are exact.

Length

1 km = 0.6215 mi

1 mi = 1.609 km

1 m = 1.0936 yd = 3.281 ft = 39.37 in

*1 in = 2.54 cm

*1 ft = 12 in = 30.48 cm

*1 yd = 3 ft = 91.44 cm

1 lightyear = 1 $c \cdot y$ = 9.461 × 10^{15} m

*1 Å = 0.1 nm

Area

*1 m^2 = 10^4 cm^2

1 km^2 = 0.3861 mi^2 = 247.1 acres

*1 in^2 = 6.4516 cm^2

1 ft^2 = 9.29 × 10^{-2} m^2

1 m^2 = 10.76 ft^2

*1 acre = 43,560 ft^2

1 mi^2 = 640 acres = 2.590 km^2

Volume

*1 m^3 = 10^6 cm^3

*1 L = 1000 cm^3 = 10^{-3} m^3

1 gal = 3.786 L

1 gal = 4 qt = 8 pt = 128 oz = 231 in^3

1 in^3 = 16.39 cm^3

1 ft^3 = 1728 in^3 = 28.32 L = 2.832 × 10^4 cm^3

Time

*1 h = 60 min = 3.6 ks

*1 d = 24 h = 1440 min = 86.4 ks

1 y = 365.24 d = 31.56 Ms

Speed

1 km/h = 0.2778 m/s = 0.6215 mi/h

1 mi/h = 0.4470 m/s = 1.609 km/h

1 mi/h = 1.467 ft/s

Angle and Angular Speed

*π rad = 180°

1 rad = 57.30°

1° = 1.745 × 10^{-2} rad

1 rev/min = 0.1047 rad/s

1 rad/s = 9.549 rev/min

Mass

*1 kg = 1000 g

*1 tonne = 1000 kg = 1 Mg

1 u = 1.6606 × 10^{-27} kg

1 kg = 6.022 × 10^{23} u

1 slug = 14.59 kg

1 kg = 6.852 × 10^{-2} slug

1 u = 931.50 MeV/c^2

Density

*1 g/cm^3 = 1000 kg/m^3 = 1 kg/L

(1 g/cm^3)g = 62.4 lb/ft^3

Force

1 N = 0.2248 lb = 10^5 dyn

1 lb = 4.4482 N

(1 kg)g = 2.2046 lb

Pressure

*1 Pa = 1 N/m^2

*1 atm = 101.325 kPa = 1.01325 bars

1 atm = 14.7 lb/in^2 = 760 mmHg

   = 29.9 inHg = 33.8 ftH_2O

1 lb/in^2 = 6.895 kPa

1 torr = 1 mmHg = 133.32 Pa

1 bar = 100 kPa

Energy

*1 kW·h = 3.6 MJ

*1 cal = 4.1840 J

1 ft·lb = 1.356 J = 1.286 × 10^{-3} Btu

*1 L·atm = 101.325 J

1 L·atm = 24.217 cal

1 Btu = 778 ft·lb = 252 cal = 1054.35 J

1 eV = 1.602 × 10^{-19} J

1 u·c^2 = 931.50 MeV

*1 erg = 10^{-7} J

Power

1 horsepower = 550 ft·lb/s = 745.7 W

1 Btu/min = 17.58 W

1 W = 1.341 × 10^{-3} horsepower

   = 0.7376 ft·lb/s

Magnetic Field

*1 G = 10^{-4} T

*1 T = 10^4 G

Thermal Conductivity

1 W/m·K = 6.938 Btu·in/h·ft^2·F°

1 Btu·in/h·ft^2·F° = 0.1441 W/m·K

B

Numerical Data

Terrestrial Data

Acceleration of gravity g	9.80665 m/s^2
Standard value	32.1740 ft/s^2
At sea level, at equator†	9.7804 m/s^2
At sea level, at poles†	9.8322 m/s^2
Mass of earth M_E	$5.98 \times 10^{24} \text{ kg}$
Radius of earth R_E, mean	$6.37 \times 10^6 \text{ m}; 3960 \text{ mi}$
Escape speed $\sqrt{2R_E g}$	$1.12 \times 10^4 \text{ m/s}; 6.95 \text{ mi/s}$
Solar constant‡	1.35 kW/m^2
Standard temperature and pressure (STP):	
Temperature	273.15 K
Pressure	$101.325 \text{ kPa}; 1.00 \text{ atm}$
Molar mass of air	28.97 g/mol
Density of air (STP), ρ_{air}	1.293 kg/m^3
Speed of sound (STP)	331 m/s
Heat of fusion of H_2O (0°C, 1 atm)	333.5 kJ/kg
Heat of vaporization of H_2O (100°C, 1 atm)	2.257 MJ/kg

†Measured relative to the earth's surface.
‡Average power incident normally on 1 m^2 outside the earth's atmosphere at the mean distance from the earth to the sun.

Astronomical Data

Earth	
Distance to moon†	$3.844 \times 10^8 \text{ m}; 2.389 \times 10^5 \text{ mi}$
Distance to sun, mean†	$1.496 \times 10^{11} \text{ m}; 9.30 \times 10^7 \text{ mi}; 1.00 \text{ AU}$
Orbital speed, mean	$2.98 \times 10^4 \text{ m/s}$
Moon	
Mass	$7.35 \times 10^{22} \text{ kg}$
Radius	$1.738 \times 10^6 \text{ m}$
Period	27.32 d
Acceleration of gravity at surface	1.62 m/s^2
Sun	
Mass	$1.99 \times 10^{30} \text{ kg}$
Radius	$6.96 \times 10^8 \text{ m}$

† Center to center.

Physical Constants

Gravitational constant	G	$6.672\,6 \times 10^{-11}\,\text{N·m}^2/\text{kg}^2$
Speed of light	c	$2.997\,924\,58 \times 10^8\,\text{m/s}$
Fundamental charge	e	$1.602\,177\,33 \times 10^{-19}\,\text{C}$
Avogadro's number	N_A	$6.022\,136\,7 \times 10^{23}\,\text{particles/mol}$
Gas constant	R	$8.314\,51\,\text{J/mol·K}$
		$1.987\,22\,\text{cal/mol·K}$
		$8.205\,78 \times 10^{-2}\,\text{L·atm/mol·K}$
Boltzmann's constant	$k = R/N_A$	$1.380\,658 \times 10^{-23}\,\text{J/K}$
		$8.617\,385 \times 10^{-5}\,\text{eV/K}$
Unified mass unit	$u = (1/N_A)\,\text{g}$	$1.660\,540 \times 10^{-24}\,\text{g}$
Coulomb constant	$k = 1/4\pi\epsilon_0$	$8.987\,551\,788 \times 10^9\,\text{N·m}^2/\text{C}^2$
Permittivity of free space	ϵ_0	$8.854\,187\,817 \times 10^{-12}\,\text{C}^2/\text{N·m}^2$
Permeability of free space	μ_0	$4\pi \times 10^{-7}\,\text{N/A}^2$
		$1.256\,637 \times 10^{-6}\,\text{N/A}^2$
Planck's constant	h	$6.626\,075\,5 \times 10^{-34}\,\text{J·s}$
		$4.135\,669\,2 \times 10^{-15}\,\text{eV·s}$
	$\hbar = h/2\pi$	$1.054\,572\,66 \times 10^{-34}\,\text{J·s}$
		$6.582\,122\,0 \times 10^{-16}\,\text{eV·s}$
Mass of electron	m_e	$9.109\,389\,7 \times 10^{-31}\,\text{kg}$
		$510.999\,1\,\text{keV}/c^2$
Mass of proton	m_p	$1.672\,623\,1 \times 10^{-27}\,\text{kg}$
		$938.272\,3\,\text{MeV}/c^2$
Mass of neutron	m_n	$1.674\,929 \times 10^{-27}\,\text{kg}$
		$939.565\,6\,\text{MeV}/c^2$
Bohr magneton	$m_B = e\hbar/2m_e$	$9.274\,015\,4 \times 10^{-24}\,\text{J/T}$
		$5.788\,382\,63 \times 10^{-5}\,\text{eV/T}$
Nuclear magneton	$m_n = e\hbar/2m_p$	$5.050\,786\,6 \times 10^{-27}\,\text{J/T}$
		$3.152\,451\,66 \times 10^{-8}\,\text{eV/T}$
Magnetic flux quantum	$\phi_0 = h/2e$	$2.067\,834\,6 \times 10^{-15}\,\text{T·m}^2$
Quantized Hall resistance	$R_K = h/e^2$	$2.581\,280\,7 \times 10^4\,\Omega$
Rydberg constant	R_H	$1.097\,373\,153\,4 \times 10^7\,\text{m}^{-1}$
Josephson frequency–voltage quotient	$2e/h$	$4.835\,979 \times 10^{14}\,\text{Hz/V}$
Compton wavelength	$\lambda_C = h/m_e c$	$2.426\,310\,58 \times 10^{-12}\,\text{m}$

For additional data, see the last four pages in the book and the following tables in the text.

Periodic Table of Elements

1																	18
1 **H** 1.00797	**2**												**13**	**14**	**15**	**16**	**17**	**2** **He** 4.003
3 **Li** 6.941	**4** **Be** 9.012												**5** **B** 10.81	**6** **C** 12.011	**7** **N** 14.007	**8** **O** 15.9994	**9** **F** 19.00	**10** **Ne** 20.179
11 **Na** 22.990	**12** **Mg** 24.31	**3**	**4**	**5**	**6**	**7**	**8**	**9**	**10**	**11**	**12**		**13** **Al** 26.98	**14** **Si** 28.09	**15** **P** 30.974	**16** **S** 32.064	**17** **Cl** 35.453	**18** **Ar** 39.948
19 **K** 39.102	**20** **Ca** 40.08	**21** **Sc** 44.96	**22** **Ti** 47.88	**23** **V** 50.94	**24** **Cr** 52.00	**25** **Mn** 54.94	**26** **Fe** 55.85	**27** **Co** 58.93	**28** **Ni** 58.69	**29** **Cu** 63.55	**30** **Zn** 65.38		**31** **Ga** 69.72	**32** **Ge** 72.59	**33** **As** 74.92	**34** **Se** 78.96	**35** **Br** 79.90	**36** **Kr** 83.80
37 **Rb** 85.47	**38** **Sr** 87.62	**39** **Y** 88.906	**40** **Zr** 91.22	**41** **Nb** 92.91	**42** **Mo** 95.94	**43** **Tc** (98)	**44** **Ru** 101.1	**45** **Rh** 102.905	**46** **Pd** 106.4	**47** **Ag** 107.870	**48** **Cd** 112.41		**49** **In** 114.82	**50** **Sn** 118.69	**51** **Sb** 121.75	**52** **Te** 127.60	**53** **I** 126.90	**54** **Xe** 131.29
55 **Cs** 132.905	**56** **Ba** 137.33	**57–71** **Rare Earths**	**72** **Hf** 178.49	**73** **Ta** 180.95	**74** **W** 183.85	**75** **Re** 186.2	**76** **Os** 190.2	**77** **Ir** 192.2	**78** **Pt** 195.09	**79** **Au** 196.97	**80** **Hg** 200.59		**81** **Tl** 204.37	**82** **Pb** 207.19	**83** **Bi** 208.98	**84** **Po** (210)	**85** **At** (210)	**86** **Rn** (222)
87 **Fr** (223)	**88** **Ra** (226)	**89–103** **Actinides**	**104** **Rf** (261)	**105** **Ha** (260)	**106** (263)	**107** (262)	**108** (265)	**109** (266)										

Rare Earths (Lanthanides)	**57** **La** 138.91	**58** **Ce** 140.12	**59** **Pr** 140.91	**60** **Nd** 144.24	**61** **Pm** (147)	**62** **Sm** 150.36	**63** **Eu** 152.0	**64** **Gd** 157.25	**65** **Tb** 158.92	**66** **Dy** 162.50	**67** **Ho** 164.93	**68** **Er** 167.26	**69** **Tm** 168.93	**70** **Yb** 173.04	**71** **Lu** 174.97
Actinides	**89** **Ac** 227.03	**90** **Th** 232.04	**91** **Pa** 231.04	**92** **U** 238.03	**93** **Np** 237.05	**94** **Pu** (244)	**95** **Am** (243)	**96** **Cm** (247)	**97** **Bk** (247)	**98** **Cf** (251)	**99** **Es** (252)	**100** **Fm** (257)	**101** **Md** (258)	**102** **No** (259)	**103** **Lr** (260)

The 1–18 group designation has been recommended by the International Union of Pure and Applied Chemistry (IUPAC).

Atomic Numbers and Atomic Masses

Name	Symbol	Atomic Number	Mass	Name	Symbol	Atomic Number	Mass
Actinium	Ac	89	227.03	Mercury	Hg	80	200.59
Aluminum	Al	13	26.98	Molybdenum	Mo	42	95.94
Americium	Am	95	(243)	Neodymium	Nd	60	144.24
Antimony	Sb	51	121.75	Neon	Ne	10	20.179
Argon	Ar	18	39.948	Neptunium	Np	93	237.05
Arsenic	As	33	74.92	Nickel	Ni	28	58.69
Astatine	At	85	(210)	Niobium	Nb	41	92.91
Barium	Ba	56	137.3	Nitrogen	N	7	14.007
Berkelium	Bk	97	(247)	Nobelium	No	102	(259)
Beryllium	Be	4	9.012	Osmium	Os	76	190.2
Bismuth	Bi	83	208.98	Oxygen	O	8	15.9994
Boron	B	5	10.81	Palladium	Pd	46	106.4
Bromine	Br	35	79.90	Phosphorus	P	15	30.974
Cadmium	Cd	48	112.41	Platinum	Pt	78	195.09
Calcium	Ca	20	40.08	Plutonium	Pu	94	(244)
Californium	Cf	98	(251)	Polonium	Po	84	(210)
Carbon	C	6	12.011	Potassium	K	19	39.102
Cerium	Ce	58	140.12	Praseodymium	Pr	59	140.91
Cesium	Cs	55	132.905	Promethium	Pm	61	(147)
Chlorine	Cl	17	35.453	Protactinium	Pa	91	231.04
Chromium	Cr	24	52.00	Radium	Ra	88	(226)
Cobalt	Co	27	58.93	Radon	Rn	86	(222)
Copper	Cu	29	63.55	Rhenium	Re	75	186.2
Curium	Cm	96	(247)	Rhodium	Rh	45	102.905
Dysprosium	Dy	66	162.50	Rubidium	Rb	37	85.47
Einsteinium	Es	99	(252)	Ruthenium	Ru	44	101.1
Erbium	Er	68	167.26	Rutherfordium	Rf	104	(261)
Europium	Eu	63	152.0	Samarium	Sm	62	150.36
Fermium	Fm	100	(257)	Scandium	Sc	21	44.96
Fluorine	F	9	19.00	Selenium	Se	34	78.96
Francium	Fr	87	(223)	Silicon	Si	14	28.09
Gadolinium	Gd	64	157.25	Silver	Ag	47	107.870
Gallium	Ga	31	69.72	Sodium	Na	11	22.990
Germanium	Ge	32	72.59	Strontium	Sr	38	87.62
Gold	Au	79	196.97	Sulfur	S	16	32.064
Hafnium	Hf	72	178.49	Tantalum	Ta	73	180.95
Hahnium	Ha	105	(260)	Technetium	Tc	43	(98)
Helium	He	2	4.003	Tellurium	Te	52	127.60
Holmium	Ho	67	164.93	Terbium	Tb	65	158.92
Hydrogen	H	1	1.00797	Thallium	Tl	81	204.37
Indium	In	49	114.82	Thorium	Th	90	232.04
Iodine	I	53	126.90	Thulium	Tm	69	168.93
Iridium	Ir	77	192.2	Tin	Sn	50	118.69
Iron	Fe	26	55.85	Titanium	Ti	22	47.88
Krypton	Kr	36	83.80	Tungsten	W	74	183.85
Lanthanum	La	57	138.91	Uranium	U	92	238.03
Lawrencium	Lr	103	(260)	Vanadium	V	23	50.94
Lead	Pb	82	207.19	Xenon	Xe	54	131.29
Lithium	Li	3	6.941	Ytterbium	Yb	70	173.04
Lutetium	Lu	71	174.97	Yttrium	Y	39	88.906
Magnesium	Mg	12	24.31	Zinc	Zn	30	65.38
Manganese	Mn	25	54.94	Zirconium	Zr	40	91.22
Mendelevium	Md	101	(258)				

ILLUSTRATION CREDITS

Part Openers

Part VI p. 1142 Courtesy AT&T Archives

Chapter 36

Opener p.1143 Lawrence Livermore/Photo Researchers; **p.1155** **(right)** Education Development Center; **(left)** Education Development Center; **p.1158 (left)** R. Trump/IBM Research; **(right)** Courtesy of E. D. Williams; **(bottom)** Courtesy E. D. Williams; **p.1159 (top left)** D. M. Eigler, and E. K. Schweizer, IBM Almaden Research Center; **(top right)** D. M. Eigler, and E. K. Schweizer, IBM Almaden Research Center; **(bottom left)** D. M. Eigler, and E. K. Schweizer, IBM Almaden Research Center; **(bottom right)** D. M. Eigler, and E. K. Schweizer, IBM Almaden Research Center.

Chapter 37

Opener p.1169 Dr. Bruce Schardt/Courtesy Digital Instruments, Inc.; **p.1170** **(Figure 37–01) (a)** Adapted from Eastman Kodak and Wabash Instrument Corporation; **(b)** Adapted from Eastman Kodak and Wabash Instrument Corporation; **p.1188 (left)** A. Jayaraman/AT&T Bell Labs; **(right)** A. Jayaraman/AT&T Bell Labs; **p.1189 (all images)** David Parker/Photo Researchers; **p.1193** © Robert Landau/Westlight.

Chapter 38

Opener p.1203 Courtesy AT&T Archives; **p.1210 (Figure 38–07)** Will and Deni McIntyre/Photo Researchers; **p.1217 (Figure 38–14)** **(a)** Courtesy Dr. J. A. Marquisee; **(b)** Courtesy Dr. J. A. Marquisee; **p.1221 (top left)** Richard Walters 2/89, p.52/*Discover*; **(top right)** Dr. Jeremy Burgess/Science Photo Library/Photo Researchers; **(bottom left)** Thomas R. Taylor /Photo Researchers; **(bottom right)** Courtesy AT&T Archives; **p.1224 (top left)** Chris Kovach 3/91, p.69/*Discover*; **(top right)** Srinivas Manne, University of California-Santa Barbara; **(bottom left)** Dr. F. A. Quiocho and J. C. Spurlino/Howard Hughes Medical Institute, Baylor College of Medicine; **(bottom right)** W. Krätschmer/Max-Planck Institute for Nuclear Physics; **p.1225** Museum of Modern Art; **p.1229** C. Falco/Photo Researchers; **p.1232 (top)** Dr. Jeremy Burgess/Science Photo Library/Photo Researchers; **(left)** Courtesy

AT&T Archives; **(right)** AT&T Archives; **p.1233 (top)** Dr. Jeremy Burgess/Science Photo Library /Photo Researchers; **(left)** AT&T Archives; **(right)** AT&T Archives; **p.1234 (bottom)** Courtesy of Texas Instruments; **(top)** Courtesy of David Sarnoff Research Center; **p.1235 (top left)** Images by Patterson Electronics and computer processing by John Sanford; **(top right)** Images by Patterson Electronics and computer processing by John Sanford; **(bottom left)** Images by Patterson Electronics and computer processing by John Sanford; **(bottom right)** Images by Patterson Electronics and computer processing by John Sanford.

Chapter 39

Opener p.1243 Albert Einstein Archives/AIP Niels Bohr Library; **p.1244** Courtesy NRAO/AUI; **p.1265** C. Powell, P. Fowler & D. Perkins/Science Photo Library/Photo Researchers; **p.1272** © Michael Freeman; **p.1273** N.A.S.A.

Chapter 40

Opener p.1284 Hans Wolf /The Image Bank; **p.1297 (left)** © 1991 by the Metropolitan Museum of Art; **(center)** Courtesy of Paintings Conservation Dept., Metropolitan Museum of Art; **(right)** Paintings Conservation Dept., Metropolitan Museum of Art; **p.1300** Jerry Mason/Photo Researchers; **p.1305 (left)** Courtesy of Princeton Plasma Physics Laboratory; **(right)** Courtesy of Princeton Plasma Physics Laboratory; **p.1306 (left)** Courtesy of Lawrence Livermore National Laboratory/U. S. Department of Energy; **(right)** Courtesy of Lawrence Livermore National Laboratory/U. S. Department of Energy.

Chapter 41

Opener p.1313 Lawrence Livermore Lab/Science Photo Library/Photo Researchers; **p.1315** ICRR (Institute for Cosmic Ray Research), The University of Tokyo; **p.1316 (top)** Science Photo Library /Photo Researchers; **(bottom)** Lawrence Berkeley Laboratory/Science Photo Library/Photo Researchers; **p.1317** Lawrence Berkeley Lab/Science Photo Library/Photo Researchers; **p.1318 (Figure 41–01)** Richard Ehrlich; **(bottom)** CERN; **p.1319** CERN; **p.1321** Lawrence Berkeley Lab/Science Photo Library/Photo Researchers.

ANSWERS

Problem answers are calculated using $g = 9.81$ m/s^2 unless otherwise specified in the Problem. Differences in the last figure can easily result from differences in rounding the input data and are not important.

To help you master the techniques in Examples and to solve the intermediate-level problems at the end of each chapter, the problem maps preceding the answers for the chapters indicate which Examples and odd-numbered intermediate-level Problems deal with similar material.

Chapter 36

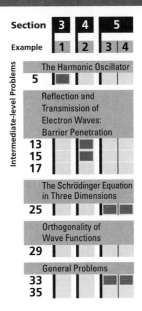

Section 3 4 5
Example 1 2 3 4

Intermediate-level Problems

The Harmonic Oscillator
5

Reflection and Transmission of Electron Waves: Barrier Penetration
13
15
17

The Schrödinger Equation in Three Dimensions
25

Orthogonality of Wave Functions
29

General Problems
33
35

1. True

3. (a)

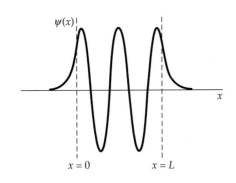

(b)

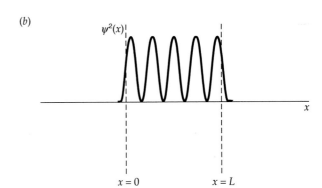

5. Answer given in the problem.

7. $[(2/\sqrt{\pi})(m\omega_0/\hbar)^{3/2}]^{1/2}$

9. Answer given in the problem.

11. $m\omega_0\hbar/2$

13. (a) $k_1/\sqrt{2}$, $(2mU_0)^{1/2}/\hbar$; (b) 0.0294; (c) 0.971; (d) 9.71×10^5 particles. Classically, 100% of the particles are transmitted.

15. (a) $0.1k_1$, $(0.02mU_0)^{1/2}/\hbar$; (b) 0.671; (c) 0.329; (d) 3.29×10^5 particles. Classically, 100% of the particles are transmitted.

17. 0.342

19. (1, 1, 1), (1, 1, 2), (1, 2, 1), (1, 1, 3), (1, 2, 2), (1, 2, 3), (1, 1, 4), (1, 3, 1), (1, 3, 2), (1, 2, 4)

21. (a) (1, 1, 1), (1, 1, 2), (1, 1, 3), (1, 2, 1), (1, 1, 4), (1, 2, 2), (1, 2, 3), (1, 1, 5), (1, 2, 4), (1, 3, 1), (1, 1, 6), (1, 3, 2); (b) (1, 1, 4) and (1, 2, 2); (1, 1, 6) and (1, 3, 2)

23. (a) $\psi(x, y, z) = (8/L^3)^{1/2} \cos(\pi x/L) \sin(\pi y/L) \sin(\pi z/L)$; (b) The energy levels are the same.

25. $50h^2/8mL^2$

27. $10E_1$

29. Answer given in the problem.

31. Answer given in the problem.

33. $20\,\hbar^2\pi^2/mL^2$, $21\hbar^2\pi^2/mL^2$, $21\hbar^2\pi^2/mL^2$

35. (a)

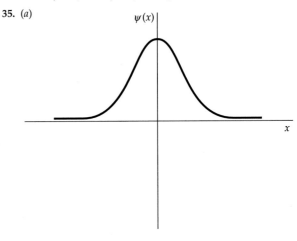

(b)

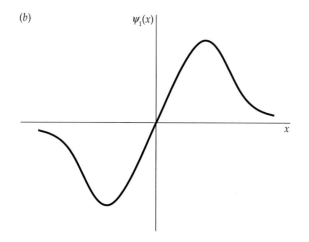

37. Answer given in the problem.
39. $\sqrt{2}\,[m\omega_0/(\pi\hbar)]^{1/4}$
41. Answer given in the problem.

Chapter 37

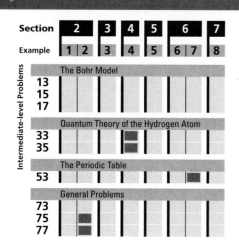

1. decrease
3. (a)
5. (d)
7. Answer given in the problem.
9. 1.89 eV, 656 nm; 2.55 eV, 486 nm; 2.86 eV, 434 nm
11. (a) 0.85 eV, 1459 nm; (b) 4052 nm, 2627 nm, 2168 nm
13. (a) 13.6 eV; (b) 54.4 eV; (c) 122.4 eV
15. Answer given in the problem.
17. (a) Answer given in the problem; (b) $R_H = 1.096\,776 \times 10^7\ \text{m}^{-1}$, $R_\infty = 1.097\,373 \times 10^7\ \text{m}^{-1}$; (c) 0.0544%
19. (c)
21. (a) $(2\sqrt{3})\hbar$; (b) $-3, -2, -1, 0, 1, 2, 3$;

(c)

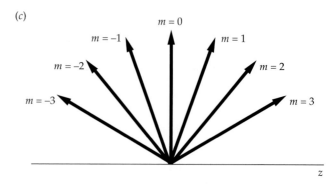

23. (a) 0, 1, 2; (b) 0; $-1, 0, 1$; $-2, -1, 0, 1, 2$; (c) 18
25. (a) 45°; (b) 26.6°; (c) 8.05°
27. (a) $n \geq 1$, $\ell \geq 0$; (b) $n \geq 2$, $\ell \geq 1$; (c) $n \geq 3$, $\ell \geq 2$
29. (a) 4; (b) $n = 2$, $\ell = 0$, $m = 0$; $n = 2$, $\ell = 1$, $m = 1$; $n = 2$, $\ell = 1$, $m = 0$; $n = 2$, $\ell = 1$, $m = -1$
31. (a) $0.0605a_0^{-3/2}$; (b) $0.00366a_0^{-3}$; (c) $0.046a_0^{-1}$
33. (a) 0.00092; (b) 0
35. $(3/2)a_0$
37. 0.323
39. $\ell = 0$ or $\ell = 1$
41. 5/2 or 3/2
43. In sodium, the screening of the nuclear charge by the inner electrons is less for the 3s state than for the 3p state; there is no screening in hydrogen.
45. Iron and cobalt each have two valence electrons in the 4s shell so they have similar properties. They differ only in that cobalt has one more electron in the 3d shell. Neon has a closed shell structure so it is inert. Sodium has one more electron, which is in the outer 3s shell, so its properties are very different.
47. (a) silicon; (b) calcium
49. (c)
51. (a) $1s^2\,2s^2\,2p^6\,3s^2\,3p^1$; (b) $1s^2\,2s^2\,2p^6\,3s^2\,3p^6\,3d^5\,4s^1$
53. 3.4
55. (a) $2s^1$ or $2p^1$; (b) $1s^2\,2s^2\,2p^6\,3p^1$; (c) $1s^1 2s^1$
57. (a) 0.0611 nm, 0.0580 nm; (b) 0.0543 nm
59. zirconium
61. (a) 1.01 nm; (b) 0.155 nm
63. (c)
65. These atoms have a single outer p-shell electron, which is shielded from the nucleus more than an s-shell electron.
67. $n = 4$ to $n = 1$.
69. $n = 3$ to $n = 2$, $n = 9$ to $n = 3$, $n = 7$ to $n = 4$
71. (a) 1.61774 eV, 1.61041 eV; (b) 7.33×10^{-3} eV; (c) 63.3 T
73. Answer given in the problem.
75. (a) $R_H = 1.096\,776 \times 10^{-7}\ \text{m}^{-1}$, $R_D = 1.097\,074 \times 10^{-7}\ \text{m}^{-1}$; (b) 0.178 nm
77. (a) $R_T = 1.097\,174 \times 10^{-7}\ \text{m}^{-1}$; (b) 0.0598 nm, 0.238 nm
79. Answer given in the problem.

Chapter 38

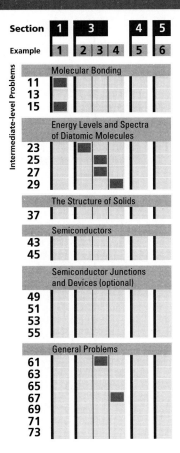

1. polar

3. Ne

5. (a) covalent; (b) ionic; (c) metallic

7. releases energy, 3.02 eV

9. 43.6%

11. 0.499 eV

13.

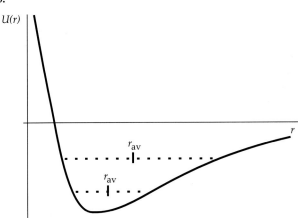

From the graph we see that r_{av} increases with increasing vibration energy. As a result, such a molecule expands when heated.

15. (a) −6.64 eV; (b) 5.70 eV; (c) 0.64 eV

17. (a) 3.1 eV; (b) $C = 1.37 \times 10^{-13}$ eV·nm^n, $n = 19.7$

19. The force constant is similar to that of a stiff ordinary spring.

21. 0.110 nm

23. (a) 0.504 u; (b) 7.00 u; (c) 6.86 u; (d) 0.980 u

25. 0.00314 eV

27. 1.58×10^{-5} eV

29. 477 N/m

31. (a) ^{35}Cl: 0.9722 u, ^{37}Cl: 0.9737 u, $\Delta\mu/\mu = 0.00153$ u; (b) $\Delta f/f = -0.00153$

33. (a) cube with sides of length 2R; (b) 0.524

35. 2.07 g/cm^3

37. 0.740

39. (d)

41. (a) p-type; (b) n-type

43. 3.17 nm, 8.46 nm

45. 342×10^{-10} m

47. (b)

49.

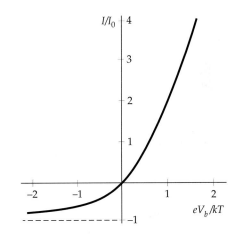

51. 250

53.

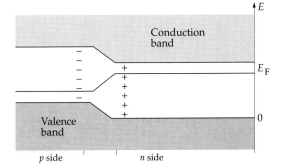

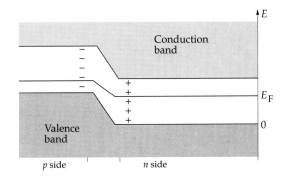

55. p-type, 10^{23} holes/m^3

57. (a) ionic; (b) covalent; (c) metallic

59. 2.63×10^{-29} C·m

61. 9.55×10^{-4} eV

63. Answer given in the problem.

65. 1.1 eV

67. 1551 N/m

69. $r_0 = a$, $U_{min} = -U_0$, $r_0 = 0.074$ nm, $U_0 = 4.52$ eV

71. $F_x \propto 1/x^4$

73. (a) $I = 1.45 \times 10^{-46}$ kg·m^2, $E_{0r} = 2.39 \times 10^{-4}$ eV

(b)

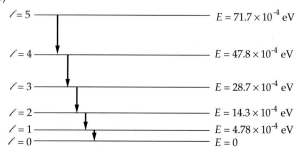

(c), (d) Transitions that obey $\Delta\ell = -1$ are indicated by the arrows in the diagram.

$\Delta\ell$	ΔE	λ
$\ell = 1 \rightarrow \ell = 0$	4.78×10^{-4} eV	2.60 mm
$\ell = 2 \rightarrow \ell = 1$	9.55×10^{-4} eV	1.30 mm
$\ell = 3 \rightarrow \ell = 2$	14.4×10^{-4} eV	0.862 mm
$\ell = 4 \rightarrow \ell = 3$	19.1×10^{-4} eV	0.650 mm
$\ell = 5 \rightarrow \ell = 4$	23.9×10^{-4} eV	0.519 mm

These photons fall in the microwave and short radio wave portion of the electromagnetic spectrum.

75. $f = 1.9 \times 10^{13}$ Hz, $\lambda = 1.6 \times 10^{-5}$ m

Chapter 39

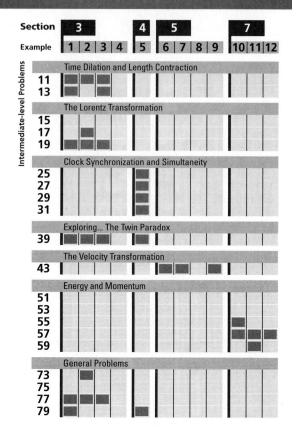

1. The friend in the car.

3. (a) 6.63 m; (b) 12.6 m

5. (a) 599 m; (b) 13.4 km

7. (a) 130 y; (b) 88.1 y

9. (a) 0.6 m; (b) 2.5 ns

11. $0.8c$

13. (a) 978 m; (b) 9.57×10^7 m; (c) 1.02×10^{-7} m

15. Answer given in the problem.

17. $0.141c$

19. (a) 2.56×10^{-9} s; (b) 8×10^8 particles/cm^2·s; (c) 2.4×10^8 particles/cm^2·s

21. Answers given in the problem.

23. Yes

25. 80 c·min

27. Answer given in the problem.

29. $L_p V/c^2 = 60$ min, which is the same as the time interval found in Problem 28.

31. 4.39 μs

33. 0.0637

35. $0.696c$

37. (a) 11.3 y; (b) 40 y

39. (a) 50 signals/y; (b) 533.3 signals; (c) 1067 signals; (d) 50 signals/y; (e) 267 signals; (f) 1333 signals; (g) A by 2.67 y

41. $0.994c$

43. (a) $0.976c$; (b) $0.997c$

45. (a) 1.005; (b) 1.15; (c) 1.67; (d) 7.09

47. (a) $0.155E_0$; (b) $1.29E_0$; (c) $6.09E_0$

49. 2.97 GeV

51. Answer given in the problem.

53. Answer given in the problem.

55. (a) 1625 MeV/c; (b) $0.866c$

57. 608 MeV, 389 MeV

59. (a) 290 MeV; (b) 625 MeV

61. (a) $0.943c$; (b) 3 MeV; (c) 2.83 MeV/c; (d) 0.877 MeV;
(e) 4.12 MeV/c^2

63. Answers given in the problem.

65. (b)

67. (a) 5.33 y; (b) 3.53 y

69. $0.999c$

71. $0.866c$

73. 1.85×10^4 y

75. (a) 2.87×10^9 MeV; (b) 0.133 s

77. (a) $0.625c$; (b) 31.2 y

79. (a) 4.17 μs; (b) 7.71 μs; (c) 2.5 μs

81. (a) 2.1 μs; (b) 2.59 μs; (c) 0.493 μs; (d) 2.59 μs; (e) 4.36 h; (f) 19 h

83. (a) $c/3$; (b) 20 m; (c) 0.20 μs

85. Answer given in the problem.

87. Answer given in the problem.

89. Answer given in the problem.

91. Answer given in the problem.

Chapter 40

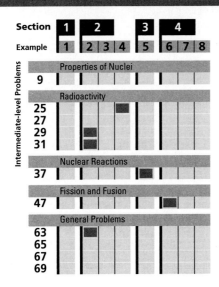

1. (a) ^{15}N, ^{13}N; (b) ^{57}Fe, ^{58}Fe; (c) ^{117}Sn, ^{119}Sn

3. (a) 31.99 MeV, 5.332 MeV/nucleon;
(b) 333.7 MeV, 8.557 MeV/nucleon;
(c) 1636 MeV, 7.868 MeV/nucleon

5. (a) $3C/4\pi R_0^3$; (b) 1.17×10^{14} g/cm^3

7. 1.17×10^{14} g/cm^3

9. (a) 7.81 fm, 6.77 fm; (b) 199 MeV

11. α decay is often followed by β^- decay (not β^+ decay) because the daughter nucleus is neutron rich.

13. It would cause inaccuracies since we assume the initial ratio of ^{14}C/^{12}C was the same as today's ratio.

15. (a) 500 counts/s; (b) 250 counts/s; (c) 125 counts/s

17. 3.61×10^{10} disintegrations/s

19. (a) 4.868 MeV; (b) 6.999 MeV

21. 3350 y

23. (a) 13,950 y; (b) Yes

25. Answer given in the problem.

27. (a) Answer given in the problem. (b) 6.76×10^{-3} s^{-1}, 103 s

29. (a) 4.16×10^{-9}/s^{-1}; (b) 5.28 y

31. (a) 156 h; (b) 551 h

33. Answer given in the problem.

35. (a) 4.032 MeV; (b) 18.35 MeV; (c) 4.785 MeV

37. (a) 1.20 MeV; (b) The mass of $^{13}_{6}$C includes one less electron than that of $^{13}_{7}$N so one electron mass plus the mass of the emitted electron must be added.

39. The neutrons emitted in fission are fast (energy of the order of 1 MeV), whereas the probability of the capture varies inversely with neutron speed, so a moderator is needed to slow down the emitted neutrons to thermal energies (of the order of 0.02 eV).

41. They escape from the reactor or are captured by nuclei other than ^{235}U.

43. 1.56×10^{19} fissions/s

45. 1.16×10^8 K

47. 3.20×10^{10} J

49. (a), (b) Answer given in the problem. (c) 3.74×10^{38} protons/s, 5.07×10^{10} y

51. Pressure changes affect the electron clouds surrounding the nucleus and therefore affect the spacing of the nuclei slightly, and temperature changes affect the vibration energy of nuclei, but neither affects the internal structure of the nucleus responsible for radioactivity.

53. The temperature of the sun is so great that the thermal energy of positively charged nuclei such as ^{1}H and ^{2}He can overcome the Coulomb repulsion and bring the nuclei close enough to fuse. On earth, the temperature is much too small for thermal energy to overcome the Coulomb energy of repulsion.

55. $\lambda = 0.0693$ s^{-1}

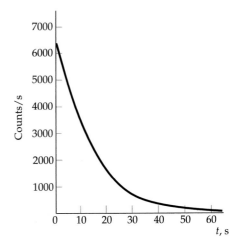

55. $\lambda = 0.0693 \text{ s}^{-1}$

57. 0.156 MeV

59. Yes, 1h

61. 6.60×10^3 decays/s

63. 6.3 L

67. (a)

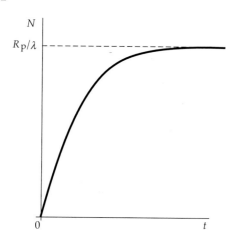

65. (a) 22.96 MeV; (b) 4179 MeV; (c) 1286 MeV

(b) 8.66×10^4 nuclei

69. (a) 4 fm; (b) 310 MeV/c; (c) Answer given in the problem. (d) 310 MeV

71. (a) 1.19 MeV/c; (b) 0.752 keV; (c) 0.0962%

73. Answer given in the problem. (b) 55 collisions

(d)

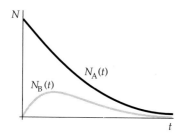

75. (a)–(c) Answer given in the problem.

77. Answer given in the problem.

Chapter 41

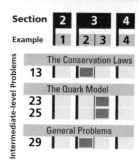

1. Both are hadrons, that is, they interact via strong nuclear interactions and are composed of quarks. Baryons have $\frac{1}{2}$-integral spins, whereas mesons have zero or integral spin. Mesons are less massive.

3. from the lifetime

5. False

7. (a) 279.2 MeV; (b) 1876.6 MeV; (c) 211.4 MeV

9. (a) $\Delta S = +1$, weak interaction; (b) $\Delta S = +2$, not allowed; (c) $\Delta S = +1$, weak interaction

11. (a) $\Delta S = +2$, not allowed; (b) $\Delta S = +1$, weak interaction

13. (a) No, $n \rightarrow p + e^- + \bar{\nu}_e$; (b) $\Omega^- \rightarrow p + 3e^- + 3\bar{\nu}_e + 2\bar{\nu}_\mu + 2\nu_\mu + e^+ + \nu_e$ (c) Only strangeness is not conserved.

15. Two quarks make a meson and three quarks make a baryon.

17. (a) $+1, +1, 0, p^+$; (b) $+1, 0, 0, n$; (c) $+1, +1, -1, \Sigma^+$; (d) $+1, -1, -1, \Sigma^-$; (e) $+1, 0, -2, \Xi^0$; (f) $+1, -1, -2, \Xi^-$

19. *uuu*

21. (a) $c\bar{d}$; (b) $\bar{c}d$

23. (a) *uds*; (b) $\bar{u}\,\bar{u}\,\bar{d}$; (c) *dds*

25. (a) *sss*; (b) *dss*

27. (a) False; (b) True; (c) False; (d) True; (e) True

29. (a) Yes; (b) $\Xi^0 \rightarrow p + e^- + \bar{\nu}_e + \nu_\mu + \bar{\nu}_\mu + 2\gamma$; (c) Only strangeness is not conserved. (d) No, the rest mass of $(\Lambda^0 + \pi^0)$ is greater than that of Σ^0.

31. (a) 38.1 MeV; (b) 6.72; (c) $K_{proton} = 4.93$ MeV, $K_{pion} = 33.2$ MeV

33. (a), (b) Answer given in the problem. (c) $(7.5 \times 10^{-12})c$, 40.3 s; (d) $(30 \times 10^{-12})c$, 161 s

INDEX

Numbers in **bold** indicate additional display material, such as diagrams; *n* indicates a footnote; AP indicates material in the Appendixes.

Pages 1–655 are found in Volume 1; pages 656–1141 are found in Volume 2; pages 1142–1335 are found in Volume 3.

Aberrations in optical images, 1073, 1090
Absolute temperature scale, 544–546, 551, 613–614
Absorption spectra of diatomic molecules, 1212, 1217–1219
Absorption spectrum, 1031
ac circuits [*see* Alternating current (ac)]
ac Josephson effect, 845
Accelerated reference frame, 357–358, 418, 1243
Acceleration (*see also* Velocity)
 angular, 257–260
 average and instantaneous, 27, 62
 and center of mass, 218–221
 centripetal, 125–126
 constant, 29–38, 149–150
 definition, 27–28
 electric force, 674
 and gravity, 30, 87, 89
 and harmonic motion, 404, 405
 and Lorentz transformation, 1247
 Newton's laws and, 83–84, 85, 88–89
 and rotation, 275
 tangential, 258
 vectors, 59–62
Accelerators of particles, 1294–1295, 1313, **1316**
Acceptor levels in semiconductors, 1226
Accommodation, eye, 1091
Action at a distance, 92, 667
Action-reaction pair, 84, 89–90, 100
Adiabatic process, 276, 588–590
Allowed energy, 524
Alpha decay, 1293–1294
Alternating current (ac) [*see also* Direct current (dc); *RLC* circuit]
 in capacitors, conductors and inductors, 964–968
 vs. direct current, 959
 and electric motors, 985–988
 in generators, 960–961, 972, 973–981
 LC circuits, 969–971
 phasors, 968–969
 in resistors, 961–964
 root-mean-square (rms) values, 962–964

transformer, 982–984
Altitude and pressure, 380
Ammeter, 809–810, 896, 962
Ammonia (NH_3) molecule, 1211
Amorphous solid, 1219–1220
Ampère, André-Marie, 883, 895, 1017
Ampere (A)
 and Coulomb, 659
 definition, 895–897
 unit of current, 4, 787, AP1
Ampère's law, 898–901, 1000–1001, 1003
Amperian current, 904
Amplifier, *pnp* transistor as, **1230**, 1231
Amplitude, 404, 427, 448, 482
Analyzer, polarization, 1054
Anderson, Carl, 1317
Angle of incidence, refraction, 1041, 1042
Angle of minimum deviation, rainbows, 1052
Angles, conversion factors for, AP2
Angular acceleration, 257–259
Angular displacement, 258–259
Angular frequency, 405–407, 449
Angular magnification, 1094
Angular momentum (*see also* Conservation of momentum; Momentum)
 atomic spin, 1183–1184, 1316–1319
 conservation of, 304–309, 1179
 definition, 297–300
 of doublet, 1193–1194
 gyroscope, 303
 vs. magnetic moment, 906
 particle, 297
 quantization, 309–311, 1213
 quantum numbers, 1178
 and rotation, 210, 295–297, 312
 spin, 300, 310, 312
 and torque, 300–302
 unit of, 310, 312
 z component of, 1177
Angular speed, 258, AP2
Angular velocity, 258, 297
Anisotropic material, 1057
Anode in battery, 798
Antenna, electric dipole, 1005–1007
Antiderivative, 39
Antinode, waves, **488**, 489, 492
Antiparticles (*see also* Particles)
 antiproton/proton collision, **239**
 creation, 1317, 1318, 1330
 electric charge, 1328
 leptons, 1314
 mass of, 1314–1315, 1328
 neutrino, 1291, 1314–1315, 1320
 quarks, 1323, **1324**
 spin, 1316–1319, 1328

Antiquarks, **1324**
Aphelion, 322
Apparent depth, lenses, 1080
Apparent weight, 88
Arc discharge, 740
Archimedes, 381, **382**
Archimedes' principle, 380–382
Area, 7, AP2, AP14–15
Aristotle, 2
Armature, electric motor, 985
Astigmatism, 1090, 1091
Aston, Francis William, 865
Astronomical numerical data, AP3
Astronomical telescope, 1094, 1096–1099, **1098**, 1128
Astronomical unit (AU), 322
Asymmetric wave function, 1206–1208
Atmosphere (*see also* Pressure)
 escape speed, 330, 331, 558
 law of, 380
 units of, 376
Atomic magnetic moments, 906–908
Atomic mass, **1287**
Atomic number, 1169
Atomic orbitals, 1211
Atomic spectra, 1170–1171
Atomic theory of matter, 1313
Atoms (*see also* Electrons; Elementary particles; Molecular bonding; Neutrons; Nuclear physics; Protons)
 Bohr model of hydrogen, 1169–1176
 as elementary particles, 1313
 fine structure, 1183–1185
 nuclear, 1170–1171
 nuclei, properties of, 1284–1288
 optical atomic spectra, 1192–1194
 periodic table, 1185–1192
 plum pudding model, **1171**
 polyatomic molecules, 1210–1212
 quantization, 515
 quantum theory of atoms, 1176–1178
 quantum theory of hydrogen atom, 1178–1183
 rest energies of, **1266**
 selection rules, 1179
 spin, 1316–1319
 spin-orbit effect, 1183–1185
 X-ray spectra, 1194–1195
Atwood's machine, 111
AU (astronomical unit), 322
Avalanche breakdown current, 1228
Average acceleration, 27
Average force, 226–227
Average power, 962–963, 974–975
Average speed, 21
Average velocity, 19–20, 29

conservative force, 178–179
definition, 162
electric dipole in electric field, 677
electrostatic, 753–755, 758, 1207–1208
and equilibrium, 166–168
gravitational, 164, 216
and harmonic motion, 410, **531**
and harmonic oscillator, 1148–1150
of ion in crystal, 1220, 1222
magnetic dipole in magnetic field, 870–871
vs. separation in ionic bonding, 1205
of a spring, 165–166, 416
and three dimensional Schrödinger equations, 1160–1162
Potential-energy curve, 167
Potential-energy function, 163
Pound (lb), unit of force, 5, 89
Power
conversion factors for, AP2
and energy, 159–162
and kinetic energy of rotation, 274–275
and torque, 274
unit of, 4, 159, AP1
Power, electric
average, 962–963, 974–975
dissipated in conductors, 794
RLC circuit, 974–975
transformers, 982–984
Power, optical instruments
lens, 1084
magnifying power, 1094, 1095, 1097
resolving power, 1128, 1132
Power factor, *RLC* circuit, 975
Powers of *x* and 10 (exponents), **5**, 8–9, 11, AP12–14
Poynting, Sir John, 1008
Poynting vector, 1008
Prefixes for SI units, 4–5
Pressure
altitude vs., 380
conversion factors for, AP2
definition, 376
fluid, 376–380
gas, 548, 550–551
gauge, 379
ideal gas and, 617–618
PV diagrams, 576–578
radiation, 1008–1011
standard conditions, 548
unit of, 376, 379
Principal quantum number, 1177
Principal rays
for mirror, 1075
for negative lens, 1086
for positive lens, 1085
Principia (Newton), 1018
Principle of equivalence, 329, 1270–1273
Principle of Newtonian reality, 1244
Prism, light spectra, 1028, **1048**
Probability, 520–521, 623–625
Probability densities, 1180–1183
Projectile motion, 64–72
Propagation of light, 1028, 1040–1041, 1244–1245
Proper length, Lorentz transformation, 1250

Proportion review, direct and inverse, AP9–10
Propulsion, 241–243
Protons (*see also* Atoms)
accelerated, 1294–1295, 1313, **1316**
antiproton, 1317
in atom, 1169
bubble chamber tracks, **238, 239, 1313**
electric charge quantization, 658–659
as elementary particle, 1313, 1320, 1321
N and *Z* numbers, 1285–1286
nuclei, 1284–1285
rest energies of, **196**
sharing in hydrogen bond, 1209–1210
Pseudoforces, 128–129
P-type semiconductor, 1226, 1227
PV (pressure/volume) diagrams, gas, 576–578
Pythagorean theorem, 57, 238

Q factor (quality factor), 424, 975
Q value, nuclear reaction, 1295
QCD (quantum chromodynamics), 1326, 1327–1328
Quadratic equation review, AP9–10, AP12
Quality factor (*Q* factor), 424, 975
Quantization
of angular momentum, 309–311
of electric charge, 658–659
of magnetic flux, 949
Quantization of energy
in atoms, 515
in hydrogen atom, 531–532
in microscopic systems, 199–200
particle in a box, 1145–1146
and standing waves, 519–520
Quantum chromodynamics (QCD), 1326, 1327–1328
Quantum electrodynamics, 1325
Quantum gravity, theory of, 1330
Quantum Hall effect, 873–874
Quantum mechanics (*see also* Elementary particles)
and classical physics, 2–3
Schrödinger and, 520
bound-state problem, 1151
and covalent bonds, 1206
polyatomic molecules, 1210–1212
and spin-orbit effect, 1184
and wave packet representation, 1152–1153
Quantum number
angular momentum, 1178, 1183–1185, 1316
covalent bonds, 1206
definition, 526
and energy quantization, 200
hydrogen atom, 1177–1178
magnetic, 1177
orbital, 1177
principal, 1177
rotational, 1213
in spherical coordinates, 1177–1178
vibrational, 1215
Quantum states, 1161, 1204–1205

Quantum theory
of atoms, 1176–1178
electrical conduction, 839–840
for hydrogen atom, 1178–1183
and Schrödinger, 520–521, 525
superconductivity, 843–844
Quantum-mechanical system and molecular bonding, 1203–1210
Quark confinement, 1325
Quarks (*see also* Elementary particles)
and the Big Bang, 1330
as elementary particles, 1323–1325, 1327
and field quanta, 1326
hadrons composed of, 1314
in Standard Model, 659*n*, 1323, 1327–1328
Quarter-wave plate, 1057
Quasar, **196**
Quasi-static adiabatic process, 576, 588

R factor, 644
Radial electric field, 728
Radial equation, 1177
Radial probability density, 1181
Radial ray, 1075
Radiation (*see also* Radioactivity)
background and black-body, 1329–1330
Cerenkov, 468, **1098**
electric-dipole, 1005–1007
electromagnetic, 200
and light sources, 1031–1033
molecular absorption and emittance, 1212
muons, 1251–1252, 1291
nuclear, 2
from photons, 511–515
synchrotron, 1005
thermal energy, 640, 646–648
Radiation era, 1331
Radiation frequency, 1173
Radiation pressure, 1008–1011
Radio waves, 1005
Radioactive carbon dating, 1292
Radioactivity
alpha decay, 1293–1294
beta decay, 1291–1292
gamma decay, 1293
rate of, 1288–1291
Radius of allowed orbit, 1173
Radius of curvature, plane mirror, 1076, 1080
Rainbows, 1048–1052
Raman, C. V., 1031
Raman scattering, 1031
Range of projectile, 64, 66
Ranger and monkey problem, 70–71
Rayleigh, Lord, 1031 (*see also* Strutt, John William)
Rayleigh scattering, 1031
Rayleigh's criterion for resolution, 1128
Rays
approximation in diffraction, 463
diagrams for lenses and mirrors, 1075–1077, 1085–1087
principal types, 1075, 1085, 1086
wavefronts of, 454

Prefixes for Powers of 10	Multiple	Prefix	Abbreviation
	10^{18}	exa	E
	10^{15}	peta	P
	10^{12}	tera	T
	10^9	giga	G
	10^6	mega	M
	10^3	kilo	k
	10^2	hecto	h
	10^1	deka	da
	10^{-1}	deci	d
	10^{-2}	centi	c
	10^{-3}	milli	m
	10^{-6}	micro	μ
	10^{-9}	nano	n
	10^{-12}	pico	p
	10^{-15}	femto	f
	10^{-18}	atto	a

Some Physical Data			
Acceleration of gravity at earth's surface	g	$9.81 \text{ m/s}^2 = 32.2 \text{ ft/s}^2$
Radius of earth	R_E	$6370 \text{ km} = 3960 \text{ mi}$
Mass of earth	M_E	$5.98 \times 10^{24} \text{ kg}$
Mass of sun		$1.99 \times 10^{30} \text{ kg}$
Mass of moon		$7.36 \times 10^{22} \text{ kg}$
Escape speed at earth's surface		$11.2 \text{ km/s} = 6.95 \text{ mi/s}$
Standard temperature and pressure (STP)		$0°C = 273.15 \text{ K}$ $1 \text{ atm} = 101.3 \text{ kPa}$
Earth–moon distance		$3.84 \times 10^8 \text{ m} = 2.39 \times 10^5 \text{ mi}$
Earth–sun distance (mean)		$1.50 \times 10^{11} \text{ m} = 9.30 \times 10^7 \text{ mi}$
Speed of sound in dry air (at STP)		331 m/s
Density of air		1.29 kg/m^3
Density of water		1000 kg/m^3
Heat of fusion of water	L_f	333.5 kJ/kg
Heat of vaporization of water	L_v	2.257 MJ/kg

The Greek Alphabet

Alpha	A	α	Iota	I	ι	Rho	P	ρ
Beta	B	β	Kappa	K	κ	Sigma	Σ	σ
Gamma	Γ	γ	Lambda	Λ	λ	Tau	T	τ
Delta	Δ	δ	Mu	M	μ	Upsilon	Y	υ
Epsilon	E	ϵ	Nu	N	ν	Phi	Φ	ϕ
Zeta	Z	ζ	Xi	Ξ	ξ	Chi	X	χ
Eta	H	η	Omicron	O	o	Psi	Ψ	ψ
Theta	Θ	θ	Pi	Π	π	Omega	Ω	ω

Abbreviations for Units

A	ampere		lb	pound
Å	angstrom (10^{-10} m)		L	liter
atm	atmosphere		m	meter
Btu	British thermal unit		MeV	mega-electron volt
Bq	becquerel		Mm	megameter (10^6 m)
C	coulomb		mi	mile
°C	degree Celsius		min	minute
cal	calorie		mm	millimeter
Ci	curie		ms	millisecond
cm	centimeter		N	newton
dyn	dyne		nm	nanometer (10^{-9} m)
eV	electron volt		pt	pint
°F	degree Fahrenheit		qt	quart
fm	femtometer, fermi (10^{-15} m)		rev	revolution
ft	foot		R	roentgen
Gm	gigameter (10^9 m)		Sv	seivert
G	gauss		s	second
Gy	gray		T	tesla
g	gram		u	unified mass unit
H	henry		V	volt
h	hour		W	watt
Hz	hertz		Wb	weber
in	inch		y	year
J	joule		yd	yard
K	kelvin		μm	micrometer (10^{-6} m)
kg	kilogram		μs	microsecond
km	kilometer		μC	microcoulomb
keV	kilo-electron volt		Ω	ohm

Some Conversion Factors

$1 \text{ m} = 39.37 \text{ in} = 3.281 \text{ ft} = 1.094 \text{ yd}$

$1 \text{ m} = 10^{15} \text{ fm} = 10^{10} \text{ Å} = 10^9 \text{ nm}$

$1 \text{ km} = 0.6215 \text{ mi}$

$1 \text{ mi} = 5280 \text{ ft} = 1.609 \text{ km}$

$1 \text{ lightyear} = 1 \ c \cdot \text{y} = 9.461 \times 10^{15} \text{ m}$

$1 \text{ in} = 2.540 \text{ cm}$

$1 \text{ L} = 10^3 \text{ cm}^3 = 10^{-3} \text{ m}^3 = 1.057 \text{ qt}$

$1 \text{ h} = 3.6 \text{ ks}$

$1 \text{ y} = 365.24 \text{ d} = 3.156 \times 10^7 \text{ s}$

$1 \text{ km/h} = 0.278 \text{ m/s} = 0.6215 \text{ mi/h}$

$1 \text{ ft/s} = 0.3048 \text{ m/s} = 0.6818 \text{ mi/h}$

$1 \text{ rev} = 2\pi \text{ rad} = 360°$

$1 \text{ rad} = 57.30°$

$1 \text{ rev/min} = 0.1047 \text{ rad/s}$

$1 \text{ slug} = 14.59 \text{ kg}$

$1 \text{ tonne} = 10^3 \text{ kg} = 1 \text{ Mg}$

$1 \text{ atm} = 101.3 \text{ kPa} = 1.013 \text{ bar} = 76.00 \text{ cmHg} = 14.70 \text{ lb/in}^2$

$1 \text{ N} = 10^5 \text{ dyn} = 0.2248 \text{ lb}$

$1 \text{ lb} = 4.448 \text{ N}$

$1 \text{ Pa} \cdot \text{s} = 10 \text{ poise}$

$1 \text{ J} = 10^7 \text{ erg} = 0.7373 \text{ ft} \cdot \text{lb} = 9.869 \times 10^{-3} \text{ L} \cdot \text{atm}$

$1 \text{ kW} \cdot \text{h} = 3.6 \text{ MJ}$

$1 \text{ cal} = 4.184 \text{ J} = 4.129 \times 10^{-2} \text{ L} \cdot \text{atm}$

$1 \text{ L} \cdot \text{atm} = 101.3 \text{ J} = 24.22 \text{ cal}$

$1 \text{ eV} = 1.602 \times 10^{-19} \text{ J}$

$1 \text{ Btu} = 778 \text{ ft} \cdot \text{lb} = 252 \text{ cal} = 1054 \text{ J}$

$1 \text{ horsepower} = 550 \text{ ft} \cdot \text{lb/s} = 746 \text{ W}$

$1 \text{ W/m} \cdot \text{K} = 6.938 \text{ Btu} \cdot \text{in/h} \cdot \text{ft}^2 \cdot °\text{F}$

$1 \text{ T} = 10^4 \text{ G}$

$1 \text{ kg weighs about } 2.205 \text{ lb}$

Some Physical Constants

Avogadro's number	N_A		$6.022\ 136\ 7 \times 10^{23}$ particles/mol
Boltzmann's constant	k		$1.380\ 658 \times 10^{-23}$ J/K
Bohr magneton	$m_B = e\hbar/2m_e$		$9.274\ 015\ 4 \times 10^{-24}$ J/T
Coulomb constant	$k = 1/4\pi\epsilon_0$		$8.987\ 551\ 788 \times 10^9$ N·m^2/C^2
Compton wavelength	$\lambda_C = h/2e$		$2.426\ 310\ 58 \times 10^{-12}$ m
Fundamental charge	e		$1.602\ 177\ 33 \times 10^{-19}$ C
Gas constant	$R = N_A k$		$8.314\ 51$ J/mol·K $= 1.987\ 22$ cal/mol·K $= 8.205\ 78 \times 10^{-2}$ L·atm/mol·K
Gravitational constant	G		$6.672\ 6 \times 10^{-11}$ N·m^2/kg^2
Mass, of electron	m_e		$9.109\ 389\ 7 \times 10^{-31}$ kg $= 510.999\ 1$ keV/c^2
of proton	m_p		$1.672\ 623\ 1 \times 10^{-27}$ kg $= 938.272\ 3$ MeV/c^2
of neutron	m_n		$1.674\ 929 \times 10^{-27}$ kg $= 939.565\ 6$ MeV/c^2
Permeability of free space	μ_0		$4\pi \times 10^{-7}$ N/A^2
Planck's constant	h		$6.626\ 075\ 5 \times 10^{-34}$ J·s $= 4.135\ 669\ 2 \times 10^{-15}$ eV·s
	$\hbar$		$1.054\ 572\ 66 \times 10^{-34}$ J·s $6.582\ 122\ 0 \times 10^{-16}$ eV·s
Speed of light	c		$2.997\ 924\ 58 \times 10^8$ m/s
Unified mass unit	u		$1.660\ 540 \times 10^{-27}$ kg $= 931.494\ 32$ MeV/c^2

Mathematical Symbols

$=$	is equal to	Δx	change in x
$\neq$	is not equal to	$\lvert x \rvert$	absolute value of x
$\approx$	is approximately equal to	$n!$	$n(n-1)(n-2)\cdots 1$
$\sim$	is of the order of	Σ	sum
$\propto$	is proportional to	lim	limit
$>$	is greater than	$\Delta t \to 0$	Δt approaches zero
$\geq$	is greater than or equal to	$\dfrac{dx}{dt}$	derivative of x with respect to t
$\gg$	is much greater than		
$<$	is less than	$\dfrac{\partial x}{\partial t}$	partial derivative of x with respect to t
$\leq$	is less than or equal to		
$\ll$	is much less than	$\int$	integral